Edward A. (Edward Albert) Bowser

Academic Algebra

For the Use of Common and High Schools and Academies

Edward A. (Edward Albert) Bowser

Academic Algebra
For the Use of Common and High Schools and Academies

ISBN/EAN: 9783337159429

Printed in Europe, USA, Canada, Australia, Japan

Cover: Foto ©berggeist007 / pixelio.de

More available books at **www.hansebooks.com**

ACADEMIC ALGEBRA;

FOR THE USE OF

COMMON AND HIGH SCHOOLS AND ACADEMIES.

WITH NUMEROUS EXAMPLES.

BY

EDWARD A. BOWSER, LL.D.,

PROFESSOR OF MATHEMATICS AND ENGINEERING IN RUTGERS COLLEGE.

OAKLAND HIGH SCHOOL,

BOSTON, U.S.A.:

D. C. HEATH & CO., PUBLISHERS.

1895.

Norwood Press:
Berwick & Smith, Boston, U.S.A.

PREFACE.

THIS work is designed as a text-book for Common and High Schools and Academies, and to prepare students for entering Colleges and Scientific Schools. The book is a complete treatise on Algebra up to and through the Progressions, and including Permutations and Combinations and the Binomial Theorem.

The aim has been to explain the principles concisely and clearly, bestowing great care upon the explanations and proofs of the fundamental operations and rules. Copious illustrations have been given to make the work intelligible and interesting to young students; and numerous explanatory notes have been all along inserted, to guard the pupil against the errors which experience shows to be almost universal among beginners. Thoroughness has been aimed at, rather than multiplicity of subjects. If a student has not time to master a complete course, it is better for him to omit entirely subjects that are less necessary, than to go rapidly over too many things.

In the earlier chapters, some of the most interesting practical applications of the subject have been introduced. Thus, a chapter on easy equations and problems precedes the chapters on Factoring and Fractions. By this course the beginner soon becomes acquainted with the ordinary Algebraic processes without encountering too many of their difficulties; and he is learning at the same time something of the more attractive parts of the subject. Nothing is more pleasing to a young student than to see and feel that he can *use* his knowledge to some practical end.

Throughout the book are numerous examples fully worked out, to illustrate the most useful applications of important rules, and to exhibit the best methods of arranging the work. No principle is well learned by a pupil and thoroughly fixed in his mind till he can use it. For this purpose a large number of examples is given at the ends of the chapters. These examples have been selected and arranged so as to illustrate and enforce every part of the subject. Each set has been carefully graded, commencing with some which are very easy, and proceeding to others which are more difficult. Complicated examples have been excluded, because they consume time and energy which may be spent more profitably on other branches of mathematics.

The chief sources from which I have derived assistance in preparing this work are the treatises of Wood, De Morgan, Serret, Todhunter, Colenzo, Hall and Knight, Smith, and Chrystal.

My thanks are due to those of my friends who have kindly assisted me in reading the MS., correcting the proof-sheets, and verifying copy.

E. A. B.

RUTGERS COLLEGE,
NEW BRUNSWICK, N.J., *June*, 1888.

TABLE OF CONTENTS.

CHAPTER I.

FIRST PRINCIPLES.

CHAPTER III.

SUBTRACTION.

CHAPTER IV.

MULTIPLICATION.

CHAPTER V.

DIVISION.

CHAPTER VI.

SIMPLE EQUATIONS OF ONE UNKNOWN QUANTITY.

CHAPTER VII.

FACTORING — GREATEST COMMON DIVISOR — LEAST COMMON MULTIPLE.

GREATEST COMMON DIVISOR.

CHAPTER XI.

INVOLUTION AND EVOLUTION.

EVOLUTION.

CHAPTER XII.

THE THEORY OF EXPONENTS — SURDS.

SURDS (RADICALS).

CHAPTER XIII.

QUADRATIC EQUATIONS OF ONE UNKNOWN QUANTITY.

CHAPTER XIV.

SIMULTANEOUS QUADRATIC EQUATIONS.

CHAPTER XV.

CHAPTER XVI.

CHAPTER XVII.

PERMUTATIONS AND COMBINATIONS — BINOMIAL THEOREM.

PERMUTATIONS AND COMBINATIONS.

BINOMIAL THEOREM.

ALGEBRA.

CHAPTER I.

FIRST PRINCIPLES.

1. Quantity and its Measure. — Quantity is any thing that is capable of increase, diminution, and measurement; as time, space, motion, weight, and area.

To *measure* a quantity is to find how many times it contains another quantity of the same kind, taken as a standard of comparison. This standard is called a *unit*.

For example, if we wish to determine the quantity of a *weight*, we must take a *unit of weight*, such as a pound, or an ounce, and observe how many times it is contained in the quantity to be measured. If we wish to measure *area*, we must take a *unit of area*, as a square foot, square yard, or acre, and see how many times it is contained in the area to be measured. So also, if we wish to measure the value of a sum of money, or any portion of time, we must take a unit of value, as a dollar or a sovereign, or a unit of time, as a day or a year, and see how many times it is contained in the quantity to be measured.

2. Number. — The relation between any quantity and its unit is always expressed by a *number; a number* therefore simply shows *how many times* any quantity to be measured contains another quantity, arbitrarily assumed as the unit. All quantities, therefore, can be expressed by numbers.

All numbers are *concrete* or *abstract*.

A Concrete Number is one in which the kind of quantity which it measures is expressed or understood; as 6 books, 10 men, 4 days.

An Abstract Number is one in which the kind of quantity which it measures is not expressed ; as 6, 10, 4.

The word *quantity* is often used with the same meaning as *number*. Numbers may be either whole or fractional. The word *integer* is often used instead of *whole number*.

3. Mathematics. — Mathematics is the science which treats of the measurement and relations of quantities. It is divided into two parts, *Pure Mathematics* and *Mixed Mathematics.*

Pure Mathematics consists of the four branches, *Arithmetic, Algebra, Geometry,* and *Calculus.*

Mixed Mathematics is the application of Pure Mathematics to the *Mechanic Arts.*

4. Algebra. — Algebra is that branch of Mathematics in which we reason about numbers by means of *symbols.* The different symbols used represent the numbers themselves, the manner in which they are related to one another, and the operations performed on them.

In Arithmetic, numbers are represented by ten characters, called figures, which are variously combined according to certain rules, and which have but one single definite value. In Algebra, on the contrary, numbers are represented either by figures, as in Arithmetic, or by symbols which may have any value we choose to assign to them.

5. Algebraic Symbols. — The *symbols* employed in Algebra are of four kinds : symbols of *quantity*, symbols of *operation*, symbols of *relation*, and symbols of *abbreviation.*

6. Symbols of Quantity. — The *symbols of quantity* may be any characters whatever, but those that are most commonly used are figures and the letters of the alphabet ; and as in the simplest mathematical problems there are certain quantities given, in order to determine other quantities which are unknown, it is usual to represent the known quantities by figures and by the first letters of the alphabet, a, b, c, etc. ; a', b', c', etc., read *a prime, b prime, c prime,* etc. ; a_1, b_1, c_1, etc., read *a one, b one, c one,* etc. ; while the unknown quantities are represented by the final letters of the alphabet, v, x, y, z, v', x', y', z', etc.

Known Quantities are those whose values are given.

Unknown Quantities are those whose values are required.

Since all quantities can be expressed by *numbers* (Art. 2), it is only these *numbers* with which we are concerned, and the symbols of quantity, whether figures or letters, always represent *numbers*.

In Arithmetic a character has but one definite and invariable value, while in Algebra a symbol may stand for any quantity we choose to assign to it (Art. 4); but while there is no restriction as to the numerical values a symbol may represent, it is understood that in the same piece of work it keeps the same value throughout. Thus, when we say "let $a = 2$," we do not mean that a must have the value 2 always, but only in the particular example we are considering. Also, we may operate with symbols without assigning to them any particular value at all; and it is with such operations that Algebra is chiefly concerned.

7. Symbols of Operation. — The *symbols of operation* are the same in Algebra as in Arithmetic, or in any other branch of Mathematics, and are the following :

8. The Sign of Addition, $+$, is called *plus*. When placed before a number it denotes that the number is to be *added*. Thus, $6 + 3$, read 6 *plus* 3, means that 3 is to be added to 6 ; $a + b$, read a *plus* b, denotes that the number represented by b is to be added to the number represented by a ; or, more briefly, it denotes that b is to be added to a. If a represent 8, and b represent 5, then $a + b$ represents 13.

Similarly $a + b + c$, read a *plus* b *plus* c, denotes that we are to add b to a, and then add c to the result.

9. The Sign of Subtraction, $-$, is called *minus*. When placed before a number it denotes that the number is to be *subtracted*. Thus, $a - b$, read a *minus* b, denotes that the number represented by b is to be subtracted from the number represented by a ; or. more briefly. that b is to be subtracted from a. If a represent 8, and b represent 5, then $a - b$ represents 3.

Similarly $a - b - c$, read a *minus* b *minus* c, denotes that we are to subtract b from a, and then subtract c from the result.

If neither $+$ nor $-$ stands before a quantity, $+$ is always understood; thus a means $+ a$.

Quantities which have the same sign, either $+$ or $-$, are said to have like signs. Thus, $+ a$ and $+ b$ have like signs, also $- a$ and $- b$; but $+ a$ and $- b$ have unlike signs.

Note. — Although there are many signs used in Algebra, when the *sign* of a quantity is spoken of, it means the $+$ or $-$ sign which is prefixed to it; and when we speak of *changing the signs* of an expression, it means that we are to change $+$ to $-$ and $-$ to $+$ wherever they occur.

The sign $\sim$ is sometimes used to denote the difference of two numbers when it is not known which of them is the greater. Thus, $a \sim b$ denotes the difference of the numbers represented by a and b; and is equal to $a - b$, or $b - a$, according as a is greater or less than b; but this symbol $\sim$ is very rarely required.

10. The Sign of Multiplication, $\times$, is read *into*, or *times*, or *multiplied by*. When placed between two numbers it denotes that they are to be *multiplied* together. Thus, $a \times b$, read a *into* b, denotes that the number represented by a is to be multiplied by the number represented by b, or, more briefly, that a is to be multiplied by b, or that the two are to be multiplied together. The numbers to be multiplied together are called *factors*, and the result of the multiplication is called a *product*. Thus 5, a, and b are the factors of the product $5 \times a \times b$. If a represent 8, and b represent 4, then $a \times b$ represents 32; a and b are the factors of the product $a \times b$, or 8 and 4 are the factors of 32. Similarly $a \times b \times c$ denotes the product of the numbers a, b, and c. If a represent 6, b represent 8, and c represent 10, then $a \times b \times c$ represents 480, and $5 \times a \times b \times c$ represents 2400.

Sometimes a *point* is used instead of the sign $\times$; or, still more commonly, one number is placed close after the other without any sign between them. Thus, $a \times b$, $a \cdot b$, and ab all mean the same thing, viz., the product of a and b; also, $a \times b \times c$, or $a \cdot b \cdot c$, or abc, denotes the product of the numbers a, b, and c. If a, b, and c represent 2, 5, and 10 respectively, then abc represents 100.

If one factor of a product is equal to 0, the whole product must be equal to 0, *whatever values the other factors may have.* A factor 0 is sometimes called a "zero factor." *

The sign of multiplication must not be omitted when numbers are expressed in the ordinary way by figures. Thus 23 cannot be used to represent the product of 2 and 3, because 23 is used to mean the number *twenty-three*. Nor can the product of 2 and 3 be represented by 2.3, because 2.3 is used to mean *two and three-tenths*. We must therefore represent the product of 2 and 3 by placing the sign of multiplication between them, as follows: 2×3. When the numbers to be multiplied together are represented by letters, or by letters and a figure, it is usual to omit the sign of multiplication for the sake of brevity, and write them in succession close to each other; thus, the product of the numbers 7, a, b, c, and d would be written $7abcd$, instead of $7 \times a \times b \times c \times d$, or $7 \cdot a \cdot b \cdot c \cdot d$, and would have the same meaning.

11. The Sign of Division, $\div$, is read *divided by*, or simply *by*. When placed between two numbers, it denotes that the number which precedes it is to be divided by the number which follows it. Thus, $a \div b$, read a *divided by* b, or a *by* b, denotes that the number represented by a is to be divided by the number represented by b, or, more briefly, that a is to be divided by b. If a represent 8, and b represent 2, then $a \div b$ represents 4. Most frequently, to express division, the number to be divided is placed over the other

---

* It is a common mistake of beginners to say that an Algebraic expression like $a \times 0$ or $0 \times a$ is equal to a, by supposing it to mean a not multiplied at all; whereas $a \times 0$ or $0 \times a$ signifies 0 taken a times, or a taken 0 times, and is therefore equal to 0.

with a horizontal line between them, in the manner of a fraction in Arithmetic. Thus, $\frac{a}{b}$ is used instead of $a \div b$, and has the same meaning. Also, the sign of division may be replaced by a vertical line, straight or curved. Thus, $a \lfloor b$, or $b)a$ is used instead of $a \div b$, and has the same meaning.

NOTE. — It is important for the student to notice the *order* of the operations in such expressions as $a + b \times c$ and $a - b \div c$. The former means that b is first to be multiplied by c, and the result added to a. The latter means that b is first to be divided by c, and the result subtracted from a.

12. The Exponential Sign. — This sign is a small figure or letter written at the right of and above a number to show how many times the number is taken as a factor, and is called an *exponent*. Thus, a^2 is used to denote $a \times a$, or that a is taken twice as a factor; a^3 is used to denote $a \times a \times a$, or that a is taken three times as a factor; a^4 is used to denote $a \times a \times a \times a$, or that a is taken four times as a factor; and a^n is used to denote $a \times a \times a \times a$, etc., to n factors, or that a is taken n times as a factor. Similarly $a^2 b^4 cd^3$ is used to denote $aabbbbcddd$, and $7a^3cd^2$ is used for $7aaacdd$.

If a factor be multiplied by itself any number of times the product is called a *power* of that factor. Thus,

$a \times a$ is called the second power of a, and is written a^2;

$a \times a \times a$ is called the third power of a, and is written a^3;

$a \times a \times a \times a$ is called the fourth power of a, and is written a^4;

and so on. Similarly $aaabbc$ is called the product of the third power of a, the second power of b, and c, and is written $a^3 b^2 c$.

The second power of a, i.e., a^2, is usually read *a to the second power*, or *a square*. The third power of a, i.e., a^3, is usually read *a to the third power*, or *a cube*. There are no such words in use for the higher powers; the fourth power of a, i e., a^4, is usually read *a to the fourth power*, or briefly, *a fourth power*; and so on. When the exponent is unity it is omitted. Thus we do not write a^1, but simply a, which is the same as a^1, and means *a to the first power.*

13. The Radical Sign, $\sqrt{}$. — A *root* of a quantity is a factor, which, multiplied by itself a certain number of times, will produce the given quantity. The *square root* of a quantity is that quantity whose square or second power is equal to the given quantity. Thus the square root of 16 is 4, because 4^2 is equal to 16; the square root of a^2 is a, of 81 is 9.

The square root of a is denoted by $\sqrt[2]{a}$, or more simply $\sqrt{a}$.

Similarly the *cube, fourth, fifth,* etc., *root* of any quantity is that quantity whose third, fourth, fifth, etc., power is equal to the given quantity.

The roots are denoted by the symbols $\sqrt[3]{}$, $\sqrt[4]{}$, $\sqrt[5]{}$, etc.; thus, $\sqrt[3]{27a^3}$ denotes the cube root of $27a^3$, which is $3a$, because $3a$ to the third power is $27a^3$. Similarly $\sqrt[5]{32}$ is 2. The small figure placed on the left side of the symbol is called the *index* of the root. Thus 2 is the index of the square root, 3 of the cube root, 4 of the fourth root, and so on; the index, however, is generally omitted in denoting the *square* root; thus $\sqrt{a}$ is written instead of $\sqrt[2]{a}$.

The symbol $\sqrt{}$ is sometimes called the *radical sign.* When this sign with the proper index on the left side of it is placed over a quantity it denotes that some root of the quantity is to be extracted.

14. Symbols of Relation. — The *symbols of relation* are the following:

The *sign of equality*, =, is read *equals*, or *is equal to.* When placed between two numbers, it denotes that they are equal to each other. Thus $a = b$, read *a equals b*, or *a is equal to b*, denotes that the number represented by a is equal to the number represented by b: or, more briefly, that a equals b. And $a + b = c$ denotes that the sum of the numbers a and b is equal to the number c; so that if a represent 8 and b represent 4, then c must represent 12.

The *signs of inequality,* $>$ and $<$, are read *is greater than,* and *is less than,* respectively. When either is placed between two numbers it denotes that they are unequal to each other, the opening of the angle in both cases being turned towards the greater number. Thus $a > b$, read *a is greater than b,* denotes that the number a is greater than the number b, and $b < a$, read *b is less than a,* denotes that the number b is less than the number a.

The *sign of ratio,* $:$, is read *is to* or *to.* When placed between two numbers it denotes their ratio. Thus $a : b$, read *a is to b,* or *the ratio of a to b,* denotes the ratio of the number a to the number b. A proportion, or two equal ratios, is expressed by writing the sign $=$ or the sign $::$ between two equal ratios. Thus

$$a : b = c : d, \text{ or } a : b :: c : d,$$

read *a is to b as c is to d,* or *the ratio of a to b equals the ratio of c to d.*

The *sign of variation,* $\propto$, is read *varies as.* When placed between two numbers it denotes that they increase and decrease together, in the same ratio. Thus $x \propto y$, read *x varies as y,* denotes that x and y increase and decrease together.

15. Symbols of Abbreviation. — The *symbols of abbreviation* are the following:

The *signs of deduction,* $\therefore$ is read *hence* or *therefore,* and $\because$ is read *since* or *because.*

The *signs of aggregation* are the *bar* $|$, the *parenthesis* $(\ )$, the *bracket* $[\ ]$, the *brace* $\{\ \}$, and the *vinculum* $\overline{}$. These are employed to connect two or more numbers which are to be treated as if they formed one number. Thus, suppose we have to denote that the sum of a and b is to be multiplied by c; we denote it thus $(a + b) \times c$ or $\{a + b\} \times c$, or simply $(a + b) c$ or $\{a + b\} c$; here we mean that the *whole* of $a + b$ is to be multiplied by c. If we omit the parenthesis, or brace, we have $a + bc$, and this denotes that b *only* is to be multiplied by c and the result added to a.

Also $(a + b + c) \times (d + e)$ denotes that the result expressed by $a + b + c$ is to be multiplied by the result expressed by $d + e$. This may also be denoted simply thus $(a + b + c)(d + e)$, just as $a \times b$ is shortened into ab. If we omit the parenthesis we have $a + b + cd + e$, and this denotes that c *only* is to be multiplied by d *only*, and the result added to $a + b + e$.

Also $\sqrt{}\ (a + b + c)$ denotes that we are to obtain the result expressed by $a + b + c$, and then take the square root of this result.

Also $(ab)^2$ denotes $ab \times ab$; and $(ab)^3$ denotes $ab \times ab \times ab$.

Also $(a + b + c) \div (d + e)$ denotes that the result expressed by $a + b + c$ is to be divided by the result expressed by $d + e$. This may also be expressed by the bracket thus $[a + b + c] \div [d + e]$, or the brace $\{a + b + c\} \div \{d + e\}$, or the vinculum $\overline{a + b + c} \div \overline{d + e}$, or $\dfrac{a + b + c}{d + e}$, where the line between the numerator and denominator acts as a vinculum.

The *signs of continuation* are dots . . . . . , or dashes – – – – – –, and are read *and so on*.

16. Algebraic Expressions. — The four kinds of symbols which have been explained are called *Algebraic symbols* (Art. 5). Any collection of Algebraic symbols is called an *Algebraic expression*, or briefly, an *expression*. Thus $4a + 5b - c + x$ is an expression; $3b + 4c$ is the Algebraic expression for 3 times the number b increased by 4 times the number c.

The *numerical value* of an expression is the number obtained by giving a particular value to each letter, and then performing the operations indicated.

We shall now give some examples in finding the numerical values of expressions, as an exercise in the use of the symbols which have been explained.

EXAMPLES.

If $a = 1$, $b = 2$, $c = 3$, $d = 4$, $e = 5$, find the numerical values of the following expressions:

1. $9a + 2b + 3c - 2d$.

Here we have $9a + 2b + 3c - 2d =$
$$9 \times 1 + 2 \times 2 + 3 \times 3 - 2 \times 4 =$$
$$9 + 4 + 9 - 8 = 14 \quad Ans.$$

2. $7ae + 3bc + 9d$.

Here we have $7ae + 3bc + 9d =$
$$7 \times 1 \times 5 + 3 \times 2 \times 3 + 9 \times 4 = 89 \ Ans.$$

3. $abcd + abce + abde + acde + bcde$.   *Ans.* 274.

4. $\dfrac{4ac}{b} + \dfrac{8bc}{d} - \dfrac{5cd}{e}$.   6.

5. $\dfrac{cde}{ab} + \dfrac{5bcd}{ae} - \dfrac{6ade}{bc}$.   34.

If $a = 1$, $b = 3$, $c = 5$, and $d = 0$, find the numerical values of the following:

6. $a^2 + 2b^2 + 3c^2 + 4d^2$.   *Ans.* 94.

7. $a^4 - 4a^3b + 6a^2b^2 - 4ab^3 + b^4$.   16.

8. $\dfrac{12a^3 - b^2}{3a^2} + \dfrac{2c^2}{a + b^2} - \dfrac{a + b^2 + c^3}{5b^3}$.   5.

If $a = 1$, $b = 2$, $c = 3$, $d = 5$, and $e = 8$, find the numerical values of the following:

9. $b^2(a^2 + e^2 - c^2)$.   *Ans.* 224.

10. $\sqrt{(2b + 4d + 5e)}$.   8

11. $(a^2 + b^2 + c^2)(e^2 - d^2 - c^2)$.   420

12. $e - \{\sqrt{(e + 1)} + 2\} + (e - \sqrt[3]{e})\sqrt{(e - 4)}$.   15.

13. Find the value of $x^2 - 2x - 9$ when $x = 5$.

EXPLANATION. — If $x =$ any number, as for example, 5, then x^2 (which $= x \cdot x$) $= 5x$, x^3 (which $= x \cdot x^2$) $= 5x^2$, x^4 (which $= x \cdot x^3$) $= 5x^3$, etc. Hence examples like 13 may be solved as follows:

EXPLANATION.

$x^2 - 2x - 9$ when $x = 5$. $x^2 = 5x$

$\underline{x^2 = 5x}$ $5x - 2x = 3x$

$\quad\quad \underline{3x = 15}$ $3x = 15$

$\quad\quad\quad\quad 6 = result.$ $3x - 9 = 15 - 9 = 6.$

14. Find the value of $x^5 - 50x^4 - 49x^3 - 100x^2 - 101x - 50$, when $x = 51$.

These examples may be conveniently solved as follows:

$$
\begin{array}{c|cccccc}
 & x^5 & - 50x^4 & - 49x^3 & - 100.c^2 & - 101x & - 50 \\
51 & & + 51 & + 51 & + 102 & + 102 & + 51 \\
\hline
 & & + x^4 & + 2x^3 & + 2x^2 & + x & + 1 \\
\end{array}
$$

∴ result is 1.

15. Find the value of $x^4 - 11x^3 - 11x^2 - 11x - 11$ for $x = 12$. *Ans.* 1.

16. Find the value of $x^4 - 8x^3 - 19x^2 - 9x - 8$ for $x = 10$. *Ans.* 2.

17. Factor — Coefficient. — When two or more numbers are multiplied together the result is called the product, and each of the numbers multiplied together to form the product is called a *factor* of the product (Art. 10). Thus, $3 \times 4 \times 5 = 60$, and each of the numbers 3, 4, and 5 is a factor of the product 60. Factors expressed by letters are called *literal* factors; factors expressed by figures are called *numerical* factors. Thus, in the product $4ab$, 4 is called a *numerical* factor, while a and b are called *literal* factors.

The proof is given in Arithmetic that it is immaterial in what order the factors of a product are written; it is usual, however, to arrange them in alphabetic order.

The numerical factor is called the *coefficient* of the remaining factors. Thus in the expression $4ab$, 4 is the *coefficient*, and denotes that ab is taken 4 times. But it is sometimes convenient to consider any factor, or factors, of a product as the coefficient of the remaining factors. Thus, in the product $5abc$, $5a$ may be appropriately called the coefficient of bc, or $5ab$ the coefficient of c.

The coefficient is called *numerical* or *literal*, according as it is a number, or one or more letters. Thus, in the quantities $5x$ and mx, 5 is a *numerical* and m a *literal* coefficient.

When no numerical coefficient is expressed, 1 is always understood. Thus, a is the same as $1a$.

A coefficient placed before any parenthesis indicates that

every term of the expression within the parenthesis is to be multiplied by that coefficient.

Care must be taken to distinguish between a *coefficient* and an *exponent*. Thus $4a$ means *four times* a, or $a + a + a + a$; here 4 is a *coefficient*. But a^4 means a *times* a *times* a *times* a, or $a \times a \times a \times a$, or $aaaa$ (Art. 12). That is, if $a = 4$,

$$4a = 4 \times a = 4 \times 4 = 16,$$

but $a^4 = a \times a \times a \times a = 4 \times 4 \times 4 \times 4 = 256.$

18. A Term, its Dimensions, and Degree — Homogeneous — Similar. — A *term* is an Algebraic expression in which no two of the parts are connected by the sign of addition or subtraction. Thus $4a$, $5a^2bc$, and $4xy \div 5ab$ are terms. $2a$, $4c^2d$, and $-5a^3d$ are the terms of the expression $2a + 4c^2d - 5a^3d$.

Each of the literal factors of a term is called a *dimension* of the term, and the number of the literal factors or dimensions is called the *degree* of the term. Thus a^2b^3c or $aabbbc$ is said to be of six dimensions or of the sixth degree, because it contains six literal factors, viz., a twice, b three times, and c once. A numerical coefficient is not counted; thus a^2b^3 and $5a^2b^3$ are of the same degree, i.e., the fifth degree, since there are five literal factors, viz., a twice and b three times.

It is clear that the *degree* of a term, or the *number of its dimensions*, is the *sum of the exponents of its literal factors*, provided we remember that if no exponent be expressed 1 must be understood (Art. 12). Thus $a^3b^4c^2$ is of the ninth degree, since $3 + 4 + 2 = 9$.

Terms are *homogeneous* when they are of the same degree. Thus a^3b, a^2b^2, $3ab^3$, are homogeneous.

Terms are *similar* or *like* when they have the same literal part, i.e., when they have the same letters and the corresponding letters affected with the same exponents. Otherwise they are said to be *unlike*. Thus, the terms $3a^3b^2$, $5a^3b^2$, and $-a^3b^2$ are *similar*, or *like*; but the terms a^2b and

ab^2 are *unlike*, since, although the letters are the same, they are not raised to the same power.

19. Simple and Compound Expressions. — A *simple expression* consists of only *one* term, as $5ab$, and is called a *monomial*.

A *compound expression* consists of *two or more* terms, and is called a *polynomial*, or *multinomial*.

A *binomial* is a polynomial of two terms. Thus, $ab^2 + 2ac$ is a *binomial*.

A *trinomial* is a polynomial of three terms. Thus, $a + b - c$ is a *trinomial*.

A polynomial is said to be *homogeneous* when all its terms are of the same degree. Thus $5ab^2 + 7a^2b + 9b^3$ is homogeneous, for each term is of the third degree.

When a polynomial consists of several terms of different degrees, the degree of the polynomial is that of its highest term.

A polynomial is said to be *arranged* according to the powers of any letter it contains when the exponents of that letter occur in the order of their magnitudes, either *increasing* or *decreasing*. Thus, $a^4 + 4a^3b + 6a^2b^2 + 4ab^3$ is arranged according to the descending powers of a, and $4ab^3 + 6a^2b^2 + 4a^3b + a^4$ is arranged according to the ascending powers of a.

The *reciprocal* of a number is 1 divided by that number. Thus, the reciprocal of a is $\frac{1}{a}$. If the product of two numbers is 1, each number is the reciprocal of the other.

20. Positive and Negative Quantities. — In Arithmetic we deal with numbers connected by the signs $+$ and $-$, and in finding the value of an expression such as $1 + 3 - 2 + 4$ we understand that the numbers to which the sign $+$ is prefixed are *additive*, and those to which the sign $-$ is prefixed are *subtractive*, while the first term, 1, to which no sign is prefixed, is counted among the additive terms. The same thing is true in Algebra; thus in the expression $5a + 7b - 3c - 2d$ we understand the symbols $5a$ and $7b$ to be additive, while $3c$ and $2d$ are subtractive.

But in Arithmetic the sum of the additive terms is always greater than the sum of the subtractive terms, i.e., we are always required to subtract a smaller number from a greater; if the reverse were the case the result would have no Arithmetic meaning, i.e., we could not in Arithmetic subtract a greater number from a smaller. In Algebra, however, not only may the sum of the subtractive terms exceed that of the additive, but a subtractive term may stand alone, and yet have a meaning quite intelligible. It is therefore usual to divide all Algebraic quantities into *positive quantities* and *negative quantities*, according as they are preceded by the sign $+$ or the sign $-$; and this is quite irrespective of any actual process of addition and subtraction.

ILLUSTRATION (1) Suppose a ship were to start from the equator and sail northward 100 miles and then southward 80 miles, the Algebraic statement would be

$$100 - 80 = +20.$$

Here the positive sign of the result indicates that the ship is 20 miles *north* of the equator. But if the ship first sailed 80 miles northward and then southward 100 miles, the Algebraic statement would be

$$80 - 100 = -20.$$

Here the negative sign of the result indicates that the ship is 20 miles *south* of the equator.

(2) Suppose a man were to gain $40 and then lose $36, his total *gain* would be $4. But if he first gained $36 and then lost $40, he sustained a *loss* of $4.

The corresponding Algebraic statements would be

$$\$40 - \$36 = +\$4,$$
$$\$36 - \$40 = -\$4.$$

Here the negative quantity in the second case is interpreted as a *debt*, i.e., a sum of money opposite in character to the positive quantity or *gain* in the first case. In Arithmetic we would call it a debt or loss of $4. In Algebra we make the equivalent statement that it is a gain of $-$4.

(3) Suppose a man starts at a certain point and walks 100 yards to the right in a straight line, and then walks back 70 yards, he will be 30 yards to the *right* of his starting point. If he first walks from the same point 70 yards to the right and then walks back 70 yards, he will be at the point from which he started. But if he first walks to the right 70 yards and then walks back 100 yards, he will be 30 yards to the *left* of his starting point. The corresponding Algebraic statements are

$$100 \text{ yards} - 70 \text{ yards} = 30 \text{ yards}$$
$$70 \text{ '' } - 70 \text{ '' } = 0 \text{ ''}$$
$$70 \text{ '' } - 100 \text{ '' } = -30 \text{ '' } .$$

Here we see that the negative sign may be taken as indicating *a reversal of direction.* In Arithmetic we would say the man was 30 yards to the left of his starting point. In Algebra we say he was -30 yards to the right of his starting point.

There are numerous instances like the preceding in which it is convenient for us to be able to represent not only the *magnitude* but the *nature* or *quality* of the things about which we are reasoning. As in the preceding cases, in a question of position we may have to distinguish a distance measured to the *north* of the equator from a distance measured to the *south* of it; or a distance measured to the *right* of a certain starting point from a distance measured to the *left* of it; or we may have to distinguish a sum of money *gained* from a sum of money *lost;* and so on. These pairs of related quantities the Algebraist distinguishes by means of the signs $+$ and $-$. Thus if the things to be distinguished are gain and loss, he may denote by 4 or $+4$ a *gain,* and then he will denote by -4 a *loss* of the same extent. In this way we can conceive the possibility of the independent existence of negative quantities. The signs $+$ and $-$, therefore, are used to indicate the *nature* of quantities as *positive* or *negative,* as well as to indicate *addition* and *subtraction* (Arts. 8 and 9).

In Arithmetic we are concerned only with the numbers which begin at 0 and are represented by the symbols 0, 1,

2, 3, 4, etc. without limit, and intermediate fractions. But the quantities which we usually measure by numbers in Algebra do not really begin at any point, but extend in opposite directions without limit. In order therefore to measure such quantities on a uniform system, the symbols of Algebra are considered as increasing from 0 in two opposite directions; i.e., besides the symbols used in Arithmetic, we consider another set -1, -2, -3, -4, etc. without limit, and intermediate fractions. Symbols in one direction are preceded by the sign $+$, and are called *positive;* and those in the other direction are preceded by the sign $-$, and are called *negative.* Symbols without a sign prefixed are considered to have $+$ prefixed.

These two sets of symbols may be illustrated as follows :

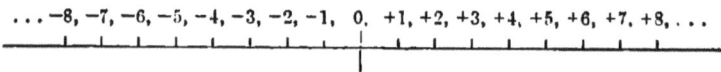

$$\ldots -8, -7, -6, -5, -4, -3, -2, -1, \quad 0. \quad +1, +2, +3, +4, +5, +6, +7, +8, \ldots$$

the positive being those in the right direction from zero, and the negative those in the left direction from the same point.

Thus, if 4 represent a distance of 4 miles measured to the *right* of a certain point, -4 will represent a distance of 4 miles measured to the *left* of the same point. If $+4$ represent 4 degrees *above* zero, -4 will represent 4 degrees *below* zero. If $+4$ represent 4 years *after* Christ, -4 will represent 4 years *before* Christ. If $+4$ represent a *fall* of four feet, -4 will represent a *rise* of 4 feet. If $+4$ represent a *gain* of $4, -4 will represent a *loss* of $4. In general, when we have to consider quantities the exact reverse of each other in their nature or quality, we may regard the quantities of either quality as positive, and those of the opposite quality as negative. It matters not which quality we take as the positive one so long as we take the opposite one as negative; but having assumed at the commencement of an investigation a certain quality as positive, the important point is to use it uniformly and consistently throughout.

The *absolute value* of any quantity is the number represented by this quantity taken independently of the sign which precedes the number. Thus, 3 and -3 have the same absolute value.

Negative quantities are often spoken of as *less than zero*. For example, if a man's debts exceed his assets by $4, it is said that "he is worth $4 less than nothing." In the language of Algebra it would be said "he is worth −$4."

A negative number is said to be *Algebraically greater* than another when it is *numerically less*, or *when it has the smaller absolute value.* Thus $-3 > -6$, since -3 is only 3 less than 0 while -6 is 6 less than 0, or as a person who owes $3 is better off than one who owes $6; or in the case of the thermometer, when the mercury is at 10° below 0 (marked $-10°$) at one hour, and at $-5°$ at another hour, the temperature is said to be increasing; i.e., $-5° > -10°$. Also, in Algebra, *zero is greater than any negative quantity*, as a man who has no property or debt is considered better off than one who is in debt. Thus it is easy to see that in the series on page 16 each number is greater by unity than the one immediately to the left of it.

21. Additions and Multiplications may be Made in any Order. — (1) When a number of terms are connected by the signs + and −, the value of the result is the same in whatever order the terms are taken; thus $6 + 5$ and $5 + 6$ give the same result viz., 11; and so also $a + b$ and $b + a$ give the same result, viz., the sum of the numbers which are represented by a and b. We may express this fact Algebraically thus, $a + b = b + a$. Similarly $a - b + c = a + c - b$, for in the first of the two expressions b is taken from a, and c added to the result; in the second c is added to a, and b taken from the result.

Similar reasoning applies to all Algebraic expressions. Hence we may write the terms of an expression in any order we please, provided each has its proper sign.

Thus it appears that $a - b$ may be written in the equivalent form $- b + a$. As an illustration we may suppose that a represents a gain of a pounds. and $-b$ a loss of b pounds; it is clearly immaterial whether the gain precedes the loss, or the loss precedes the gain

(2) When one number, whether integral or fractional, is multiplied by a second, the result is the same as when the second is multiplied by the first.

The proof for whole numbers is as follows: Write down a rows of units, putting b units in each row, thus:

$$| \; | \; | \; | \; | \ldots \ldots b \text{ in a row,}$$
$$| \; | \; | \; | \; | \ldots \ldots$$
$$| \; | \; | \; | \; | \ldots \ldots a \text{ rows.}$$

Then counting by rows there will be b units in a row repeated a times, i.e., $b \times a$ units. Counting by columns there will be a units in a column repeated b times, i.e., $a \times b$ units.

$$\therefore \; ba = ab.$$

These two laws are together called the *Commutative Law*, or *Law of Commutation*.

22. Suggestions for the Student in Solving Examples. — In solving examples the student should clearly explain how each step follows from the one before it; for this purpose short verbal explanations are often necessary.

The sign "$=$" should never be used except to connect quantities which are equal. Beginners should be particularly careful not to employ the sign of equality in any vague and inexact sense. The signs of equality, in the several steps of the work, should be placed one under the other, unless the expressions are very short.

In elementary work too much importance cannot be attached to *neatness of style and arrangement*. The beginner should remember that *neatness* is in itself conducive to *accuracy*.

EXAMPLES.

Find the numerical value of the following expressions, when $a = 1$, $b = 2$, $c = 3$, $d = 4$, and $e = 5$.

1. $a^2 + b^2 + c^2 + d^2 + e^2$.   *Ans.* 55.
2. $abc^2 + bcd^2 - dea^2$.   94.
3. $c^4 + 6c^2b^2 + b^4 - 4c^3b - 4cb^3$.   81.

4. $\dfrac{b^2c^2}{4a} + \dfrac{de}{b^2} - \dfrac{32}{b^4}.$ *Ans.* 12.

5. $\dfrac{a^4 + 4a^3b + 6a^2b^2 + 4ab^3 + b^4}{a^3 + 3a^2b + 3ab^2 + b^3}.$ 3

6. $(a+b)(b+c) - (b+c)(c+d) + (c+d)(d+e).$ 43

7. $(a - 2b + 3c)^2 - (b - 2c + 3d)^2 + (c - 2d + 3e)^2.$ 72

8. $\sqrt{(4c^2 + 5d^2 + e)}.$ 11.

9. $\sqrt{(e^2 + d^2 + c^2 - a^2)}.$ 7.

If $a = 8, b = 6, c = 1, x = 9, y = 4,$ find the value of

10. $\sqrt[3]{\left(\dfrac{6cy^4}{x^2}\right)} + 2\sqrt{\left(\dfrac{3a^3}{4b^3}\right)}.$ *Ans.* $5\frac{1}{3}$.

11. Find the difference between abx and $a + b + x$, when $a = 5, b = 7,$ and $x = 12.$ *Ans.* 396.

12. When $a = 3$, find the difference between a^2 and $2a$, a^3 and $3a$, a^4 and $4a$, a^5 and $5a$, a^6 and $6a$.

Ans. 3, 18, 69, 228, and 711.

13. Find the value of $3\sqrt{c} + 2a\sqrt{(2a + b - x)}$, when $a = 6, b = 5, c = 4, x = 1.$ *Ans.* 54.

14. Find the value of $(9 - y)(x + 1) + (x + 5)(y + 7)$ $- 112$, when $x = 3$ and $y = 5.$ *Ans.* 0.

Find the value of

15. $x^4 - 11x^3 - 11x^2 - 13x + 11$ for $x = 12.$ $-1.$

16. $x^4 - x^3 - 4x^2 - 3x - 5$ for $x = 3.$ 4.

17. $x^5 - 3x^2 - 8$ for $x = 4.$ 968.

18. $3x^4 - 60x^3 + 54x^2 + 60x + 58$ for $x = 19.$ 115.

Express the following in Algebraic symbols:

19. Seven times a, plus the third power of b. $7a + b^3.$

20. Six times the cube of a multiplied by the square of b, diminished by the square of c multiplied by the fourth power of d. *Ans.* $6a^3b^2 - c^2d^4.$

21. 3 into x minus m times y, divided by m minus n.

Ans. $(3x - my) \div (m - n).$

22. Four times the fourth power of a, diminished by six times the cube of a into the cube of b, and increased by four times the fourth power of b. *Ans.* $4a^4 - 6a^3b^3 + 4b^4.$

CHAPTER II.

ADDITION.

23. Addition — Algebraic Sum. — Addition in Algebra is the process of finding the *Algebraic sum* of several quantities. The Algebraic sum of several quantities is their aggregate value, and it is usual to find the simplest equivalent expression for it.

It is convenient to make *three cases* in Addition ; (1) when the terms to be added are *like* (Art. 18), and have *like signs;* (2) when they are *like* but have *unlike signs;* and (3) when they are *unlike.*

24. Case 1. To Add Terms which are Like and have Like Signs. — Let it be required to add $8x^2y$, $4x^2y$, and $7x^2y$.

Here $8x^2y$ is x^2y taken 8 times, $4x^2y$ is x^2y taken 4 times, and $7x^2y$ is x^2y taken 7 times; therefore x^2y is taken in all $8 + 4 + 7 = 19$ times, and hence the sum is $19x^2y$.

The truth of this will be evident to the beginner when he remembers that the three quantities 8 lbs., 4 lbs., and 7 lbs., added together, give 19 lbs.

Similarly $12ab + 3ab + 5ab + ab = 21ab$.

Let it be required to add $-3ab$, $-7ab$, and $-9ab$.

Here $-3ab$ is ab taken -3 times, $-7ab$ is ab taken -7 times, and $-9ab$ is ab taken -9 times ; therefore ab is taken in all -19 times, and hence the sum is $-19ab$.

The truth of this will be evident from the consideration that, if a sum of money be diminished successively by $3, $7, and $9, it is diminished altogether by $19.

Therefore, *to add like terms which have the same sign, add the numerical coefficients, prefix the common sign, and annex the common symbols.*

For example, $6a + 3a + a + 7a = 17a$, and $-2ab - 7ab$
$-9ab = -18ab$.

25. Case 2. To Add Terms which are Like, but have Unlike Signs. — Let it be required to add $9a$ and $-4a$.

Here $-4a$ destroys 4 of the 9 time , a, and gives when added to it, $5a$. This is usually expressed by saying $-4a$ *will cancel* $+4a$ *in the term* $9a$, and leave $+5a$ for the aggregate or sum of the two terms.

For if $9a$ denote \$9 which a man has in his possession, and $-4a$ denote a debt of \$4, then the aggregate value of his money is \$5.

In like manner if it be required to add $8a$, $-9a$, $-a$, $3a$, $4a$, $-11a$, a, we find the sum of the positive terms to be $16a$, and the sum of the negative terms to be $-21a$; now $+16a$ will cancel $-16a$ in the term $-21a$, which leaves $-5a$ for the aggregate or sum of the terms.

Therefore, *to add like terms which have not all the same sign, add all the positive numerical coefficients into one sum, and all the negative numerical coefficients into another; take the difference of these two sums, prefix the sign of the greater, and annex the common symbols.*

For example $7a - 3a + 11a + a - 5a - 2a = 19a - 10a = 9a$, and $5ab - 6ab + 2ab - 7ab - 3ab + 4ab = 11ab - 16ab = -5ab$.

We need not, however, strictly adhere to this rule, for since terms may be added or subtracted in any order (Art. 21), we may choose the order we find most convenient.

Thus, in the last example, we may say $5ab$ added to $-6ab$ gives $-ab$; adding $-ab$ to $+2ab$ gives $+ab$; adding $+ab$ to $-7ab$ gives $-6ab$; adding $-6ab$ to $-3ab$ gives $-9ab$; adding $-9ab$ to $+4ab$ gives $-5ab$, for the sum, which is the same as was found by the rule.

26. Case 3. To Add Terms which are not all Like Terms. — Let it be required to add $4a + 5b - 7c + 3d$, $3a - b + 2c + 5d$, $9a - 2b - c - d$, and $-a + 3b + 4c - 3d + e$.

It is convenient to arrange the terms in columns, so that like terms shall stand in the same column ; and then add each column, beginning with that on the left, as follows :

$$4a + 5b - 7c + 3d$$
$$3a -\ b + 2c + 5d$$
$$9a - 2b -\ c -\ d$$
$$-a + 3b + 4c - 3d + e$$
$$\overline{15a + 5b - 2c + 4d + e}$$

Here the terms $4a$, $3a$, $9a$, and $-a$ are all like terms ; the sum of the positive terms is $16a$; there is one negative term, viz., $-a$, so that the sum of the terms in the first column by Art. 25 is $+15a$; the sign $+$ may be omitted by Art. 9. Similarly $5b - b - 2b + 3b = 5b$, $-7c + 2c - c + 4c = -2c$, and so on ; there being no term similar to e, it is connected to the other terms by its proper sign.

Therefore, *to add terms which are not all like terms, add together the terms which are like terms, by the rule in Case 2, and set down the other terms each preceded by its proper sign.*

In the two following examples the terms are arranged suitably in columns.

$$x^3 + 2x^2 -\ 3x + 1 \qquad a^2 +\ ab +\ b^2 - c$$
$$4x^3 + 7x^2 +\ x - 9 \qquad 3a^2 - 3ab - 7b^2$$
$$-2x^3 +\ x^2 -\ 9x + 8 \qquad 4a^2 + 5ab + 9b^2$$
$$-3x^3 -\ x^2 + 10x - 1 \qquad a^2 - 3ab - 3b^2$$
$$\overline{\qquad 9x^2 -\ x - 1 \qquad 9a^2 \qquad\qquad - c}$$

In the first example we have in the first column $x^3 + 4x^3 - 2x^3 - 3x^3 = 5x^3 - 5x^3 = 0$; this is usually expressed by saying *the terms which involve x^3 cancel each other.*

Similarly, in the second example, the terms which involve ab cancel each other ; and also those which involve b^2 cancel each other.

27. Remarks on Addition. — We have seen that when two or more *like* terms are to be added together they may be collected into a single term (Arts. 24 and 25). If, however, the terms are *unlike* they cannot be collected. Thus we write the sum of a and b in the form $a + b$, and the sum of a and $-b$ in the form $a - b$.

From the foregoing examples it will be observed that in Algebra the word *sum* is used in a wider sense than in Arithmetic. Thus, in the language of Arithmetic, $a - b$ signifies that b is to be subtracted from a, and has no other meaning; but in Algebra it also means the sum of the two quantities a and $-b$ without any regard to the relative magnitudes of a and b.

When quantities are connected by the signs $+$ and $-$, the resulting expression is called their *Algebraic sum*. Thus the Algebraic sum of $12a - 29a + 14a$ is $-3a$.

In Algebra, wherever the word *sum* is used without an adjective, the *Algebraic sum* is understood.

EXAMPLES.

1.	2.	3.	4.
$4ax$	$6ab$	$2bx^2$	$2a^2b$
$5ax$	$-7ab$	$-3bx^2$	$-a^2b$
$-3ax$	$-3ab$	$-9bx^2$	$11a^2b$
$2ax$	$5ab$	$7bx^2$	$-5a^2b$
$-7ax$	$-9ab$	$2bx^2$	$4a^2b$
ax	$2ab$	$-4bx^2$	$-9a^2b$
$2ax$	$-6ab$	$-5bx^2$	$2a^2b$

5.

$$7x^2 - 3xy \qquad + x$$
$$3x^2 \qquad - y^2 + 3x - y$$
$$-2x^2 + 4xy + 5y^2 - x - 2y$$
$$\qquad -7xy - y^2 + 9x - 5y$$
$$4x^2 \qquad + 4y^2 - 2x$$
$$\overline{12x^2 - 6xy + 7y^2 + 10x - 8y}$$

Add together the following expressions:

6. $a+2b-3c$, $-3a+b+2c$, $2a-3b+c$.   *Ans.* 0.

7. $3a+2b-c$, $-a+3b+2c$, $2a-b+3c$. $4a+4b+4c$.

8. $-3x+2y+z,\ x-3y+2z,\ 2x+y-3z.$ *Ans.* 0.

9. $-x+2y+3z,\ 3x-y+2z,\ 2x+3y-z.$ $4x+4y+4z.$

10. $4a+3b+5c,\ -2a+3b-8c,\ a-b+c.$ $3a+5b-2c.$

11. $-15a-19b-18c,\ 14a+14b+8c,\ a+5b+9c.$ $-c.$

12. $25a-15b+c,\ 13a-10b+4c,\ a+20b-c.$ $39a-5b+4c.$

13. $-16a-10b+5c,\ 10a+5b+c,\ 6a+5b-c.$ $5c.$

In adding together several expressions containing terms with different powers of the same letter, it will be found convenient to arrange all the expressions in *ascending* or *descending* powers of that letter (Art. 19).

14. $3x^3+7-5x^2,\ 2x^2-8-9x,\ 4x-2x^3+3x^2.$

Arranging the terms in the *descending* powers of x, we have

$$
\begin{array}{rrr}
3x^3 & -5x^2 & +7 \\
& 2x^2-9x & -8 \\
-2x^3 & +3x^2+4x & \\
\hline
x^3 & & -5x-1 \ \ \textit{Ans.}
\end{array}
$$

15. $3ab^2-2b^3+a^3,\ 5a^2b-ab^2-3a^3,\ 8a^3+5b^3,\ 9a^2b-2a^3 +ab^2.$ *Ans.* $3b^3+3ab^2+14a^2b+4a^3.$

It will be observed that this answer is arranged according to *descending* powers of b, and *ascending* powers of a.

16. $2x^2-2xy+3y^2,\ 4y^2+5xy-2x^2,\ x^2-2xy-6y^2.$
                     Ans. $x^2+xy+y^2.$

17. $a^3-a^2+3a,\ 3a^3+4a^2+8a,\ 5a^3-6a^2-11a.$
                     Ans. $9a^3-3a^2.$

18. $x^3+3x^2y+3xy^2,\ -3x^2y-6xy^2-x^3,\ 3x^2y+4xy^2.$
                     Ans. $3x^2y+xy^2.$

19. $x^3-2ax^2+a^2x+a^3,\ x^3+3ax^2,\ 2a^3-ax^2-2x^3.$
                     Ans. $a^2x+3a^3.$

20. $2ab-3ax^2+2a^2x,\ 12ab+10ax^2-6a^2x,\ -8ab+ax^3 -5a^2x.$ *Ans.* $6ab-9a^2x+7ax^2+ax^3.$

21. $x^2+y^4+z^3,\ -4x^2-5z^3,\ 8x^2-7y^4+10z^3,\ 6y^4-6z^3.$
                     Ans. $5x^2.$

22. $x^4-4x^3y+6x^2y^2-4xy^3+y^4,\ 4x^3y-12x^2y^2+12xy^3-4y^4,$ $6x^2y^2-12xy^3+6y^4,\ 4xy^3-4y^4,\ y^4.$ *Ans.* $x^4.$

CHAPTER III.

SUBTRACTION.

28. Subtraction — Algebraic Difference. — Subtraction in Algebra is the process of finding the difference between two Algebraic quantities.

The *Algebraic Difference* of two quantities is the number of units which must be added to one in order to produce the other. Thus, what is the difference between 2 and 6 means " how many units added to 2 will make 6 " ? The *Difference* is sometimes called the *Remainder.*

The *Subtrahend* is the quantity to be subtracted; or it is the one *from* which we measure. Thus, 2 is the subtrahend in the above example.

The *Minuend* is the quantity from which the subtrahend is taken ; or it is the one *to* which we measure. Thus, 6 is the minuend in the above example.

If the minuend is Algebraically greater than the subtrahend, the difference is positive (Art. 20).

If the minuend is Algebraically less than the subtrahend, the difference is negative.

In Arithmetic we cannot subtract a greater number from a less one, because subtraction in Arithmetic means taking a *less* number from a *greater.* But in Algebra there is no such restriction, because Algebraic subtraction means *finding a difference.*

29. Rule for Algebraic Subtraction. — Let distances to the right of the zero point be called positive, and those to the left of the same point be called negative (Fig. 1, Art. 20). Also call measuring toward the right from *any* point positive, and measuring toward the left from any point negative.

Then the difference between 2 and 6 means either *how many units must we measure, and in what direction,* in order to pass from 2 to 6 or to pass from 6 to 2. In the first case we begin at 2 and measure four units to the right and say 2 from 6 is +4. In the second case we begin at 6 and measure four units to the left and say 6 from 2 is −4. That is, if we subtract 2 from 6 the difference is 4 ; but if we subtract 6 from 2 the difference is −4.

Also to find the difference between −1 and +1, we may begin at −1 and measure 2 units to the right and get +2, or we may begin at +1 and measure 2 units to the left and get −2 ; i.e., if we subtract −1 from +1 the difference is +2, but if we subtract +1 from −1 the difference is −2.

Similarly the difference between −2 and −7 is −5 or +5, according as we measure from −2 toward the left to −7 or from −7 toward the right to −2 ; i.e., if we subtract −2 from −7 the remainder is −5, but if we subtract −7 from −2 the remainder is +5. And also, the difference between −6 and +7 is +13 or −13 according as we measure from −6 to +7 or from +7 to −6 ; i.e., if we subtract −6 from +7 the difference is 13, but if we subtract 7 from −6 the difference is −13.

Hence we see that the remainder in each case is found by changing the Algebraic sign of the subtrahend, and then adding it Algebraically to the minuend.

Otherwise thus. Suppose we have to take 9 + 5 from 16 ; the result is the same as if we first take 9 from 16, and then take 5 from the remainder ; that is, the result is denoted by 16 − 9 − 5.

Thus $$16 - (9 + 5) = 16 - 9 - 5.$$

Here we enclose 9 + 5 in parenthesis in the first expression, because we are to take the *whole* of 9 + 5 from 16 (Art. 15).

Suppose we have to take 9 − 5 from 16. If we take 9 from 16, we obtain 16 − 9 ; but we have thus taken too much from 16, for we had to take, not 9, but 9 diminished by 5. Hence we must increase the result by 5 ; and thus we obtain $16 - (9 - 5) = 16 - 9 + 5.$

Similarly, $16 - (6 + 4 - 1) = 16 - 6 - 4 + 1.$

In like manner suppose we have to subtract $b - c$ from a. If we subtract b from a, we obtain $a - b$; but we have thus taken too much from a, for we are required to take, not b, but b diminished by c. Hence we must increase the result $a - b$ by c; and thus we obtain $a - (b - c) = a - b + c$ for the true remainder.

Similarly, $a - (b + c - d) = a - b - c + d.$

Suppose we have to subtract $b - c + d - e$ from a. This is the same thing as subtracting $b + d - c - e$ from a (Art. 21). If we subtract $b + d$ from a, we obtain $a - b - d$; but we have thus taken too much from a, for we were to take, not $b + d$, but $b + d$ diminished by c and e. Hence we must increase the result by $c + e$, and thus obtain

$$a - (b - c + d - e) = a - b - d + c + e = a - b + c - d + e.$$

From considering each of these examples, it is evident that *subtracting a positive number is the same thing as adding an equal negative number*, and also that *subtracting a negative number is the same thing as adding an equal positive number*. Therefore, Algebraic subtraction is equivalent to the Algebraic addition of a number with the opposite Algebraic sign. Hence for subtraction we have the following

RULE.

Change the signs of all the terms in the subtrahend, and then add the result to the minuend.

EXAMPLES.

1. Let it be required to subtract $3x - y + z$ from $4x - 3y + 2z$.

Changing the signs of all the terms in the subtrahend, it stands as follows: $-3x + y - z$. Then collecting as in addition, we have

$$4x - 3y + 2z - 3x + y - z = x - 2y + z.$$

2. From $3x^4 + 5x^3 - 6x^2 - 7x + 5$

take $2x^4 - 2x^3 + 5x^2 - 6x - 7$.

Changing the signs of all the terms in the subtrahend, and proceeding as in addition, we have

$$3x^4 + 5x^3 - 6x^2 - 7x + 5$$
$$-2x^4 + 2x^3 - 5x^2 + 6x + 7$$
$$\overline{x^4 + 7x^3 - 11x^2 - x + 12}$$

REM. — The beginner may solve a few examples by *actually* changing the signs of the subtrahend and going through the operation as fully as we have done in these two examples; but he may gradually accustom himself to perform the subtraction without *actually* changing the signs, but merely changing them *mentally*, as in the following example.

3. From $8ab + 7ac + 2c^2$ take $5ab - 4ac + 3c^2 - d$.

Writing the subtrahend under the minuend so that similar terms shall fall in the same column, for convenience (Art. 26), we have

$$8ab + 7ac + 2c^2$$
$$5ab - 4ac + 3c^2 - d$$
$$\overline{3ab + 11ac - c^2 + d}$$

Changing the sign of $5ab$ from $+$ to $-$ and adding it to $8ab$, we have $3ab$; in like manner, changing the sign of $-4ac$ from $-$ to $+$ and adding it to $7ac$, we have $11ac$; also changing the sign of $+3c^2$ from $+$ to $-$ and adding it to $2c^2$, we have $-c^2$; changing the sign of $-d$ and adding it, we have $+d$.

Every example in subtraction may be verified by adding the remainder to the subtrahend; the sum will be equal to the minuend.

30. Remarks on Addition and Subtraction. — In Arithmetic addition always produces *increase* and subtraction *decrease;* but in Algebra addition may produce *decrease* and subtraction may produce *increase.* Thus in Algebra we may add $-4a$ to $8a$ and obtain the Algebraic sum $4a$, which is smaller than $8a$; or we may subtract $-3a$ from $5a$ and obtain the Algebraic difference $8a$, which is larger than $5a$.

EXAMPLES.

1. From $\quad\quad 5x^2 + xy - 3y^2$
Subtract $\quad\quad 2x^2 + 8xy - 7y^2$

Remainder $\quad\quad 3x^2 - 7xy + 4y^2$

2. From $\quad x^4 - 2x^3 \quad\quad\quad - 9x + 4$
Subtract $\quad 2x^4 \quad\quad\quad - 3x^2 + 7x - 8$

Remainder $\quad - x^4 - 2x^3 + 3x^2 - 16x + 12$

From

3. $15x + 10y - 18z$ subtract $2x - 8y + z$.
$\quad\quad\quad\quad\quad\quad\quad$ *Ans.* $13x + 18y - 19z$.

4. $x - y - z$ subtract $-10x - 14y + 15z$.
$\quad\quad\quad\quad\quad\quad\quad$ *Ans.* $11x + 13y - 16z$.

5. $25a - 16b - 18c$ take $4a - 3b + 15c$.
$\quad\quad\quad\quad\quad\quad\quad$ *Ans.* $21a - 13b - 33c$.

6. $yz - zx + xy$ take $-xy + yz - zx$. $\quad\quad 2xy$.

7. $-2x^3 - x^2 - 3x + 2$ take $x^3 - x + 1$.
$\quad\quad\quad\quad\quad\quad\quad$ *Ans.* $-3x^3 - x^2 - 2x + 1$.

8. $4x^2 - 3x + 2$ take $-5x^2 + 6x - 7$. $\quad 9x^2 - 9x + 9$.

9. $x^3 + 11x^2 + 4$ take $8x^2 - 5x - 3$. $\quad x^3 + 3x^2 + 5x + 7$.

10. $-8a^2x^2 + 5x^2 + 15$ take $9a^2x^2 - 8x^2 - 5$.
$\quad\quad\quad\quad\quad\quad\quad$ *Ans.* $-17a^2x^2 + 13x^2 + 20$.

11. $\frac{1}{2}x^2 - \frac{1}{3}xy - \frac{3}{2}y^2$ take $-\frac{3}{2}x^2 + xy - y^2$.
$\quad\quad\quad\quad\quad\quad\quad$ *Ans.* $2x^2 - \frac{4}{3}xy - \frac{1}{2}y^2$.

12. $\frac{2}{3}a^2 - \frac{5}{6}a - 1$ take $-\frac{2}{3}a^2 + a - \frac{1}{2}$. $\quad \frac{4}{3}a^2 - \frac{7}{6}a - \frac{1}{2}$.

13. $\frac{1}{3}x^2 - \frac{1}{3}x + \frac{1}{6}$ take $\frac{1}{3}x - 1 + \frac{1}{2}x^2$. $\quad -\frac{1}{6}x^2 - \frac{5}{6}x + \frac{7}{6}$.

14. $\frac{3}{8}x^2 - \frac{2}{3}ax$ take $\frac{1}{3} - \frac{1}{4}x^2 - \frac{5}{6}ax$. $\quad \frac{5}{8}x^2 + \frac{1}{6}ax - \frac{1}{3}$.

15. $\frac{3}{4}x^3 - \frac{1}{3}xy^2 - y^2$ take $\frac{1}{2}x^2y - \frac{5}{6}y^2 - \frac{1}{2}xy^2$.
$\quad\quad\quad\quad\quad\quad\quad$ *Ans.* $\frac{3}{4}x^3 - \frac{1}{2}x^2y - \frac{1}{6}y^2$.

31. The Use of Parentheses.*—A *parenthesis* indicates that the terms enclosed within it are to be considered as one quantity (Art. 15). On account of the extensive use which

---

* As the bracket, brace, bar, and vinculum all have the same significance as the parenthesis (Art. 15), the rules for their removal or introduction are the same.

is made of parentheses in Algebra, it is necessary that the student should become acquainted with the rules for their removal or introduction.

32. Plus Sign before the Parenthesis. — *When a parenthesis is preceded by the sign* +, *the parenthesis can be removed without making any change in the expression within the parenthesis.*

This rule has already been illustrated in Arts. 25 and 26; it is in fact the *rule for addition.*

7 + (12 + 4) means that 12 and 4 are to be added and their sum added to 7. It is clear that 12 and 4 may be added separately or together without altering the result.

Thus $7 + (12 + 4) = 7 + 12 + 4 = 23.$

Also $a + (b + c)$ means that b and c are to be added together and their sum added to a.

Thus $a + (b + c) = a + b + c.$

7 + (12 − 4) means that to 7 we are to add the excess of 12 over 4; now if we add 12 to 7, we have added 4 too much, and must therefore take 4 from the result.

Thus $7 + (12 - 4) = 7 + 12 - 4 = 15.$

Similarly $a + (b − c)$ means that to a we are to add b diminished by c.

Thus $a + (b - c) = a + b - c.$

Therefore Conversely : *Any part of an expression may be enclosed within a parenthesis and the sign* + *placed before it, the sign of every term within the parenthesis remaining unaltered.*

Thus, the expression $a − b + c − d + e$ may be written in any of the following ways :

$$a-b+c+(-d+e),\ a-b+(c-d+e),\ a+(-b+c-d+e),$$

and so on.

33. Minus Sign before the Parenthesis. — *When a parenthesis is preceded by the sign* −, *the parenthesis may be removed if the sign of every term within the parenthesis be changed.*

This rule has already been illustrated in Art. 29 ; it is in fact the *rule for subtraction*. The rule is evident, because the sign — before a parenthesis shows that the whole expression within the parenthesis is to be subtracted, and the subtraction is effected by changing the signs of all the terms of the expression to be subtracted.

Thus $\qquad a - (b + c) = a - b - c.$

Also $\qquad a - (b - c) = a - b + c.$

Therefore Conversely : *Any part of an expression may be enclosed within a parenthesis and the sign — placed before it, provided the sign of every term within the parenthesis be changed.* The proof of this operation is to clear the parenthesis introduced, and thus obtain the original expression.

Thus $a - b + c + d - e$ may be written in the following ways :

$$a-b+c-(-d+e),\ a-b-(-c-d+e),\ a-(b-c-d+e),$$

and so on.

34. Compound Parentheses. — Expressions may occur with more than one pair of parentheses ; these parentheses may be removed in succession by the preceding rules.

We may either begin with the outside parenthesis and go inward, or begin with the inside parenthesis and go outward. It is usually best to begin with the inside parenthesis. The beginner is recommended always to remove *first the inside pair*, next the inside of all that remain, and so on. Thus for example ;

$$a + \{b + (c - d)\} = a + \{b + c - d\} = a + b + c - d$$
$$a + \{b - (c - d)\} = a + \{b - c + d\} = a + b - c + d.$$
$$a - \{b + (c - d)\} = a - \{b + c - d\} = a - b - c + d.$$
$$a - \{b - (c - d)\} = a - \{b - c + d\} = a - b + c - d.$$

It will be seen in these examples that, to prevent confusion between different pairs of parentheses, we employ those of different *forms;* and hence we use, besides the parenthesis, the brace, the bracket, and sometimes the vinculum (Art. 15).

Thus, for example,

$$a - [b - \{c - (d - \overline{e - f})\}] = a - [b - \{c - (d - e + f)\}]$$
$$= a - [b - \{c - d + e - f\}] = a - [b - c + d - e + f]$$
$$= a - b + c - d + e - f.$$

Also

$$\iota - 2b - [4a - 6b - \{3a - c + (5a - 2b - \overline{3a - c + 2b})\}]$$
$$= a - 2b - [4a - 6b - \{3a - c + (5a - 2b - 3a + c - 2b)\}]$$
$$= a - 2b - [4a - 6b - \{3a - c + 5a - 2b - 3a + c - 2b\}]$$
$$= a - 2b - [4a - 6b - 3a + c - 5a + 2b + 3a - c + 2b]$$
$$= a - 2b - 4a + 6b + 3a - c + 5a - 2b - 3a + c - 2b$$
$$= 2a, \text{ by collecting like terms.}$$

EXAMPLES.

Simplify the following expressions by removing the parentheses and collecting like terms.

1. $a - (b - c) + a + (b - c) + b - (a + c)$. *Ans.* $a + b - c$.
2. $a - [b + \{a - (b + a)\}]$. a.
3. $a - [2a - \{3b - (4c - 2a)\}]$. $a + 3b - 4c$.
4. $\{a - (b - c)\} + \{b - (c - a)\} - \{c - (a - b)\}$. $3a - b - c$.
5. $2a - (5b + [3c - a]) - (5a - [b + c])$. $-2a - 4b - 2c$.
6. $-[a - \{b - (c - a)\}] - [b - \{c - (a - b)\}]$. $b - a$.
7. $-(-(-(-x))) - (-(-y))$. $x - y$.
8. $-[5x - (11y - 3x)] - [5y - (3x - 6y)]$. $-5x$.
9. $-[15x - \{14y - (15z + 12y) - (10x - 15z)\}]$.
                                        Ans. $-25x + 2y$.
10. $8x - \{16y - [3x - (12y - x) - 8y] + x\}$. $11x - 36y$.
11. $-[x - \{z + (x - z) - (z - x) - z\} - x]$. $2x - 2z$.
12. $-[a + \{a - (a - x) - (a + x) - a\} - a]$. $2a$.
13. $-[a - \{a + (x - a) - (x - a) - a\} - 2a]$. a.
14. $2a - [2a - \{2a - (2a - \overline{2a - a})\}]$. a.
15. $16 - x - [7x - \{8x - (9x - \overline{3x - 6x})\}]$. $16 - 12x$.
16. $2x - [3y - \{4x - (5y - \overline{6x - 7y})\}]$. $12x - 15y$.
17. $2a - [3b + (2b - c) - 4c + \{2a - (3b - \overline{c - 2b})\}]$. $4c$.
18. $a - [5b - \{a - (5c - 2c - b - 4b) + 2a - (a - 2b + c)\}]$.
                                        Ans. $3a - 2c$.

19. $x^4 - [4x^3 - \{6x^2 - (4x-1)\}] - (x^4 + 4x^3 + 6x^2 + 4x + 1)$.

Ans. $-8x^3 - 8x$.

When the beginner has had a little practice the number of steps may be considerably diminished; he may begin at the outside and remove two or more parentheses at once, as follows:

20. $a - [2b + \{3c - 3a - (a + b)\} + 2a - (b + 3c)]$

$= a - 2b - 3c + 3a + a + b - 2a + b + 3c$

$= 3a$.

21. $a - (b - c) - [a - b - c - 2\{b + c - 3(c - a) - d\}]$.

Ans. $6a + 2b - 2c - 2d$.

22. $2x - (3y - 4z) - \{2x - (3y + 4z)\} - \{3y - (4z + 2x)\}$.

Ans. $2x - 3y + 12z$.

23. $-20(a - d) + 3(b - c) - 2[b + c + d - 3\{c + d$

$- 4(d - a)\}]$. *Ans.* $4a + b + c$.

24. $-4(a + d) + 24(b - c) - 2[c + d + a - 3\{d + a$

$- 4(b + c)\}]$. *Ans.* $-50c$.

25. $2(3b - 5a) - 7[a - 6\{2 - 5(a - b)\}]$.

Ans. $-227a + 216b + 84$.

26. $-10\{a - 6[a - (b - c)]\} + 60\{b - (c + a)\}$. $-10a$.

27. $-3\{-2[-4(-a)]\} + 5\{-2[-2(-a)]\}$. $4a$.

28. $-2\{-[-(x - y)]\} + \{-2[-(x - y)]\}$. 0.

NOTE. — The line between the numerator and denominator of a fraction is a kind of vinculum. Thus $\dfrac{x - 3}{2}$ is equivalent to $\frac{1}{2}(x - 3)$.

29. $\dfrac{1}{4}\{a - 5(b - a)\} - \dfrac{3}{2}\left\{\dfrac{1}{3}\left(b - \dfrac{a}{3}\right) - \dfrac{2}{9}\left[a - \dfrac{3}{4}\left(b - \dfrac{4a}{5}\right)\right]\right\}$.

Ans. $\tfrac{11}{5}a - 2b$.

30. $35\left[\dfrac{3x - 4y}{5} - \dfrac{1}{10}\left\{3x - \dfrac{5}{7}(7x - 4y)\right\}\right] + 8(y - 2x)$.

Ans. $12x - 30y$.

31. $\dfrac{3}{8}\left\{\dfrac{4}{3}(a - b) - 8(b - c)\right\} - \left\{\dfrac{b - c}{2} - \dfrac{c - a}{3}\right\}$

$- \dfrac{1}{2}\left\{c - a - \dfrac{2}{3}(a - b)\right\}$. *Ans.* $a - \tfrac{13}{3}b + \tfrac{10}{3}c$.

32. $\frac{1}{2}x - \frac{1}{2}\left(\frac{2}{3}y - \frac{1}{2}z\right) - \left[x - \left\{\frac{1}{2}x - \left(\frac{1}{3}y - \frac{1}{4}z\right)\right\} - \left(\frac{2}{3}y - \frac{1}{2}z\right)\right].$

Ans. 0.

The terms of an expression can be placed in parentheses in various ways (Arts. 32 and 33). Thus,

33. $ax - bx + cx - ay + by - cy$

may be written

$$(ax - bx) + (cx - ay) + (by - cy),$$

or $\qquad (ax - bx + cx) - (ay - by + cy),$

or $\qquad (ax - ay) - (bx - by) + (cx - cy).$

Whenever a factor is common to every term within a parenthesis, it may be placed outside of the parenthesis as a multiplier of the expression within. Thus,

34. $\qquad ax^3 + 7 - cx - dx^2 - c + bx - dx^3 + bx^2 - 2x$

$= (ax^3 - dx^3) + (bx^2 - dx^2) + (bx - cx - 2x) + (7 - c)$

$= (a - d)x^3 + (b - d)x^2 + (b - c - 2)x + (7 - c).$

In this result, $(a - d)$, $(b - d)$, $(b - c - 2)$ are regarded as the coefficients of x^3, x^2, and x, respectively (Art. 17). Hence we have here placed together in parentheses the coefficients of the different powers of x so as to have the sign $+$ before each parenthesis.

35. $\qquad -a^2x - 7a + a^2y + 3 - 2x - ab$

$= -(a^2x - a^2y) - (7a + ab) - (2x - 3)$

$= -(x - y)a^2 - (7 + b)a - (2x - 3).$

We have here placed together in parentheses the coefficients of the different powers of a so as to have the sign $-$ before each parenthesis.

In the following four examples place together in parentheses the coefficients of the different powers of x so that the sign $+$ will be before all the parentheses.

36. $ax^4 + bx^2 + 5 + 2bx - 5x^2 + 2x^4 - 3x.$

Ans. $(a + 2)x^4 + (b - 5)x^2 + (2b - 3)x + 5.$

37. $3bx^2 - 7 - 2x + ab + 5ax^3 + cx - 4x^2 - bx^3.$

Ans. $(5a - b)x^3 + (3b - 4)x^2 + (c - 2)x + ab - 7.$

38. $2 - 7x^3 + 5ax^2 - 2cx + 9ax^3 + 7x - 3x^2.$

Ans. $(9a - 7)x^3 + (5a - 3)x^2 + (7 - 2c)x + 2.$

39. $2cx^5 - 3abx + 4dx - 3bx^4 - a^2x^5 + x^4$.

Ans. $(2c - a^2)x^5 + (1 - 3b)x^4 + (4d - 3ab)x$.

In the following four examples place together in parentheses the coefficients of the different powers of x so that the sign $-$ will be before all the parentheses.

40. $ax^2 + 5x^3 - a^2x^4 - 2bx^3 - 3x^2 - bx^4$.

Ans. $-(a^2 + b)x^4 - (2b - 5)x^3 - (3 - a)x^2$.

41. $7x^3 - 3c^2x - abx^5 + 5ax + 7x^5 - abcx^3$.

Ans. $-(ab - 7)x^5 - (abc - 7)x^3 - (3c^2 - 5a)x$.

42. $ax^2 + a^2x^3 - bx^2 - 5x^2 - cx^3$.

Ans. $-(c - a^2)x^3 - (b + 5 - a)x^2$.

43. $3b^2x^4 - bx - ax^4 - cx^4 - 5c^2x - 7x^4$.

Ans. $-(a + c + 7 - 3b^2)x^4 - (b + 5c^2)x$.

Simplify the following expressions, and in each result place together in parentheses the coefficients of the different powers of x. This is known as *re-grouping the terms according to the powers of* x.

44. $ax^3 - 2cx - [bx^2 - \{cx - dx - (bx^3 + 3cx^2)\} - (cx^2 - bx)]$.

Ans. $(a - b)x^3 - (b + 2c)x^2 - (b + c + d)x$.

45. $ax^2 - 3\{-ax^3 + 3bx - 4[\frac{1}{6}cx^3 - \frac{2}{3}(ax - bx^2)]\}$.

Ans. $(3a + 2c)x^3 + (a + 8b)x^2 - (8a + 9b)x$.

46. $x^5 - 4bx^4 - \frac{1}{6}\left[12ax - 4\left\{3bx^4 - 9\left(\frac{cx}{2} - bx^5\right) - \frac{3}{2}ax^4\right\}\right]$.

Ans. $(6b + 1)x^5 - (a + 2b)x^4 - (2a + 3c)x$.

We shall close this chapter with a few examples in Addition and Subtraction.

47. To the sum of $2a - 3b - 2c$ and $2b - a + 7c$ add the sum of $a - 4c + 7b$ and $c - 6b$. *Ans.* $2a + 2c$.

48. Add the sum of $2y - 3y^2$ and $1 - 5y^3$ to the remainder left when $1 - 2y^2 + y$ is subtracted from $5y^3$. *Ans.* $-y^2 + y$.

49. Take $x^2 - y^2$ from $3xy - 4y^2$, and add the remainder to the sum of $4xy - x^2 - 3y^2$ and $2x^2 + 6y^2$. *Ans.* $7xy$.

50. Add together $3x^2 - 7x + 5$ and $2x^3 + 5x - 3$, and diminish the result by $3x^2 + 2$. *Ans.* $2x^3 - 2x$.

51. What expression must be subtracted from $3a - 5b + c$ so as to leave $2a - 4b + c$? *Ans.* $a - b$.

CHAPTER IV.

MULTIPLICATION.

35. Multiplication in Algebra is the process of taking any given quantity as many times as there are units in any given number.*

The *Multiplicand* is the quantity to be taken or multiplied.

The *Multiplier* is the number by which it is multiplied.

The *Product* is the result of the operation.

The multiplicand and multiplier taken together are called *Factors* of the product.

In Algebra as in Arithmetic, the product of any number of factors is the same in whatever order the factors may be taken (Art. 21). Thus, $2 \times 3 \times 5 = 2 \times 5 \times 3 = 3 \times 5 \times 2$, and so on. In like manner $abc = acb = bca$, and so on.

Also
$$2a \times 3b = 2 \times a \times 3 \times b$$
$$= 2 \times 3 \times a \times b$$
$$= 6ab.$$

36. Rule of Signs. — The rule of signs, and especially the use of the negative multiplier, usually presents some difficulty to the beginner.

(1) If $+a$ is to be multiplied by $+c$, this indicates that $+a$ is to be taken as many times as there are units in c. Now if $+a$ be taken once, the result is $+a$; if it be taken twice, the result is evidently $+2a$; if taken three times, the result is $+3a$; and so on. Therefore if $+a$ be taken c times, it is $+ca$ or $+ac$. That is
$$+a \times +c = +ac.$$

(2) If $-a$ is to be multiplied by $+c$, this indicates that $-a$ is to be taken as many times as there are units in c.

----

* This definition is true only of whole numbers.

Now if $-a$ be taken once, the result is $-a$; if it be taken twice, the result is $-2a$; if taken three times. the result is $-3a$; and so on. Therefore if $-a$ be taken c times, it is $-ca$ or $-ac$. That is,

$$-a \times +c = -ac.$$

Similarly $\quad -3 \times +4 = -3$ taken four times
$$= -3 -3 -3 -3$$
$$= -12.$$

(3) Suppose $+a$ is to be multiplied by $-c$. We have illustrated the difference between $+c$ and $-c$ (Art. 20), by supposing that $+c$ represents a line of c units measured in one direction, and $-c$ a line of c units measured in the *opposite* direction. Hence if $+a$ is to be multiplied by $-c$, this indicates that $+a$ is to be taken as many times as there are units in $+c$, and further that the direction of the line which represents the product is to be reversed.

Now $+a$ taken $+c$ times gives $+ac$; and changing the sign, which corresponds to a reversal of direction, we get $-ac$. That is

$$+a \times -c = -ac.$$

Similarly $+3 \times -4$ indicates that 3 is to be taken 4 times, and the sign changed. The first operation gives $+12$, and the second -12. That is

$$+3 \times -4 = -12.$$

(4) If $-a$ is to be multiplied by $-c$, this indicates that $-a$ is to be taken as many times as there are units in c, and then that the direction of the line which represents the product is to be reversed.

Now $-a$ taken c times gives $-ac$; and changing the sign, which corresponds to a reversal of direction, we get $+ac$. That is

$$-a \times -c = +ac.$$

Similarly -3×-4 indicates that -3 is to be taken 4 times. and the sign changed. The first operation gives -12, and the second $+12$. That is,

$$-3 \times -4 = +12.$$

(3) is sometimes expressed as follows: $+a$ multiplied by $-c$ indicates that $+a$ is to be taken as many times as there are units in c, and then the result subtracted. Now $+a$ taken $+c$ times gives $+ac$, and changing the sign, in order to subtract (Art. 29), we get $-ac$.

Similarly (4) indicates that $-a$ is to be taken as many times as there are units in c, and the result subtracted. Now $-a$ taken $+c$ times gives $-ac$, and changing the sign, in order to subtract, we get $+ac$.

Hence we have the following *Rule of Signs: The product of two terms with like signs is $+$; the product of two terms with unlike signs is $-$.*

To familiarize the beginner with the rule of signs we add a few examples in substitution, where some of the symbols denote negative quantities.

EXAMPLES.

If $a = -2$, $b = 3$, $c = -1$, $x = -5$, $y = 4$, find the value of the following:

1. $3a^2b = 3(-2)^2 \times 3 = 3 \times 4 \times 3 = 36.$
2. $-7a^3bc = -7(-2)^3 \times 3 \times (-1) = -7 \times -8 \times 3 \times -1 = -168.$

3. $8abc^2$.　　*Ans.* $-48.$
4. $6a^2c^2$.　　　$24.$
5. $-2a^4bx$.　　$480.$
6. $5c^2x^2$.　　　$500.$
7. $-7c^4xy$.　　$140.$
8. $-8ax^3$.　　$-2000.$
9. $-5a^2b^2c^2$.　$-180.$
10. $-7a^3c^3$.　　$-56.$

11. $8c^4x^3$.　*Ans.* $-1000.$
12. $3c^3x^3$.　　$375.$
13. $4c^5x^3$.　　$500.$
14. $7a^5c^4$.　　$-224.$
15. $-4a^2c^4$.　$-16.$
16. $-b^2c^2$.　　$-9.$
17. $2a^2c^3x$.　　$40.$

If $a = -4$, $b = -3$, $c = -1$, $f = 0$, $x = 4$, $y = 1$, find the value of

18. $3a^2 + bx - 4cy$.　　*Ans.* $40.$
19. $fa^2 - 2b^3 - cx^8$.　　$118.$
20. $3a^2y^3 -- 5b^2x - 2c^3$.　　$-130.$

21. $2a^3 - 3b^3 + 7cy^4.$ *Ans.* $-54.$

22. $3b^2y^4 - 4b^2f - 6c^4x.$ 3.

23. $2\sqrt{(ac)} - 3\sqrt{(xy)} + \sqrt{(b^2c^4)}.$ 1.

It is convenient to make *three cases* in Multiplication, (1) the multiplication of *monomials*, (2) the multiplication of a *polynomial by a monomial*, and (3) the multiplication of *polynomials*.

37. The Multiplication of Monomials. — Since by definition (Art. 12) we have

$$a^4 = aaaa,$$

and $$a^6 = aaaaaa,$$

$$\therefore \quad a^4 \times a^6 = aaaaaaaaaa$$
$$= a^{10}.$$

Also $$3a^2 = 3aa,$$
$$7a^3 = 7aaa.$$
$$\therefore \quad 3a^2 \times 7a^3 = 3 \times 7 \times aaaaa$$
$$= 21a^5.$$

Similarly $5a^3b^2 \times 6a^2b^4x^2 = 5aaabb \times 6aabbbbxx$
$$= 30a^5b^6x^2.$$

Also $4a^3c^2 \times 3c^3x^2 \times 3x^2 = 4aaacc \times 3cccxx \times 3xx$
$$= 36a^3c^5x^4.$$

Hence for the multiplication of monomials we have the following

RULE.

Multiply together the numerical coefficients, annex to the result all the letters, and give to each letter an exponent equal to the sum of its exponents in the factors.

For example $2x^2 \times 3x^4 \times x^6 = 6x^{2+4+6} = 6x^{12}.$

Also $5a^2b^3 \times 2b^2c^4 \times 3c^2d^4 = 30a^2b^5c^6d^4.$

NOTE. — The beginner must be careful, in applying this rule, to observe that the exponents of one letter cannot combine in any way with those of another. Thus, the expression $4a^2b^3c^4$ admits of no further simplification.

The product of three or more expressions is called *the continued product.*

38. To Multiply a Polynomial by a Monomial.
— Suppose we have to multiply $(a + b)$ by 3 ; that is, take $a + b$ 3 times. We have

$$3(a + b) = (a + b) + (a + b) + (a + b)$$
$$= (a + a + a \text{ taken three times})$$
together with $\qquad (b + b + b \text{ taken three times})$
$$= 3a + 3b.$$

Similarly $7(a + b) = 7a + 7b.$

$$m(a + b) = (a + a + a + \ldots \text{ taken } m \text{ times})$$
together with $\qquad (b + b + b + \ldots \text{ taken } m \text{ times})$
$$= ma + mb. \ldots \ldots \ldots \ldots (1)$$

Also $\qquad m(a - b) = (a + a + a + \ldots \text{ taken } m \text{ times})$
together with $\qquad (-b - b - b - \ldots \text{ taken } m \text{ times})$
$$= ma - mb. \ldots \ldots \ldots \ldots (2)$$

Similarly

$$m(a - b + c) = ma - mb + mc.$$

This is generally called the *Distributive Law.*

Hence, to multiply a polynomial by a monomial, we have the following

RULE.

Multiply each term of the polynomial separately by the monomial, and collect the results to form the complete product.

For example,

$$4(x^2 + 2xy - 4z) = 4x^2 + 8xy - 16z.$$
$$(4x^2 - 7y - 8z^3) \times 3xy^2 = 12x^3y^2 - 21xy^3 - 24xy^2z^3.$$
$$(\tfrac{2}{3}a^2 - \tfrac{1}{6}ab - b^2) \times 6a^2b^2 = 4a^4b^2 - a^3b^3 - 6a^2b^4.$$

EXAMPLES.

Multiply together

1. $4a^2b^3$ and $7a^5$.		*Ans.* $28a^7b^3$.
2. $3a^4b^7x^3$ and $5a^3bx$.		$15a^7b^8x^4$.
3. $2c^2yz^3$ and x^5y^7z.		$2x^7y^8z^4$.
4. $ab + bc$ and a^3b.		$a^4b^2 + a^3b^2c$.
5. $5x + 3y$ and $2x^2$.		$10x^3 + 6x^2y$.
6. $bc + ca - ab$ and abc.		$ab^2c^2 + a^2bc^2 - a^2b^2c$.
7. $5x^2y + xy^2 - 7x^2y^2$ and $8x^2y^3$.		$40x^4y^4 + 8x^3y^5 - 56x^4y^5$

8. $6a^3bc - 7ab^2c^2$ and a^2b^2.   *Ans.* $6a^5b^3c - 7a^3b^4c^2$.
9. $a^4b^3x - 5b^6$ and $2a^3x^5$. $2a^7b^3x^6 - 10a^3b^6x^5$.
10. $8a^2b^3 - \frac{2}{3}b^2c^3$ and $\frac{3}{2}ab^4$. $12a^3b^7 - ab^6c^3$.

39. The Multiplication of a Polynomial by a Polynomial. — Suppose we have to multiply $c + d$ by $a + b$.

Here we are required to take $c + d$ as many times as there are units in $a + b$, i.e., we are to take $c + d$ as many times as there are units in a, and then add to this product $c + d$ taken as many times as there are units in b.

Hence $(a+b)(c+d) = (c+d)$ taken a times together with $\quad\quad (c+d)$ taken b times
$$= (c+d)a + (c+d)b$$
$$= ac + ad + bc + bd\,[(1)\text{of Art. 38}].\ (1)$$

Again $(a-b)(c+d) = (c+d)$ taken a times diminished by $\quad\quad (c+d)$ taken b times
$$= (c+d)a - (c+d)b$$
$$= ac + ad - (bc + bd)\,[(1)\text{of Art. 38}]$$
$$= ac + ad - bc - bd\ (\text{Art. 33}). \quad . \quad (2)$$

Also $(a+b)(c-d) = (c-d)$ taken a times together with $\quad\quad (c-d)$ taken b times
$$= (c-d)a + (c-d)b$$
$$= ac - ad + bc - bd\,[(2)\text{of Art. 38}].\ (3)$$

Lastly $(a-b)(c-d) = (c-d)$ taken a times diminished by $\quad\quad (c-d)$ taken b times
$$= (c-d)a - (c-d)b$$
$$= ac - ad - (bc - bd)\,[(2)\text{ of Art. 38}]$$
$$= ac - ad - bc + bd\ (\text{Art. 33}). \quad . \quad (4)$$

Hence, to multiply one polynomial by another, we have the following

RULE.

Multiply each term of the multiplicand by each term of the multiplier; if the terms multiplied together have the same sign, prefix the sign + to the product, if unlike, prefix the sign —; then add the partial products to form the complete product.

If we consider each term in the second member of **(4),** and the way it was produced, we find that

$$+a \times +c = +ac.$$
$$+a \times -d = -ad.$$
$$-b \times +c = -bc.$$
$$-b \times -d = +bd.$$

These results enable us again to state the *rule of signs*, and furnish us with another proof of that rule, in addition to the one given in Art. 36. This proof of the rule of signs is perhaps a little more satisfactory than the one given in Art. 36, though it is not quite so simple.

EXAMPLES.

1. Multiply $x + 7$ by $x + 5$.

The product
$$= (x + 7)(x + 5)$$
$$= x^2 + 7x + 5x + 35$$
$$= x^2 + 12x + 35.$$

REM. — It is more convenient to write the multiplier under the multiplicand, and begin on the left and work to the right, placing like terms of the partial products in the same vertical column, as follows:

$$x + 7$$
$$x + 5$$
$$\overline{x^2 + 7x}$$
$$+ 5x + 35$$

by addition $\qquad \overline{x^2 + 12x + 35.}$

2. Multiply $3x - 4y$ by $2x - 3y$.

$$3x - 4y$$
$$2x - 3y$$
$$\overline{6x^2 - 8xy}$$
$$- 9xy + 12y^2$$

by addition $\qquad \overline{6x^2 - 17xy + 12y^2.}$

Here the first line under the multiplier is the product of the multiplicand by $2x$; the second line is the product of the multiplicand by $-3y$; like terms are placed in the same vertical column to facilitate addition.

Find the product of the following:

3. $x - 7$ and $x - 10$.	*Ans.* $x^2 - 17x + 70$.	
4. $x - 7$ and $x + 10$.	$x^2 + 3x - 70$.	
5. $x - 12$ and $x - 1$.	$x^2 - 13x + 12$.	
6. $x - 15$ and $x + 15$.	$x^2 - 225$.	
7. $-x - 2$ and $-x - 3$.	$x^2 + 5x + 6$.	
8. $-x + 5$ and $-x - 5$.	$x^2 - 25$.	
9. $x - 17$ and $x + 18$.	$x^2 + x - 306$.	
10. $-x - 16$ and $-x + 16$.	$x^2 - 256$.	
11. $2x - 3$ and $x + 8$.	$2x^2 + 13x - 24$.	
12. $3x - 5$ and $2x + 7$.	$6x^2 + 11x - 35$.	

13. $4a^2 - 5ab + 6b^2$ and $2a^2 - 3ab + 4b^2$.

$$4a^2 - 5ab + 6b^2$$
$$2a^2 - 3ab + 4b^2$$
$$\overline{}$$
$$8a^4 - 10a^3b + 12a^2b^2$$
$$- 12a^3b + 15a^2b^2 - 18ab^3$$
$$+ 16a^2b^2 - 20ab^3 + 24b^4.$$
$$\overline{}$$
$$8a^4 - 22a^3b + 43a^2b^2 - 38ab^3 + 24b^4.$$

Here the first line under the multiplier is the product of the multiplicand by $2a^2$; the second line is the product by $-3ab$; the third by $4b^2$; like terms are set down in the same vertical column to facilitate the addition.

The student will observe that both the multiplicand and multiplier are *arranged according to the descending powers of a* (Art. 19). Both factors might have been arranged according to the *ascending* powers of a. It is of no consequence which order we adopt, but we should take the *same* order for the multiplicand and multiplier.

If the multiplier and multiplicand are not arranged according to the powers of some common letter, it will be convenient to reàrrange them. Thus:

14. Multiply $3x + 4 + 2x^2$ by $4 + 2x^2 - 3x$. Arranging the factors according to the descending powers of x, the operation is as follows:

$$
\begin{array}{l}
2x^2 + 3x + 4 \\
2x^2 - 3x + 4 \\
\hline
4x^4 + 6x^3 + 8x^2 \\
 - 6x^3 - 9x^2 - 12x \\
 + 8x^2 + 12x + 16 \\
\hline
4x^4 + 7x^2 + 16.
\end{array}
$$

15. Multiply $a^2 + b^2 + c^2 - ab - bc - ca$ by $a + b + c$. Arrange according to descending powers of a.

$$
\begin{array}{l}
a^2 - ab - ac + b^2 - bc + c^2 \\
a + b + c \\
\hline
a^3 - a^2b - a^2c + ab^2 - abc + ac^2 \\
 + a^2b - ab^2 - abc + b^3 - b^2c + bc^2 \\
 + a^2c - abc - ac^2 + b^2c - bc^2 + c^3 \\
\hline
a^3 - 3abc + b^3 + c^3
\end{array}
$$

Rem. — The student should notice that he can make two exercises in multiplication from every example in which the multiplicand and multiplier are different polynomials, by changing the original multiplier into the multiplicand, and the original multiplicand into the multiplier. The result obtained should be the same in both operations. The student can therefore test the correctness of his work by interchanging the multiplicand and multiplier.

Multiply together

16. $a^2 - ab + b^2$ and $a^2 + ab + b^2$.   *Ans.* $a^4 + a^2b^2 + b^4$.

17. $x^2 + 3y^2$ and $x + 4y$. $x^3 + 4x^2y + 3xy^2 + 12y^3$.

18. $x^4 - x^2y^2 + y^4$ and $x^2 + y^2$. $x^6 + y^6$.

19. $a^2 - 2ax + 4x^2$ and $a^2 + 2ax + 4x^2$. $a^4 + 4a^2x^2 + 16x^4$.

20. $16a^2 + 12ab + 9b^2$ and $4a - 3b$. $64a^3 - 27b^3$.

21. $a^2x - ax^2 + x^3 - a^3$ and $x + a$. $x^4 - a^4$.

22. $2x^3 - 3x^2 + 2x$ and $2x^2 + 3x + 2$. $4x^5 - x^3 + 4x$.

23. $-a^5 + a^4b - a^3b^2$ and $-a - b$. $a^6 + a^3b^3$.

24. $a^3 + 2a^2b + 2ab^2$ and $a^2 - 2ab + 2b^2$. $a^5 + 4ab^4$.

Whea the coefficients are fractional we use the ordinary process of multiplication, combining the fractional coefficients by the rules of Arithmetic.*

25. Multiply $\frac{1}{3}a^2 - \frac{1}{2}ab + \frac{2}{3}b^2$ by $\frac{1}{2}a + \frac{1}{3}b$.

$$\frac{1}{3}a^2 - \frac{1}{2}ab + \frac{2}{3}b^2$$
$$\frac{1}{2}a + \frac{1}{3}b$$

$$\overline{\frac{1}{6}a^3 - \frac{1}{4}a^2b + \frac{1}{3}ab^2}$$
$$+ \frac{1}{9}a^2b - \frac{1}{6}ab^2 + \frac{2}{9}b^3$$

$$\overline{\frac{1}{6}a^3 - \frac{5}{36}a^2b + \frac{1}{6}ab^2 + \frac{2}{9}b^3.}$$

Multiply together

26. $\frac{1}{2}a^2 + \frac{1}{3}a + \frac{1}{4}$ and $\frac{1}{2}a - \frac{1}{3}$. *Ans.* $\frac{1}{4}a^3 + \frac{1}{72}a - \frac{1}{12}$.

27. $\frac{2}{3}x^2 + xy + \frac{3}{2}y^2$ and $\frac{1}{3}x - \frac{1}{2}y$. $\frac{2}{9}x^3 - \frac{3}{4}y^3$.

28. $\frac{1}{2}x^2 - \frac{2}{3}x - \frac{3}{4}$ and $\frac{1}{2}x^2 + \frac{2}{3}x - \frac{3}{4}$. $\frac{1}{4}x^4 - \frac{43}{36}x^2 + \frac{9}{16}$.

29. $\frac{2}{3}ax + \frac{2}{3}x^2 + \frac{1}{3}a^2$ and $\frac{3}{4}a^2 + \frac{3}{2}x^2 - \frac{3}{2}ax$. $\frac{1}{4}a^4 + x^4$.

It is sometimes desirable to *indicate* the product of polynomials, by enclosing each of the factors in a parenthesis, and writing them in succession. When the indicated multiplication has been actually performed, the expression is said to be *expanded*, or *developed*.

Expand the following:

30. $(2a + 3b)(a - b)$. *Ans.* $2a^2 + ab - 3b^2$.

31. $(a^2 + ax + x^2)(a^2 - ax + x^2)$. $x^4 + a^2x^2 + a^4$.

32. $(a^2 + 2ab + 2b^2)(a^2 - 2ab + 2b^2)$. $a^4 + 4b^4$.

33. $(x - 3)(x + 4)(x - 5)(x + 6)$.
  Ans. $x^4 + 2x^3 - 41x^2 - 42x + 360$

34. $(a^2 + ab + b^2)(a^3 - a^2b + b^3)(a - b)$.
  Ans. $a^6 - a^5b + a^2b^4 - b^6$.

35. $(2x^3 + 4x^2 + 8x + 16)(3x - 6)$. $6x^4 - 96$.

36. $(x^3 + x^2 + x - 1)(x - 1)$. $x^4 - 2x + 1$.

37. $(x + a)[(x + b)(x + c) - (a + b + c)(x + b) + (a^2 + ab + b^2)]$. *Ans.* $x^3 + a^3$.

---

* The student is supposed to be familiar with Arithmetic fractions, which are the only fractions that are used in this work previous to Chapter VIII.

40. Multiplication by Inspection. — Although the result of multiplying together two binomial factors can always be obtained by the methods explained in Art. 39, yet it is very important that the student should learn to write down the product rapidly *by inspection.* This is done by observing in what way the coefficients of the terms in the product arise ; thus

$$(x + 5)(x + 3) = x^2 + 5x + 3x + 15$$
$$= x^2 + 8x + 15.$$
$$(x - 5)(x + 3) = x^2 - 5x + 3x - 15$$
$$= x^2 - 2x - 15.$$
$$(x + 5)(x - 3) = x^2 + 5x - 3x - 15$$
$$= x^2 + 2x - 15.$$
$$(x - 5)(x - 3) = x^2 - 5x - 3x + 15$$
$$= x^2 - 8x + 15.$$

It will be noticed in each of these results that :

1. The product consists of three terms.

2. The first term is x^2, and the last term is the product of the second terms of the two binomial factors.

3. The middle term has for its coefficient the Algebraic sum of the second terms of the two binomial factors.

Hence the intermediate steps in the work may be omitted, and the product written down at once, as follows :

$$(x + 2)(x + 3) = x^2 + 5x + 6.$$
$$(x - 3)(x + 4) = x^2 + x - 12.$$
$$(x + 6)(x - 9) = x^2 - 3x - 54.$$
$$(x - 4y)(x - 10y) = x^2 - 14xy + 40y^2.$$
$$(x - 5y)(x + 6y) = x^2 + xy - 30y^2.$$

EXAMPLES.

Write down the values of the following products.

1	$(x + 8)(x - 5).$	*Ans.* $x^2 + 3x - 40.$
2.	$(x - 3)(x + 10).$	$x^2 + 7x - 30.$
3.	$(x + 7)(x - 9).$	$x^2 - 2x - 63.$
4.	$(x - 4)(x + 11).$	$x^2 + 7x - 44.$

5. $(x + 2)(x - 5)$.   Ans. $x^2 - 3x - 10$.
6. $(x + 9)(x - 5)$. $x^2 + 4x - 45$.
7. $(x - 8)(x + 4)$. $x^2 - 4x - 32$.
8. $(x - 6)(x + 13)$. $x^2 + 7x - 78$.
9. $(x - 11)(x + 12)$. $x^2 + x - 132$.
10. $(x - 3a)(x + 2a)$. $x^2 - ax - 6a^2$.
11. $(x - 9b)(x + 8b)$. $x^2 - bx - 72b^2$.
12. $(x - 7y)(x - 8y)$. $x^2 - 15xy + 56y^2$.

41. Special Forms of Multiplication — Formulæ.
— There are some examples in multiplication which occur
so often in Algebraic operations that they deserve especial
notice.

If we multiply $a + b$ by $a + b$ we get

$$(a + b)(a + b) = a^2 + 2ab + b^2;$$

that is $$(a + b)^2 = a^2 + 2ab + b^2. \quad . \quad . \quad . \quad (1)$$

Thus *the square of the sum of two numbers is equal to the
sum of the squares of the two numbers increased by twice their
product.*

Similarly, if we multiply $a - b$ by $a - b$ we get

$$(a - b)^2 = a^2 - 2ab + b^2. \quad . \quad . \quad . \quad (2)$$

Thus *the square of the difference of two numbers is equal
to the sum of the squares of the two numbers diminished by
twice their product.*

Also, if we multiply $a + b$ by $a - b$ we get

$$(a + b)(a - b) = a^2 - b^2. \quad . \quad . \quad . \quad (3)$$

Thus *the product of the sum and difference of two numbers
is equal to the difference of their squares.*

Because the product of two negative factors is positive
(Art. 36), it follows that the square of a negative number
is positive. For example,

$$(-a)^2 = a^2 = (+a)^2,$$

and, $$(b - a)^2 = a^2 - 2ab + b^2 = (a - b)^2.$$

Hence $a^2 - 2ab + b^2$ is the square of both $a - b$ and $b - a$.

Rem. 1. — Equations (1), (2), and (3) furnish simple examples of one of the uses of Algebra, which is to *prove general theorems respecting numbers,* and also to *express those theorems briefly.*

For example, the result $(a + b)(a - b) = a^2 - b^2$ is proved to be true, and is expressed thus by symbols more compactly than it could be by words.

A general result thus expressed by symbols is called a *formula;* hence a *formula is an Algebraic expression of a general rule.*

Rem. 2. — We may here indicate the meaning of the sign $\pm$ which is made by combining the signs $+$ and $-$, and which is called the *double sign.*

By using the double sign we may express (1) and (2) in one formula thus :

$$(a \pm b)^2 = a^2 \pm 2ab + b^2, \quad . \quad . \quad . \quad . \quad (4)$$

where $\pm$, read *plus or minus,* indicates that we may take the sign $+$ or $-$, *keeping throughout the upper sign or the lower sign.* Formulæ (1), (2), and (3) are true whatever may be the values of a and b.

The following examples will illustrate the use that can be made of formulæ (1), (2), and (3). The formulæ will sometimes be of use in Arithmetic calculations. Thus

EXAMPLES.

1.　Required the difference of the squares of 127 and 123. By formula (3) we have

$$(127)^2 - (123)^2 = (127 + 123)(127 - 123)$$
$$= 250 \times 4 = 1000.$$

2.　Required the square of 29. By formula (2)

$$(29)^2 = (30 - 1)^2 = 900 - 60 + 1 = 841.$$

3. Required the product of 53 by 47.
By formula (3)

$$53 \times 47 = (50 + 3)(50 - 3) = (50)^2 - (3)^2$$
$$= 2500 - 9 = 2491.$$

4. Required the square of 34.
By formula (1)

$$(34)^2 = (30 + 4)^2 = 900 + 240 + 16 = 1156.$$

5. Required the square of $4x + 3y$.

We can of course obtain the square by multiplying $4x + 3y$ by itself in the ordinary way. But we can obtain it by formula (1) more easily, by putting $4x$ for a and $3y$ for b. Thus

$$(4x + 3y)^2 = (4x)^2 + 2(4x \cdot 3y) + (3y)^2$$
$$= 16x^2 + 24xy + 9y^2.$$

6. Required the square of $x + y + z$.

Denote $x + y$ by a; then $x + y + z = a + z$; and by (1) we have

$$(a + z)^2 = a^2 + 2az + z^2$$
$$= (x + y)^2 + 2(x + y)z + z^2$$
$$= x^2 + 2xy + y^2 + 2xz + 2yz + z^2$$

Thus $(x + y + z)^2 = x^2 + y^2 + z^2 + 2xy + 2yz + 2xz$.

That is, *the square of the sum of three numbers is equal to the sum of the squares of the three numbers increased by twice the products of the three numbers taken two and two.*

7. Required the square of $p - q + r - s$.

Denote $p-q$ by a and $r-s$ by b; then $p-q+r-s=a+b$; and by (1) we have

$$(a+b)^2 = a^2 + 2ab + b^2 = (p-q)^2 + 2(p-q)(r-s) + (r-s)^2.$$

Then by (2) we expand $(p - q)^2$ and $(r - s)^2$.

Thus $(p - q + r - s)^2$
$$= p^2 - 2pq + q^2 + 2(pr - ps - qr + qs) + r^2 - 2rs + s^2$$
$$= p^2 + q^2 + r^2 + s^2 + 2pr + 2qs - 2pq - 2ps - 2qr - 2rs.$$

8. Required the product of $p-q+r-s$ and $p-q-r+s$.

Let $p - q = a$ and $r-s = b$; then $p-q+r-s = a+b$, and $p - q - r + s = a - b$; and by (3) we have

$$(a+b)(a-b) = a^2-b^2 = (p-q)^2 - (r-s)^2$$
$$= p^2-2pq+q^2-(r^2-2rs+s^2) \text{ by (2)}.$$

Thus $(p-q+r-s)(p-q-r+s) = p^2+q^2-r^2-s^2-2pq+2rs$.

From these examples we see that by using formulæ (1), (2), and (3), the process of multiplication may be often simplified. The student is advised first to go through the work fully as we have done; but when he becomes more familiar with this subject, he may dispense with some of the work, and thereby simplify the multiplication still more. Thus in the last example he need not substitute a and b, but apply formula (3) at once, and then (2), as follows:

$$[(p - q) + (r - s)][(p - q) - (r - s)] = (p - q)^2$$
$$- (r - s)^2 = p^2 - 2pq + q^2 - r^2 + 2rs - s^2.$$

9. Required the product of $a + b + c$, $a + b - c$, $a - b + c$, $b + c - a$.

By (3) and (1) we obtain for the product of the first two factors,

$$(a + b + c)(a + b - c) = 2ab + a^2 + b^2 - c^2. \quad (1)$$

By (3) and (2) we obtain for the product of the last two factors,

$$(a - b + c)(b + c - a) = 2ab - (a^2 + b^2 - c^2). \quad (2)$$

Multiplying together (1) and (2), we obtain

$$(2ab)^2 - (a^2 + b^2 - c^2)^2$$
$$= 2a^2b^2 + 2b^2c^2 + 2a^2c^2 - a^4 - b^4 - c^4. \quad (3)$$

Solve the following examples in multiplication by formulæ (1), (2), and (3).

10. $(15x + 14y)^2$.   *Ans.* $225x^2 + 420xy + 196y^2$.

11. $(7x^2 - 5y^2)^2$. $49x^4 - 70x^2y^2 + 25y^4$.

12. $(x^2 + 2x - 2)^2$. $x^4 + 4x^3 - 8x - 4$.

13. $(x^2 - 5x + 7)^2$. $x^4 - 10x^3 + 39x^2 - 70x + 49$.

14. $(2x^2 - 3x - 4)^2$.   *Ans.* $4x^4 - 12x^3 - 7x^2 + 24x + 16$.
15. $(x + 2y + 3z)^2$. $x^2 + 4y^2 + 9z^2 + 4xy + 6xz + 12yz$.
16. $(x^2 + xy + y^2)(x^2 + xy - y^2)$. $x^4 + 2x^3y + x^2y^2 - y^4$.
17. $(x^2 + xy + y^2)(x^2 - xy + y^2)$. $x^4 + x^2y^2 + y^4$.

42. Important Results in Multiplication. — There
are other results in multiplication which are important,
although they are not so much so as the three formulæ in
Art. 41. We place them here in order that the student may
be able to refer to them when they are wanted; they can be
easily verified by actual multiplication.

$$(a + b)(a^2 - ab + b^2) = a^3 + b^3. \quad . \quad . \quad . \quad (1)$$

$$(a - b)(a^2 + ab + b^2) = a^3 - b^3. \quad . \quad . \quad . \quad (2)$$

$$(a+b)^3 = (a+b)(a^2 + 2ab + b^2) = a^3 + 3a^2b + 3ab^2 + b^3. \quad (3)$$

$$(a-b)^3 = (a-b)(a^2 - 2ab + b^2) = a^3 - 3a^2b + 3ab^2 - b^3. \quad (4)$$

$$(a + b + c)^3 = a^3 + 3a^2(b + c) + 3a(b + c)^2 + (b + c)^3$$
$$= a^3 + b^3 + c^3 + 3a^2(b+c) + 3b^2(a+c) + 3c^2(a+b) + 6abc. \quad (5)$$

REM. — It is a useful exercise in multiplication for the student to
show that two expressions agree in giving the same result. For
example, show that
$$(a - b)(b - c)(c - a) = a^2(c - b) + b^2(a - c) + c^2(b - a).$$
Here we proceed as follows: Multiplying $(a - b)$ by $(b - c)$ we
obtain
$$(a - b)(b - c) = ab - b^2 - ac + bc;$$
then multiplying this equation by $c - a$ we obtain
$$(a - b)(b - c)(c - a) = cab - cb^2 - ac^2 + bc^2 - a^2b + ab^2 + a^2c - abc$$
$$= a^2(c - b) + b^2(a - c) + c^2(b - a) \quad . \quad . \quad . \quad (6)$$
Show that $(a - b)^2 + (b - c)^2 + (c - a)^2$
$$= 2(c - b)(c - a) + 2(b - a)(b - c) + 2(a - b)(a - c).$$
By (2) of Art. 41 we obtain
$$(a - b)^2 + (b - c)^2 + (c - a)^2$$
$$= a^2 - 2ab + b^2 + b^2 - 2bc + c^2 + c^2 - 2ac + a^2$$
$$= 2(a^2 + b^2 + c^2 - ab - bc - ca). \quad . \quad . \quad . \quad (7$$
Now $(c - b)(c - a) = c^2 - ca - cb + ab,$
$$(b - a)(b - c) = b^2 - bc - ab + ac,$$
$$(a - b)(a - c) = a^2 - ac - ab + bc;$$

therefore, by adding these three equations, we obtain

$$(c - b)(c - a) + (b - a)(b - c) + (a - b)(a - c)$$
$$= a^2 + b^2 + c^2 - ab - ac - bc; \quad (8)$$

therefore, from (7) and (8) we have

$$(a - b)^2 + (b - c)^2 + (c - a)^2$$
$$= 2(c - b)(c - a) + 2(b - a)(b - c) + 2(a - b)(a - c). \quad (9)$$

43. Results of Multiplying Algebraic Expressions.

— From an examination of the examples in multiplication, the student will recognize the truth of the following laws with respect to the result of multiplying Algebraic expressions.

(1) *In the multiplication of two polynomials, when the partial products do not contain like terms, the whole number of terms in the final product will be equal to the product of the number of terms in the multiplicand by the number of terms in the multiplier, but will be less if the partial products contain like terms, owing to the simplification produced by collecting these like terms.*

Thus as we see in Ex. 17, Art. 39, there are *two* terms in the multiplicand and *two* in the multiplier, and *four* in the product, while in Ex. 13 there are *three* terms in the multiplicand and *three* in the multiplier, and only *five* in the product.

(2) *Among the terms of the product there are always two that are unlike any other terms; these are, that term which is the product of the two terms in the factors which contain the highest power of the same letter, and that term which is the product of the two terms in the factors which contain the lowest power of the same letter.*

Thus in Ex. 13, Art. 39, there are the terms $8a^4$ and $24b^4$, and these are *unlike* any other terms; in fact, the other terms contain a raised to some power *less* than the fourth, and thus they differ from $8a^4$; and they also contain a to some power, and thus they differ from $24b^4$.

(3) *When the multiplicand and multiplier are both homogeneous* (Art. 18) *the product is homogeneous, and the degree*

of the product is the sum of the numbers which express the degrees of the multiplicand and multiplier.

Thus in Ex. 13, Art. 39, the multiplicand and multiplier are each homogeneous and of the second degree, and the product is homogeneous and of the fourth degree. In Ex. 15, Art. 39, the multiplicand is homogeneous and of the *second* degree, and the multiplier is homogeneous and of the *first* degree; the product is homogeneous and of the *third* degree. This law is of great importance, as it serves to test the accuracy of Algebraic work; the student is therefore recommended to pay great attention to the *degree* of the terms in the results which he obtains.

EXAMPLES.

Multiply

1. $4a^2 - 3b$ by $3ab$. *Ans.* $12a^3b - 9ab^2$.
2. $8a^2 - 9ab$ by $3a^2$. $24a^4 - 27a^3b$.
3. $3x^2 - 4y^2 + 5z^2$ by $2x^2y$. $6x^4y - 8x^2y^3 + 10x^2yz^2$.
4. $x^2y^3 - y^3z^4 + z^4x^2$ by $x^2y^2z^2$. $x^4y^5z^2 - x^2y^5z^6 + x^4y^2z^6$.
5. $2xy^2z^3 + 3x^2y^3z - 5x^3yz^2$ by $2xy^2z$.
  Ans. $4x^2y^4z^4 + 6x^3y^5z^2 - 10x^4y^3z^3$.
6. $-2a^2b - 4ab^2$ by $-7a^2b^2$. $14a^4b^3 + 28a^3b^4$.
7. $8xyz - 10x^3yz^3$ by $-xyz$. $-8x^2y^2z^2 + 10x^4y^2z^4$.
8. $abc - a^2bc - ab^2c$ by $-abc$. $-a^2b^2c^2 + a^3b^2c^2 + a^2b^3c^2$.
9. $x + 7$ by $x - 10$. $x^2 - 3x - 70$.
10. $x + 9$ by $x - 7$. $x^2 + 2x - 63$.
11. $2x - 3$ by $x + 8$. $2x^2 + 13x - 24$.
12. $2x + 3$ by $x - 8$. $2x^2 - 13x - 24$.
13. $x^3 - 7x + 5$ by $x^2 - 2x + 3$.
  Ans. $x^5 - 2x^4 - 4x^3 + 19x^2 - 31x + 15$.
14. $a^2 - 5ab - b^2$ by $a^2 + 5ab + b^2$.
  Ans. $a^4 - 25a^2b^2 - 10ab^3 - b^4$.
15. $x^2 - xy + x + y^2 + y + 1$ by $x + y - 1$.
  Ans. $x^3 + 3xy + y^3 - 1$.
16. $a^2 + b^2 + c^2 - bc - ca - ab$ by $a + b + c$.
  Ans. $a^3 + b^3 + c^3 - 3abc$.

17. $2ax+x^2+a^2$ by $a^2+2ax-x^2$. *Ans.* $a^4+4a^3x+4a^2x^2-x^4$.

18. $2b^2+3ab-a^2$ by $7a-5b$. $-10b^3-ab^2+26a^2b-7a^3$.

19. a^2-ab+b^2 by a^2+ab-b^2. $a^4-a^2b^2+2ab^3-b^4$.

20. $4x^2-3xy-y^2$ by $3x-2y$. $12x^3-17x^2y+3xy^2+2y^3$.

21. $x^5-x^4y+xy^4-y^5$ by $x+y$. $x^6-x^4y^2+x^2y^4-y^6$.

22. $x^4+2x^3y+4x^2y^2+8xy^3+16y^4$ by $x-2y$. x^5-32y^5.

23. $9x^2y^2+27x^3y+81x^4+3xy^3+y^4$ by $3x-y$. $243x^5-y^5$.

24. $x+2y-3z$ by $x-2y+3z$. $x^2-4y^2+12yz-9z^2$.

Write down the values of the following products by inspection.

25. $(x+7)(x+1)$.   *Ans.* x^2+8x+7.

26. $(x-7)(x+14)$. $x^2+7x-98$.

27. $(a+3b)(a-2b)$. $a^2+ab-6b^2$.

28. $(a-6)(a+13)$. $a^2+7a-78$.

29. $(2x-5)(x-2)$. $2x^2-9x+10$.

30. $(3x-1)(x+1)$. $3x^2+2x-1$.

31. $(3x+7)(2x-3)$. $6x^2+5x-21$.

Solve the following examples by formulæ (1), (2), (3) in Art. 41.

32. $(x^2+xy+y^2)(x^2-xy-y^2)$.   *Ans.* $x^4-x^2y^2-2xy^3-y^4$.

33. $(x^2+xy-y^2)(x^2-xy+y^2)$. $x^4-x^2y^2+2xy^3-y^4$.

34. $(x^3+2x^2+3x+1)(x^3-2x^2+3x-1)$. $x^6+2x^4+5x^2-1$.

35. $(x-3)^2(x^2+6x+9)$. x^4-18x^2+81.

36. $(x+y)^2(x^2-2xy-y^2)$. $x^4-4x^2y^2-4xy^3-y^4$.

Show that the following results are true:

37. $(a^2+b^2)(c^2+d^2)=(ac+bd)^2+(ad-bc)^2$.

38. $(a+b+c)^2+a^2+b^2+c^2=(a+b)^2+(b+c)^2+(c+a)^2$.

39. $(a-b)(b-c)(c-a)=bc(c-b)+ca(a-c)+ab(b-a)$.

40. $(a-b)^3+b^3-a^3=3ab(b-a)$.

41. $(a^2+ab+b^2)^2-(a^2-ab+b^2)^2=4ab(a^2+b^2)$.

42. $(a+b+c)^3-a^3-b^3-c^3=3(a+b)(b+c)(c+a)$.

43. $(a+b)^2+2(a^2-b^2)+(a-b)^2=4a^2$.

44. $(a-b)^3+(b-c)^3+(c-a)^3=3(a-b)(b-c)(c-a)$.

CHAPTER V.

DIVISION.

44. Division in Algebra is the process of finding, from a given product and one of its factors, the other factor; or it is the process of finding how many times one quantity is contained in another.

Division is therefore the converse of multiplication.

The *Dividend* is the given product; or it is the quantity to be divided.

The *Divisor* is the given factor; or it is the quantity by which we divide.

The *Quotient* is the required factor; or it is the number which shows how many times the divisor is contained in the dividend.

The above definitions may be briefly written

$$\text{quotient} \times \text{divisor} = \text{dividend},$$

or $$\text{dividend} \div \text{divisor} = \text{quotient}.$$

It is sometimes better to express this last result as a fraction; thus $$\frac{\text{dividend}}{\text{divisor}} = \text{quotient}.$$

It is convenient to make *three cases* in Division, (1) the division of one monomial by another, (2) the division of a polynomial by a monomial, (3) the division of one polynomial by another.

45. The Division of one Monomial by Another. — Since the product of 4 and x is $4x$, it follows that when $4x$ is to be divided by x the quotient is 4. Or otherwise

$$4x \div x = 4.$$

Also since the product of a and b is ab, the quotient of ab divided by a is b; that is

$$ab \div a = b.$$

Similarly $abc \div a = bc$; $abc \div b = ac$; $abc \div c = ab$; $abc \div ab = c$; $abc \div bc = a$; $abc \div ca = b$. These results may also be written

$$\frac{abc}{a} = bc; \quad \frac{abc}{b} = ac; \quad \frac{abc}{c} = ab;$$

$$\frac{abc}{ab} = c; \quad \frac{abc}{bc} = a; \quad \frac{abc}{ca} = b.$$

Also $36a^6 \div 9a^4 = \dfrac{36a^6}{9a^4} = \dfrac{36aaaaaa}{9aaaa} = 4aa$, by removing from the divisor and dividend the factors common to both, just as in Arithmetic.

Therefore $\qquad\qquad 36a^6 \div 9a^4 = 4a^2.$

Similarly $\quad 45a^4b^3c^2 \div 9a^2bc^2 = \dfrac{45aaaabbbcc}{9aabcc}$

Hence we have the following $\quad = 5a^2b^2.$

RULE.

To divide one monomial by another, divide the coefficient of the dividend by that of the divisor, and subtract the exponent of any letter in the divisor from the exponent of that letter in the dividend.

For example $72x^5y^3 \div 12x^3y^2 = 6x^{5-3}y^{3-2}$
$$= 6x^2y.$$

Also $\qquad 55a^4x^3y^5 \div 11a^2xy^2 = 5a^2x^2y^3.$

REM. — If the numerical coefficient, or the literal part of the divisor be *not* found in the dividend, we can only indicate the division. Thus if $7a$ is to be divided by $2c$, the quotient can only be indicated by $7a \div 2c$ or by $\dfrac{7a}{2c}$. In some cases, however, we may simplify the expression for the quotient by a principle already used in Arithmetic. Thus if $16a^3b^2$ is to be divided by $12abc$, the quotient is denoted by $\dfrac{16a^3b^2}{12abc}$. Here the dividend $= 4ab \times 4a^2b$ and the divisor $= 4ab \times 3c$; thus the factor $4ab$, which occurs in both dividend and divisor, may be removed in the same way as in Arithmetic, and the quotient will be denoted by $\dfrac{4a^2b}{3c}$. That is

$$\frac{16a^3b^2}{12abc} = \frac{4ab \times 4a^2b}{4ab \times 3c} = \frac{4a^2b}{3c},$$

by removing the common factor $4ab$.

N$_{\text{OTE}}$. If we apply the above rule to divide any power of a letter by the same power of the letter, we are led to a curious conclusion.

Thus by the rule

$$a^3 \div a^3 = a^{3-3} = a^0;$$

but also
$$a^3 \div a^3 = \frac{a^3}{a^3} = 1,$$

by removing the common factor a^3;

$$\therefore a^0 = 1;$$

that is, *any quantity whose exponent is 0 is equal to 1.*

The true significance of this result will be explained in Art. 115.

46. The Rule of Signs for division may be obtained from an examination of the cases which occur in multiplication, since the product of the divisor and quotient must be equal to the dividend.

Thus we have

$$+a \times (+b) = +ab, \quad \therefore +ab \div +b = +a.$$
$$-a \times (+b) = -ab, \quad \therefore -ab \div +b = -a.$$
$$+a \times (-b) = -ab, \quad \therefore -ab \div -b = +a.$$
$$-a \times (-b) = +ab, \quad \therefore +ab \div -b = -a.$$

Hence in division as well as in multiplication, *like signs produce* +, *and unlike signs produce* −.

EXAMPLES.

Divide

1. $8a^2b$ by $4ab$.   *Ans.* $2a$.
2. $-15xy$ by $3x$. $-5y$.
3. $-21a^2b^3$ by $-7a^2b^2$. $3b$.
4. $45a^6b^2x^4$ by $-9a^3bx^2$. $-5a^3bx^2$.
5. $-36a^4b^5c^3$ by $-24a^3b^3c^3$. $\frac{3}{2}ab^2$.
6. $3x^3$ by x^2. $3x$.
7. $-58a^4b^3c^2$ by $-2a^2b$. $29a^2b^2c^2$.
8. $63a^4x^5y^2$ by $-7a^4y^2$. $-9x^5$.
9. $4x^{m+n}$ by $2x^n$. $2x^m$.
10. $-77x^{m+2}$ by $11x^m$. $-7x^2$.

47. To Divide a Polynomial by a Monomial.

Since $\qquad (a - b)c = ac - bc$;

therefore $\qquad \dfrac{ac - bc}{c} = a - b.$

Also since $(a - b) \times (-c) = -ac + bc$;

therefore $\qquad \dfrac{-ac + bc}{-c} = a - b.$

Also since $(a - b + c)ab = a^2b - ab^2 + abc$;

therefore $\quad \dfrac{a^2b - ab^2 + abc}{ab} = a - b + c.$

Hence we have the following rule :

To divide a polynomial by a monomial, divide each term of the dividend separately by the divisor.

For example

$$(8a^3 - 6a^2b + 2a^2c) \div 2a^2 = 4a - 3b + c.$$
$$(9x - 12y + 3z) \div -3 = -3x + 4y - z.$$
$$(36a^3b^2 - 24a^2b^5 - 20a^4b^2) \div 4a^2b = 9ab - 6b^4 - 5a^2b.$$
$$(2x^2 - 5xy - \tfrac{3}{2}x^2y^3) \div -\tfrac{1}{2}x = -4x + 10y + 3xy^3.$$

EXAMPLES.

Divide

1. $-35x^6$ by $7x^3$. $\qquad\qquad\qquad$ *Ans.* $-5x^3.$
2. x^3y^3 by x^2y. $\qquad\qquad\qquad\qquad\qquad$ $xy^2.$
3. $4a^2b^2c^3$ by ab^2c^2. $\qquad\qquad\qquad\qquad\qquad$ $4ac.$
4. $x^2 - 2xy$ by x. $\qquad\qquad\qquad\qquad\qquad$ $x - 2y.$
5. $x^6 - 7x^5 + 4x^4$ by x^2. $\qquad\qquad$ $x^4 - 7x^3 + 4x^2.$
6. $10x^7 - 8x^6 + 3x^4$ by x^3. $\qquad$ $10x^4 - 8x^3 + 3x.$
7. $15x^5 - 25x^4$ by $-5x^3$. $\qquad\qquad$ $-3x^2 + 5x.$
8. $27x^6 - 36x^5$ by $9x^5$. $\qquad\qquad\qquad$ $3x - 4.$
9. $-24x^6 - 32x^4$ by $-8x^3$. $\qquad\qquad$ $3x^3 + 4x.$
10. $a^2 - ab - ac$ by $-a$. $\qquad\qquad$ $-a + b + c.$

48. To Divide one Polynomial by Another. — Let

it be required to divide $a^3 + 2a^2 - 3a$ by $a^2 + 3a.$

Here we are to find a quantity which when multiplied by the divisor will produce the dividend. Hence the dividend is composed of all the partial products arising from the multiplication of the divisor by each term of the quotient (Art. 39). Arranging both the dividend and divisor according to descending powers of a, we see that the first term a^3 of the dividend is the product of the first term a^2 of the divisor by the first term of the quotient (Art. 43); therefore, dividing a^3 by a^2 we obtain a for the first term of the quotient. Multiplying the whole divisor by a we obtain $a^3 + 3a^2$ for the partial product of the divisor by the first term of the quotient; subtracting this product from the dividend we obtain the first remainder $-a^2 - 3a$, which is the product of the divisor by the remaining terms of the quotient, and consequently the first term $-a^2$ of this product is the product of the first term of the divisor by the second term of the quotient. Dividing therefore this first term $-a^3$ by the first term of the divisor a^2 we obtain -1 for the second term of the quotient. Multiplying the whole divisor by -1 we obtain $-a^2 - 3a$ for the product of the divisor by the second term of the quotient; subtracting this product there is no remainder. As all the terms in the dividend have been brought down, the operation is completed. Hence $a - 1$ is the exact quotient.

The work may be arranged as follows:

$$\begin{array}{ccc} \text{Divisor.} & \text{Dividend.} & \text{Quotient.} \end{array}$$

$$+\,3a)a^3 + 2a^2 - 3a(a - 1$$
$$\underline{a^3 + 3a^2}$$
$$-\ a^2 - 3a$$
$$\underline{-\ a^2 - 3a}$$

It will be observed that in getting each term of the quotient in this example, we divide that term of the dividend containing the highest power of a by the term of the divisor containing the highest power of the same letter, and therefore, when the dividend and divisor are arranged according to descending powers of a, any term of the quotient is found by dividing the first term of the divisor

into the first term of the dividend, or into the first term of
one of the remainders.

Hence for the division of one polynomial by another, we
have the following

RULE.

*Arrange both dividend and divisor according to ascending
or descending powers of some common letter.*

*Divide the first term of the dividend by the first term of
the divisor, and write the result for the first term of the
quotient; multiply the whole divisor by this term, subtract
the product from the dividend, and to the remainder join as
many terms from the dividend, taken in order, as are required.*

*Divide the first term of the remainder by the first term of
the divisor, and write the result for the second term of the
quotient; multiply the whole divisor by this term, and subtract
the product from the last remainder.*

*Continue this operation until the remainder becomes zero,
or until the first term of the remainder will not contain the
first term of the divisor.*

This method of dividing is similar to long division in
Arithmetic, i.e., we break up the dividend into parts, and
find how often the divisor is contained in each part; and then
the sum of these partial quotients is the complete quotient.
Thus, in the example just solved, $a^3 + 2a^2 - 3a$ is divided
by the above process into two parts, viz., $a^3 + 3a^2$, and
$-a^2 - 3a$, and each of these is divided by $a^2 + 3a$, giving
for the partial quotients a and -1; thus we obtain the
complete quotient $a - 1$.

NOTE. — The divisor is often put on the *right* of the dividend and
the quotient beneath the divisor as follows:

$$
\begin{array}{ll}
\text{Dividend.} & \text{Divisor.} \\
a^2 + 2ab + b^2 \,\big|\, a + b & \\
\underline{a^2 + \ \ ab} & \ \ a + b \ \text{Quotient.} \\
\ \ \ \ \ ab + b^2 & \\
\ \ \ \ \ \underline{ab + b^2} &
\end{array}
$$

It is of great importance to arrange both dividend and divisor according to ascending or descending powers of some common letter; and to attend to this order in every part of the operation.

EXAMPLES.

1. Divide $24x^2 - 65xy + 21y^2$ by $8x - 3y$.

The operation is conveniently arranged as follows:

$$8x - 3y)24x^2 - 65xy + 21y^2(3x - 7y.$$
$$24x^2 - 9xy$$
$$\overline{\qquad\qquad}$$
$$-56xy + 21y^2$$
$$-56xy + 21y^2$$

Divide

2. $x^2 + 3x + 2$ by $x + 1$.		*Ans.* $x + 2$.
3. $x^2 - 7x + 12$ by $x - 3$.		$x - 4$.
4. $x^2 - 11x + 30$ by $x - 5$.		$x - 6$.
5. $x^2 - 49x + 600$ by $x - 25$.		$x - 24$.
6. $3x^2 + 10x + 3$ by $x + 3$.		$3x + 1$.
7. $2x^2 + 11x + 5$ by $2x + 1$.		$x + 5$.
8. $5x^2 + 11x + 2$ by $x + 2$.		$5x + 1$.
9. $2x^2 + 17x + 21$ by $2x + 3$.		$x + 7$.
10. $5x^2 + 16x + 3$ by $x + 3$.		$5x + 1$.

11. Divide $3a^4 - 10a^3b + 22a^2b^2 - 22ab^3 + 15b^4$ by $a^2 - 2ab + 3b^2$.

The operation is written as follows.

$$a^2 - 2ab + 3b^2)3a^4 - 10a^3b + 22a^2b^2 - 22ab^3 + 15b^4(3a^2 - 4ab + 5b^2$$
$$3a^4 - 6a^3b + 9a^2b^2$$
$$\overline{\qquad\qquad}$$
$$- 4a^3b + 13a^2b^2 - 22ab^3$$
$$- 4a^3b + 8a^2b^2 - 12ab^3$$
$$\overline{\qquad\qquad}$$
$$5a^2b^2 - 10ab^3 + 15b^4$$
$$5a^2b^2 - 10ab^3 + 15b^4$$

12. Divide $x^7 - 5x^5 + 7x^3 + 2x^2 - 6x - 2$ by $1 + 2x - 3x^2 + x^4$.

Arrange both dividend and divisor according to descending powers of x, and arrange the work as follows:

$$x^7 - 5x^5 + 7x^3 + 2x^2 - 6x - 2 \; \big|\; x^4 - 3x^2 + 2x + 1$$
$$\underline{x^7 - 3x^5 + 2x^4 + x^3} \overline{x^3 - 2x - 2}$$
$$\underline{- 2x^5 - 2x^4 + 6x^3 + 2x^2 - 6x}$$
$$- 2x^5 + 6x^3 - 4x^2 - 2x$$
$$\underline{}$$
$$- 2x^4 + 6x^2 - 4x - 2$$
$$- 2x^4 + 6x^2 - 4x - 2$$

We might have arranged the dividend and divisor according to *ascending* powers of x as follows:

$$-2 - 6x + 2x^2 + 7x^3 - 5x^5 + x^7 \;\big|\; 1 + 2x - 3x^2 + x^4$$
$$\underline{-2 - 4x + 6x^2 - 2x^4} \overline{-2 - 2x + x^3}$$
$$- 2x - 4x^2 + 7x^3 + 2x^4 - 5x^5$$
$$- 2x - 4x^2 + 6x^3 - 2x^5$$
$$\underline{}$$
$$x^3 + 2x^4 - 3x^5 + x^7$$
$$x^3 + 2x^4 - 3x^5 + x^7$$

We thus obtain the same quotient we had before, though the terms are in a different order.

13. Divide $a^3 + b^3 + c^3 - 3abc$ by $a + b + c$.

Arrange the dividend according to descending powers of a

$$a^3 - 3abc + b^3 + c^3 \;\big|\; a + b + c$$
$$\underline{a^3 + a^2b + a^2c} \overline{a^2 - ab - ac + b^2 - bc + c^2}$$
$$- a^2b - a^2c - 3abc$$
$$- a^2b - ab^2 - abc$$
$$\underline{}$$
$$- a^2c + ab^2 - 2abc$$
$$- a^2c - abc - ac^2$$
$$\underline{}$$
$$ab^2 - abc + ac^2 + b^3$$
$$ab^2 + b^3 + b^2c$$
$$\underline{}$$
$$- abc + ac^2 - b^2c$$
$$- abc - b^2c - bc^2$$
$$\underline{}$$
$$ac^2 + bc^2 + c^3$$
$$ac^2 + bc^2 + c^3$$

In this example we arrange the terms according to descending powers of a; then when there are two terms, such as a^2b and a^2c, which involve the same power of a, we select a new letter, as b, and put the term which contains b before the term which does not; and again of the terms ab^2 and abc, we put the former first as involving the higher power of b.

14. Divide $x^4 + 4a^4$ by $x^2 + 2ax + 2a^2$.

$$
\begin{array}{r|l}
x^4 + 4a^4 & \underline{x^2 + 2ax + 2a^2} \\
x^4 + 2ax^3 + 2a^2x^2 & x^2 - 2ax + 2a^2 \\
\hline
-2ax^3 - 2a^2x^2 & \\
-2ax^3 - 4a^2x^2 - 4a^3x & \\
\hline
\quad 2a^2x^2 + 4a^3x + 4a^4 & \\
\quad 2a^2x^2 + 4a^3x + 4a^4 & \\
\hline
\end{array}
$$

Divide

15. $x^5 - 5x^4 + 9x^3 - 6x^2 - x + 2$ by $x^2 - 3x + 2$.

$\qquad\qquad$ Ans. $x^3 - 2x^2 + x + 1$.

16. $x^5 - 2x^4 - 4x^3 + 19x^2 - 31x + 15$ by $x^3 - 7x + 5$.

$\qquad\qquad$ Ans. $x^2 - 2x + 3$.

17. $a^3 + b^3 + 3abc - c^3$ by $a + b - c$.

$\qquad\qquad$ Ans. $a^2 - ab + ac + b^2 + bc + c^2$.

18. $x^4 + 64$ by $x^2 + 4x + 8$. $\qquad x^2 - 4x + 8$.

19. $a^6 - b^6$ by $a^3 - 2a^2b + 2ab^2 - b^3$.

$\qquad\qquad$ Ans. $a^3 + 2a^2b + 2ab^2 + b^3$.

When the coefficients are fractional we may still use the ordinary process, combining the fractional coefficients by the rules of Arithmetic.

20. Divide $\frac{1}{4}x^3 + \frac{1}{72}xy^2 + \frac{1}{12}y^3$ by $\frac{1}{2}x + \frac{1}{3}y$.

$$
\begin{array}{r|l}
\frac{1}{4}x^3 + \frac{1}{72}xy^2 + \frac{1}{12}y^3 & \underline{\frac{1}{2}x + \frac{1}{3}y} \\
\frac{1}{4}x^3 + \frac{1}{6}x^2y & \frac{1}{2}x^2 - \frac{1}{3}xy + \frac{1}{4}y^2 \\
\hline
-\frac{1}{6}x^2y + \frac{1}{72}xy^2 & \\
-\frac{1}{6}x^2y - \frac{1}{9}xy^2 & \\
\hline
\quad \frac{1}{8}xy^2 + \frac{1}{12}y^3 & \\
\quad \frac{1}{8}xy^2 + \frac{1}{12}y^3 & \\
\hline
\end{array}
$$

Divide

21. $\frac{1}{8}a^3 - \frac{9}{4}a^2x + \frac{27}{2}ax^2 - 27x^3$ by $\frac{1}{2}a - 3x$.

$\qquad\qquad$ *Ans.* $\frac{1}{4}a^2 - 3ax + 9x^2$.

22. $\frac{1}{27}a^3 - \frac{1}{12}a^2 + \frac{1}{16}a - \frac{1}{64}$ by $\frac{1}{3}a - \frac{1}{4}$. $\frac{1}{9}a^2 - \frac{1}{6}a + \frac{1}{16}$.

49. Division with the Aid of Parentheses. — Sometimes it is found convenient to divide with the aid of parentheses thus :

1. Divide $x^3 - (a+b+c)x^2 + (ab+ac+bc)x - abc$ by $x-c$.

$$
\begin{array}{l}
x^3 - (a + b + c)x^2 + (ab + ac + bc)x - abc \,\big|\, \underline{x - c} \\
\underline{x^3 - cx^2} \qquad\qquad\qquad\qquad\qquad\quad x^2 - (a+b)x + ab \\
\quad -(a + b)x^2 + (ab + ac + bc)x \\
\quad \underline{-(a + b)x^2 + (a + b)cx} \\
\qquad\qquad\qquad\quad abx - abc \\
\qquad\qquad\qquad\quad \underline{abx - abc}
\end{array}
$$

Divide

2. $a^2x^4 + (2ac - b^2)x^2 + c^2$ by $ax^2 - bx + c$. *Ans.* $ax^2 + bx + c$.

3. $x^3 + (a + b + c)x^2 + (ab + ac + bc)x + abc$ by $x + b$.

$\qquad\qquad$ *Ans.* $x^2 + (a + c)x + ac$.

4. $ax^3 - (a^2 + b)x^2 + b^2$ by $ax - b$. $\qquad$ $x^2 - ax - b$.

5. $a^2(b + c) + b^2(a - c) + c^2(a - b) + abc$ by $a + b + c$.

$\qquad\qquad$ *Ans.* $a(b + c) - bc$.

50. Where the Division cannot be Exactly Performed. — In the examples given thus far the divisor has been exactly contained in the dividend. It may happen, as in Arithmetic, that the division cannot be exactly performed. In such cases it is a good exercise for the student to determine the accurate value of the remainder.

Divide $x^3 - 6x^2 + 11x + 2$ by $x - 2$.

$$
\begin{array}{l}
x^3 - 6x^2 + 11x + 2 \,\big|\, \underline{x - 2} \\
\underline{x^3 - 2x^2} \qquad\qquad\quad x^2 - 4x + \mathbf{3} \\
\quad -4x^2 + 11x \\
\quad \underline{-4x^2 + 8x} \\
\qquad\qquad 3x + 2 \\
\qquad\qquad \underline{3x - 6} \\
\qquad\qquad\quad 8
\end{array}
$$

The division can be carried no further without fractions, because x will not go into 8. We therefore express the result in the same way as in Arithmetic, that is, by adding to the quotient a fraction of which the numerator is the remainder and the denominator the divisor. Thus the result is

$$\frac{x^3 - 6x^2 + 11x + 2}{x - 2} = x^2 - 4x + 3 + \frac{8}{x - 2}.$$

EXAMPLES.

Find the remainder when

1. $x^3 - 6x^2 + 12x - 17$ is divided by $x - 3$.   *Ans.* -8.
2. $3x^3 - 7x - 9$ is divided by $x + 1$. -5.
3. $2x^3 + 5x^2 - 4x - 7$ is divided by $x + 2$. 5.
4. $4x^3 + 7x^2 - 3x - 33$ is divided by $4x - 5$. -18.
5. $27x^3 + 9x^2 - 3x - 5$ is divided by $3x - 2$. 5.
6. $16x^3 - 19 + 39x - 46x^2$ is divided by $8x - 3$. -10.
7. $8x - 8x^2 + 5x^3 + 7$ is divided by $5x - 3$. 10.

51. Important Examples in Division. — The following examples are very important; they may be easily verified, and should be carefully noticed.

$$\text{I.} \begin{cases} \dfrac{x^2 - y^2}{x - y} = x + y, \\[2mm] \dfrac{x^3 - y^3}{x - y} = x^2 + xy + y^2, \\[2mm] \dfrac{x^4 - y^4}{x - y} = x^3 + x^2y + xy^2 + y^3, \end{cases}$$

and so on; the terms in the quotient *all* being *positive*.

$$\text{II.} \begin{cases} \dfrac{x^2 - y^2}{x + y} = x - y, \\[2mm] \dfrac{x^4 - y^4}{x + y} = x^3 - x^2y + xy^2 - y^3, \\[2mm] \dfrac{x^6 - y^6}{x + y} = x^5 - x^4y + x^3y^2 - x^2y^3 + xy^4 - y^5, \end{cases}$$

and so on ; the terms in the quotient being *alternately positive and negative.*

$$\text{III.}\begin{cases}\dfrac{x^3 + y^3}{x + y} = x^2 - xy + y^2, \\[2mm] \dfrac{x^5 + y^5}{x + y} = x^4 - x^3y + x^2y^2 - xy^3 + y^4, \\[2mm] \dfrac{x^7 + y^7}{x + y} = x^6 - x^5y + x^4y^2 - x^3y^3 + x^2y^4 - xy^5 + y^6,\end{cases}$$

and so on ; the terms in the quotient being *alternately
positive and negative.*

The student can verify these results in any particular case,
and carry on these operations as far as he pleases, and he
will thus gain confidence in the truth of the following statements for which we shall hereafter give a general proof. See
College Algebra, Art. 170.

These different cases may be conveniently arranged in the
following concise statements :

$x^n - y^n$ is divisible by $x - y$ if n be *any* whole number.
(1) That is : *the difference of any two equal powers of two
numbers is always divisible by the difference of the two
numbers.*

$x^n - y^n$ is divisible by $x + y$ if n be any *even* whole
number.   (2) That is : *the difference of any two equal even
powers of two numbers is always divisible by the sum of the
numbers.*

$x^n + y^n$ is divisible by $x + y$ if n be any *odd* whole
number.   (3) That is : *the sum of any two equal odd powers
of two numbers is always divisible by the sum of the numbers.*

$x^n + y^n$ is never divisible by $x + y$ or $x - y$, when n is
an *even* whole number.

EXAMPLES.

Write the results in the following by these three statements.

1. $a^5 - b^5 \div a - b.$ *Ans.* $a^4 + a^3b + a^2b^2 + ab^3 + b^4.$
2. $x^3 - 1 \div x - 1.$ $x^2 + x + 1.$

3. $8a^3 - b^3 \div 2a - b$. $Ans.$ $4a^2 + 2ab + b^2$.

4. $8x^3 - 27y^3 \div 2x - 3y$. $4x^2 + 6xy + 9y^2$.

5. $x^4 - 16y^4 \div x + 2y$. $x^3 - 2x^2y + 4xy^2 - 8y^3$.

6. $16x^4 - y^4 \div 2x + y$. $8x^3 - 4x^2y + 2xy^2 - y^3$.

7. $x^3 + 1 \div x + 1$. $x^2 - x + 1$.

8. $a^3 + 64 \div a + 4$. $a^2 - 4a + 16$.

Divide

9. $3x^3 - 9x^2y - 12xy^2$ by $-3x$. $-x^2 + 3xy + 4y^2$.

(10.) $4x^4y^4 - 8x^3y^2 + 6xy^3$ by $-2xy$. $-2x^3y^3 + 4x^2y - 3y^2$.

11. $x^3y - 3x^2y^2 + 4xy^3$ by xy. $x^2 - 3xy + 4y^2$.

12. $-15a^3b^3 - 3a^2b^2 + 12ab$ by $-3ab$. $5a^2b^2 + ab - 4$.

13. $-3a^2 + \frac{9}{2}ab - 6ac$ by $-\frac{3}{2}a$. $2a - 3b + 4c$.

14. $\frac{9}{2}x^5y^2 - 3x^3y^4 - 6x^4y^3$ by $-\frac{3}{2}x^3y^2$. $-3x^2 + 2y^2 + 4xy$.

15. $\frac{1}{4}a^2x - \frac{1}{16}abx - \frac{3}{8}acx$ by $\frac{3}{8}ax$. $\frac{2}{3}a - \frac{1}{6}b - c$.

16. $x^2 - 11x + 30$ by $x - 5$. $x - 6$.

17. $x^2 - 7x + 12$ by $x - 3$. $x - 4$.

18. $3x^2 + x - 14$ by $x - 2$. $3x + 7$.

19. $6x^2 - 31x + 35$ by $2x - 7$. $3x - 5$.

20. $15a^2 + 17ax - 4x^2$ by $3a + 4x$. $5a - x$.

21. $60x^2 - 4xy - 45y^2$ by $10x - 9y$. $6x + 5y$.

22. $-4xy - 15y^2 + 96x^2$ by $12x - 5y$. $8x + 3y$.

(23.) $100x^3 - 3x - 13x^2$ by $3 + 25x$. $4x^2 - x$.

24. $7x^3 + 96x^2 - 28x$ by $7x - 2$. $x^2 + 14x$.

25. $x^5 + x^4y - x^3y^2 + x^3 - 2xy^2 + y^3$ by $x^2 + xy - y^2$. $x^3 + x - y$.

26. $2x^3 - 8x + x^4 + 12 - 7x^2$ by $x^2 + 2 - 3x$. $x^2 + 5x + 6$.

27. $8x^3 - 4x^2 - 128x + x^4 - 192$ by $x^2 - 16$. $x^2 + 8x + 12$.

28. $x^9 - y^9$ by $x^2 + xy + y^2$. $x^7 - x^6y + x^4y^3 - x^3y^4 + xy^6 - y^7$.

29. $2a^4 + 27ab^3 - 81b^4$ by $a + 3b$. $2a^3 - 6a^2b + 18ab^2 - 27b^3$.

30. $x^5 + x^4y + x^3y^2 + x^2y^3 + xy^4 + y^5$ by $x^3 + y^3$. $x^2 + xy + y^2$.

31. $\frac{3}{4}a^2c^3 + \frac{6}{125}a^5$ by $\frac{1}{5}a^2 + \frac{1}{2}ac$. $\frac{6}{25}a^3 - \frac{3}{5}a^2c + \frac{3}{2}ac^2$.

32. $\frac{9}{16}a^4 - \frac{3}{4}a^3 - \frac{7}{4}a^2 + \frac{4}{3}a + \frac{16}{9}$ by $\frac{3}{2}a^2 - \frac{8}{3} - a$. $\frac{3}{8}a^2 - \frac{1}{4}a - \frac{2}{3}$.

33. $x^6 - 1$ by $x - 1$. (Art. 51.) $x^5 + x^4 + x^3 + x^2 + x + 1$.

34. $x^4 - 81y^4$ by $x - 3y$. $x^3 + 3x^2y + 9xy^2 + 27y^3$.

35. $x^5 - y^5$ by $x - y$. $x^4 + x^3y + x^2y^2 + xy^3 + y^4$.

36. $a^9 - b^9$ by $a^3 - b^3$. $a^6 + a^3b^3 + b^6$.

37. $27x^3 + 8y^3$ by $3x + 2y$. $9x^2 - 6xy + 4y^2$.

. CHAPTER VI.

SIMPLE EQUATIONS OF ONE UNKNOWN QUANTITY.

52. Equations — Identical Equations. — *An Equation* is a statement in Algebraic language that two expressions are equal. Thus,

$$2x + 4 = x + 8$$

is an equation; it states that the expression $2x + 4$ is equal to the expression $x + 8$.

The two equal expressions thus connected are called *sides* or *members* of the equation. The expression to the left of the sign of equality is called the *first* side or member, and the expression to the right is called the *second* side or member. Every equation has *two* members.

An *Identical Equation*, or briefly an *Identity*, is one in which the two members are equal whatever numbers the letters represent. Thus, the following are identical equations:

$$x + 3 + x + 4 = 2x + 7,$$
$$(a + x)(a - x) - a^2 + x^2 = 0;$$

that is, these Algebraic statements are necessarily true, whatever values we assign to x and a. All the equations used in the previous chapters to express the relations of Algebraic quantities are identical equations, because they are true for all values of these quantities.

53. Equation of Condition — Unknown Quantity. — *An Equation of Condition* is one which is true only when the letters represent some particular value. For example, the equation,

$$x + 7 = 12,$$

cannot be true unless $x = 5$, and is therefore an equation of condition.

An equation of condition is called briefly an *equation*.

The letter whose value, or values, it is required to find is called the *unknown quantity*. Thus x is the unknown quantity in the above equation.

To *solve* an equation means to find the value, or values, of the unknown quantity for which 'he equation is true. These values of the unknown quantity are said to *satisfy the equation*, and are called the *roots of the equation*.

An equation which contains only one unknown quantity is called a *simple equation*, or an equation of the *first degree*, when the unknown quantity occurs only in the *first* power. It is usual to denote the unknown quantity by the letter x. The equation is said to be of the *second degree* or a *quadratic equation* when x^2 is the highest power of x which occurs, and so on.*

Thus $$2x + 6 = x + 8,$$
and $$ax + b = c$$
are simple equations.

$$x^2 - 2x = 3$$

is a quadratic equation.

54. Axioms. — *An Axiom* is a self-evident truth. The operations employed in solving equations are founded upon the following axioms :

1. If equal quantities be added to equal quantities, the sums will be equal.

2. If equal quantities be taken from equal quantities, the remainders will be equal.

3. If equal quantities be multiplied by equal quantities, the products will be equal.

4. If equal quantities be divided by equal quantities, the quotients will be equal.†

---

* The equation is supposed to be reduced to such a form that the unknown quantity is found only in the numerators of the terms, and that the exponents of its powers are expressed by positive integers.

† If the divisors are different from zero.

5. Like powers and like roots of equal quantities are equal.

These axioms may be summed up in the following one: *If the same operations be performed on equal quantities, the results will be equal.*

In the solution of equations there are two operations of frequent use. These are (1) clearing the equation of fractions, and (2) transposing the terms from one member to the other so that the unknown quantity shall finally stand alone as one member of the equation.

55. Clearing of Fractions. — Consider the equation

$$\frac{x}{2} + \frac{x}{3} + \frac{x}{6} = 2.$$

Multiplying each term by 2 × 3 × 6 (Axiom 3), we get

$$3 \times 6x + 2 \times 6x + 2 \times 3x = 2 \times 3 \times 6 \times 2,$$

or $18x + 12x + 6x = 72$;

dividing each term by 6 (Axiom 4), we get

$$3x + 2x + x = 12,$$

or $6x = 12.$

Instead of multiplying each term by 2 × 3 × 6, we might multiply each term by the least common multiple of the denominators, which is 6, and get immediately

$$3x + 2x + x = 12.$$

Hence to clear an equation of fractions, we have the following

Rule.

Multiply each term of the equation by the least common multiple of the denominators.

Clear the following equation of fractions:

$$\frac{x}{3} + \frac{x}{5} - \frac{x}{7} = 3.$$

Here the least common multiple of the denominators is the product of the denominators, 3, 5, and 7. Multiplying each term by it, we get

$$35x + 21x - 15x = 315,$$

or $41x = 315.$

Clear the following equation of fractions:

$$\frac{x}{4} + \frac{x}{6} + \frac{x}{8} + \frac{x}{12} = 4.$$

Here 24 is the least common multiple of the denominators. Multiplying each term by it, we get

$$6x + 4x + 3x + 2x = 96,$$

or $15x = 96.$

56. Transposition. — To transpose a term is to change it from one member of an equation to the other without destroying the equality of the members.

Suppose, for example, that $x - a = b.$

Add a to each member (Axiom 1) ; then we have

$$x - a + a = b + a;$$

therefore, since $-a$ and $+a$ cancel each other, we have

$$x = b + a.$$

Again, suppose that $x + b = a.$

Subtract b from each member (Axiom 2) ; then we have

$$x + b - b = a - b;$$

therefore, since $+b$ and $-b$ cancel each other, we have

$$x = a - b.$$

Here we see, in these two examples, that $-a$ has been removed from one member of the equation, and appears as $+a$ in the other; and $+b$ has been removed from one member and appears as $-b$ in the other.

It is evident that similar steps may be employed in all cases. Hence we have the following

RULE.

Any term may be transposed from one member of an equation to the other by changing its sign.

It follows from this that *the sign of every term of an equation may be changed;* for this is equivalent to transposing every term, and then making the first and second members change places. Thus, for example, suppose that

$$4x - 8 = 2x - 16.$$

Transposing every term, we have

$$-2x + 16 = -4x + 8,$$

or $$-4x + 8 = -2x + 16,$$

which is the original equation with the sign of every term changed.

This result can also be obtained by multiplying each term of the original equation by -1 (Axiom 3).

57. Solution of Simple Equations with One Unknown Quantity. — Find the value of x in the equation

$$\frac{x}{2} - \frac{18}{5} = \frac{x}{4} - \frac{x}{5}.$$

The least common multiple of the denominators is 20. Multiplying each term by 20, we get

$$10x - 72 = 5x - 4x.$$

Transposing the unknown terms to the first member, and the known terms to the second, we have

$$10x - 5x + 4x = 72.$$

Collecting the terms, we have

$$9x = 72.$$

Dividing each member by 9 (Axiom 4), we have

$$x = 8.$$

We can now give a general rule for solving any simple equation with one unknown quantity.

RULE.

Clear the equation of fractions, if necessary; transpose all the terms containing the unknown quantity to the first member of the equation, and the known quantities to the second, and collect the terms of each member. Divide both members by the coefficient of the unknown quantity, and the second member is the value required.

EXAMPLES.

1. Solve $9x + 35 = 75 + 5x$.

Here there are no fractions; transposing, we have
$$9x - 5x = 75 - 35.$$
Collecting terms, $\qquad 4x = 40.$

Dividing by 4, $\qquad\qquad x = 10.$

It is very important for the student to acquire the habit of occasionally *verifying*, that is, *proving* the truth of his results. The habit of applying such proofs tends to convince the student, and to make him self-reliant and confident in his own accuracy.

To verify the result, in the case of simple equations, we substitute the value of the unknown quantity in the original equation; if the two members are equal the result is said to be *verified*, or the equation *satisfied*.

Thus, in the last example, 10 is the *root* of the proposed equation (Art. 53). We may verify this, i.e., we may show that $x = 10$ satisfies the original equation by putting 10 for x in that equation.

Thus $\qquad 9 \times 10 + 35 = 75 + 5 \times 10,$

or $\qquad\qquad 90 + 35 = 75 + 50,$

or $\qquad\qquad\qquad 125 = 125,$

which is clearly true. Hence, since the two members are equal, $x = 10$ satisfies the equation.

2. Solve $5(x - 3) - 7(6 - x) + 3 = 24 - 3(8 - x)$

Removing parentheses,
$$5x - 15 - 42 + 7x + 3 = 24 - 24 + 3x.$$
Transposing, $\quad 5x + 7x - 3x = 24 - 24 + 15 + 42 - 3.$

Collecting terms, $\qquad\qquad 9x = 54.$

Dividing by 9, $\qquad\qquad x = 6.$

We may verify this result by putting 6 for x in the given equation.

Thus $5(6 - 3) - 7(6 - 6) + 3 = 24 - 3(8 - 6),$

or $\qquad\qquad 15 \qquad\qquad + 3 = 24 - 6,$

or $\qquad\qquad\qquad\qquad\qquad 18 = 18.$

3. Solve $5x - (4x - 7)(3x - 5) = 6 - 3(4x - 9)(x - 1)$.

Performing the multiplications indicated, we have

$$5x - (12x^2 - 41x + 35) = 6 - 3(4x^2 - 13x + 9).$$

Removing the parentheses, we have

$$5x - 12x^2 + 41x - 35 = 6 - 12x^2 + 39x - 27.$$

Erasing the term $-12x^2$ on each side, and transposing, we have

$$5x + 41x - 39x = 6 - 27 + 35.$$

Collecting terms

$$7x = 14.$$

$$\therefore \quad x = 2.$$

We may verify this result by putting 2 for x in the given equation.

The first member becomes

$$10 - (8 - 7)(6 - 5) = 10 - 1 = 9;$$

and the second member becomes

$$6 - 3(8 - 9)(2 - 1) = 6 - 3(-1) = 9.$$

Thus, since these two results are the same, $x = 2$ satisfies the equation.

NOTE. — In the first line of the solution of Ex. 3, we did not remove the parentheses until we performed the multiplications. The beginner is recommended to put down all his work as full as we have done in this example, in order to insure accuracy.

4. Solve $8x - 5[x - \{6 - 5(x - 3)\}] = 4x + 1$.

Removing parentheses, we have

$$8x - 5[x - 21 + 5x] = 4x + 1,$$

$$8x - 5x + 105 - 25x = 4x + 1.$$

$$\therefore \quad -26x = -104.$$

$$\therefore \quad x = 4.$$

Solve the following equations:

5. $2x + 7 = 3x + 3$.   *Ans.* 4.

6. $24x - 49 = 19x - 14$.   7.

7. $16x - 11 = 7x + 70$.   9.

8. $8(x - 1) + 17(x - 3) = 4(4x - 9) + 4$.   3.

9. $5x - 6(x - 5) = 2(x + 5) + 5(x - 4)$.   5.

10. $8(x - 3) - (6 - 2x) = 2(x + 2) - 5(5 - x)$.   3.

11. $3(169 - x) - (78 + x) = 29x$.   13.

12. $7x - 39 - 10x + 15 = 100 - 33x + 26$.   5.

13. $118 - 65x - 123 = 15x + 35 - 120x.$ *Ans.* 1.

14. $157 - 21(x + 3) = 163 - 15(2x - 5).$ 16.

15. $97 - 5(x + 20) = 111 - 8(x + 3).$ 30.

16. $x - [3 + \{x - (3 + x)\}] = 5.$ 5.

17. $14x - (5x - 9) - \{4 - 3x - (2x - 3)\} = 30.$ 2.

18. $5x - (3x - 7) - \{4 - 2x - (6x - 3)\} = 10.$ 1.

58. Fractional Equations. — The following are some of the most useful methods of solving *fractional equations.*

EXAMPLES.

1. Solve $\dfrac{5x + 4}{2} - \dfrac{7x + 5}{10} = 5\dfrac{3}{5} - \dfrac{x - 1}{2}.$

$5\frac{3}{5} = \frac{28}{5}$; the least common multiple of the denominators is 10; multiplying by 10, we have

$$5(5x + 4) - (7x + 5) = 56 - 5(x - 1);$$

removing parentheses,

$$25x + 20 - 7x - 5 = 56 - 5x + 5;$$

transposing, $25x - 7x + 5x = 56 + 5 - 20 + 5;$

collecting terms, $\qquad 23x = 46;$

$$\therefore \quad x = 2.$$

NOTE. — Mistakes with regard to the signs are often made in clearing an equation of fractions. In this equation the fraction $-\dfrac{x - 1}{2}$ is regarded as a single term with the minus sign before it; it is equivalent to $-\frac{1}{2}(x - 1)$. When multiplied by 10, it is well to put the result first in the form $-5(x - 1)$, and afterwards in the form $-5x + 5$, in order to secure attention to the signs.

2. Solve $4 - \dfrac{x - 9}{8} = \dfrac{x}{22} - \dfrac{1}{2}.$ *Ans.* 33.

3.   " $\frac{1}{4}(5x + 3) - \frac{1}{7}(16 - 5x) = 37 - 4x.$ 6.

4.   " $\dfrac{6x + 15}{11} - \dfrac{8x - 10}{7} = \dfrac{4x - 7}{5}.$ 3.

NOTE. — In certain cases it is more advantageous not to clear the equation of fractions at once by multiplying it throughout by the least common multiple of the denominators (Art. 55), but to clear it of fractions *partially*, and then to effect some reductions, before we remove the remaining fractions.

5. Solve $\dfrac{x-4}{3} + \dfrac{2x-3}{35} = \dfrac{5x-32}{9} - \dfrac{x+9}{28}$.

First multiply by 9, and we have

$$3x - 12 + \dfrac{18x-27}{35} = 5x - 32 - \dfrac{9x+81}{28};$$

transposing, $\dfrac{18x-27}{35} + \dfrac{9x+81}{28} = 2x - 20$.

Now clear of fractions by multiplying by the least common multiple of the denominators, which is $5 \times 7 \times 4$, or 140, and we get

$$72x - 108 + 45x + 405 = 280x - 2800;$$

transposing, $72x + 45x - 280x = -2800 + 108 - 405$;

collecting terms, $\qquad -163x = -3097$;

dividing by -163, $\qquad\qquad x = 19.$

Solve the following equations:

6. $\dfrac{x-2}{2} + \dfrac{x+10}{9} = 5.$ *Ans.* 8.

7. $\dfrac{x+19}{5} = 3 + \dfrac{x}{4}.$ 16.

8. $\dfrac{x-1}{8} = 1 + \dfrac{x+1}{18}.$ 17.

9. $\dfrac{4(x+2)}{5} = 7 + \dfrac{5x}{13}.$ 13.

10. $\dfrac{x+20}{9} + \dfrac{3x}{7} = 6.$ 7.

11. $\dfrac{x-8}{7} + \dfrac{x-3}{3} + \dfrac{5}{21} = 0.$ 4.

12. $\dfrac{x+5}{6} - \dfrac{x+1}{9} = \dfrac{x+3}{4}.$ $-\dfrac{1}{7}.$

13. $\dfrac{3(x-1)}{16} - \dfrac{5}{12}(x-4) = \dfrac{2}{5}(x-6) + \dfrac{5}{48}.$ 6.

14. $\dfrac{3x}{4} - \dfrac{6}{17}(x+10) - (x-3) = \dfrac{x-7}{51} - 4\dfrac{3}{4}.$ 7.

59. To Solve Equations whose Coefficients are Decimals, it is advisable generally to express all the decimals as common fractions, to insure accuracy, and then proceed as before; but it is often found more simple to work entirely in decimals.

EXAMPLES.

1. Solve $.\dot{6}*x + .25 - \frac{1}{9}x = 1.\dot{8} - .75x - \frac{1}{3}$.

Expressing the decimals as vulgar fractions, we have

$$\tfrac{2}{3}x + \tfrac{1}{4} - \tfrac{1}{9}x = 1\tfrac{8}{9} - \tfrac{3}{4}x - \tfrac{1}{3};$$

clearing of fractions,

$$24x + 9 - 4x = 68 - 27x - 12;$$

transposing, $24x - 4x + 27x = 68 - 12 - 9$;

$$\therefore \quad 47x = 47;$$

$$\therefore \quad x = 1.$$

2. Solve $.375x - 1.875 = .12x + 1.185$.

Transposing, $\qquad .375x - .12x = 1.185 + 1.875$;

collecting terms, $(.375 - .12)x = 3.06$;

that is $\qquad\qquad .255x = 3.06$;

dividing by .255, $\qquad\qquad x = 12$.

3. Solve $.5x - .\dot{3}x = .25x - 1$.   *Ans.* 12.

4.   " $.2x - .1\dot{6}x = .6 - .\dot{3}$.   8.

5.   " $2.25x - .125 = 3x + 3.75$. $-5\tfrac{1}{6}$.

60. Literal Equations. — A *Literal Equation* is one in which some or all the numbers are represented by letters. Thus

$$ax^2 + bx = cx + 4, \quad \text{and} \quad ax + b = cx^2 - d,$$

are literal equations. The known numbers are usually represented by the *first* letters of the alphabet, as a, b, c, etc.

---

* $.\dot{6}$ denotes the repetend .6666 etc. $= \tfrac{2}{3}$. Similarly $.\dot{8}$ denotes .888 etc. $= \tfrac{8}{9}$.

EXAMPLES.

1. Solve $ax + b^2 = bx + a^2$.

Transposing, we have
$$ax - bx = a^2 - b^2,$$
that is $\qquad (a - b)x = a^2 - b^2;$

dividing by $a - b$, the coefficient of x, we have
$$x = (a^2 - b^2) \div (a - b) = a + b.$$

2. Solve $\dfrac{x}{a} + \dfrac{x}{b} = c.$

Multiplying by ab, we have
$$bx + ax = abc,$$
that is, $\qquad (a + b)x = abc;$

dividing by $a + b$, the coefficient of x, we have
$$x = \frac{abc}{a + b}.$$

3. Solve $(a - x)(a + x) = 2a^2 + 2ax - x^2$.

$\qquad\qquad\qquad$ *Ans.* $x = -\dfrac{a}{2}.$

4. Solve $2x + bx - a = 3x - 2c.$ $\qquad x = \dfrac{a - 2c}{b - 1}.$

5. Solve $ax - bx + b^2 = a^2$. $\qquad\qquad x = a + b.$

6. Solve $(a + x)(b + x) = a(b + c) + \dfrac{a^2c}{b} + x^2$.

$\qquad\qquad\qquad$ *Ans.* $x = \dfrac{ac}{b}.$

7. Solve $a(x - a) + b(x - b) = 2ab.$ $\qquad x = a + b.$

8. Solve $2(x - a) + 3(x - 2a) = 2a.$ $\qquad x = 2a.$

9. Solve $\frac{1}{2}(x + a + b) + \frac{1}{3}(x + a - b) = b.$

$\qquad\qquad\qquad$ *Ans.* $x = b - a.$

10. Solve $(a + bx)(b + ax) = ab(x^2 - 1).$

$\qquad\qquad\qquad x = -\dfrac{2ab}{a^2 + b^2}.$

61. Problems Leading to Simple Equations. —

The preceding principles may now be employed to solve various problems.

A *Problem* is a question proposed for solution. In a problem certain quantities are given or known, and certain others which have some assigned relations to these, are required.

A *Theorem* is a truth requiring proof.

Axioms, Problems, and Theorems, are called *Propositions*.

The *Solution* of a problem by Algebra consists of two distinct parts : (1) The *Statement* of the problem, and (2) the *Solution* of the equation of the problem.

The *Statement* of the problem is the process of expressing the conditions of the problem in Algebraic language by an equation. The statement of the problem is often more difficult to beginners than the solution of the equation. No rule can be given for the statement of every particular problem. Much must depend on the skill of the student, and practice will give him readiness in this process. The following is the general plan of finding the equation :

1. Study the problem, to ascertain what quantities in it are known and what are unknown, and to understand it fully, so as to be able to prove the correctness or incorrectness of any proposed answer.

2. Represent the unknown quantity by one of the final letters of the alphabet, say x, and express in Algebraic language the relations which hold between the known and unknown quantities; an equation will thus be obtained which can be solved by the methods already given, and from which the value of the unknown quantity may be found.

Note 1. — Problems may often involve several unknown quantities, but in the present chapter we shall consider only problems in which there is *one* unknown quantity, or in which, if there are several, they are so related to one another that they can all be expressed in terms of some one of them.

EXAMPLES.

1. What number is that whose double exceeds its half by 27?

Let x represent the number;

then $2x$ represents the double of the number,

and $\dfrac{x}{2}$ represents the half of the number.

Since from the conditions of the problem the double exceeds the half by 27, we have for the equation

$$2x - \frac{x}{2} = 27.$$

Clearing of fractions,

$$4x - x = 54,$$

that is, $$3x = 54.$$

$$\therefore \; x = 18.$$

Hence the required number is 18.

Verification, $2 \times 18 - \dfrac{18}{2} = 27.$

2. The sum of two numbers is 28, and their difference is 4; find the numbers.

Let $x =$ the smaller number;

then $x + 4 =$ the greater number;

and since, from the conditions of the problem, the sum is to be equal to 28, we have for the equation

$$x + x + 4 = 28;$$

that is $$2x = 24.$$

$$\therefore \; x = 12,$$

and $$x + 4 = 16,$$

so that the numbers are 12 and 16.

Verification, $16 + 12 = 28$, and $16 - 12 = 4.$

The beginner is advised to test each solution by proving that it satisfies the data of the question.

3. A has \$80 and B has \$15. How much must A give to B in order that he may have just four times as much as B?

Let $x =$ the number of dollars that A gives to B;
then $80 - x =$ the number of dollars that A has left,
and $15 + x =$ the number of dollars that B will have after receiving x dollars from A.

But A has now four times as much as B; hence we have the equation
$$80 - x = 4(15 + x),$$
that is, $\qquad 80 - x = 60 + 4x,$
transposing and uniting, $\quad -5x = -20,$
dividing by $-5,$ $\qquad x = 4.$
Hence A must give \$4 to B.

4. A father is six times as old as his son, and in four years he will be four times as old. How old is each?

Let $x =$ the son's age in years,
then $6x =$ the father's age in years.
Also $x + 4 =$ the son's age in years, after four years,
and $6x + 4 =$ the father's age in years, after four years.
Hence, from the conditions, we have the equation
$$6x + 4 = 4(x + 4),$$
that is, $\qquad 6x + 4 = 4x + 16;$
$\therefore \ x = 6,$ the son's age,
and $\qquad\qquad 6x = 36,$ the father's age.

5. Divide 60 into two parts, so that 3 times the greater may exceed 100 by as much as 8 times the less falls short of 200.

Let $x =$ the greater part,
then $60 - x =$ the less.
Also $3x - 100 =$ the excess of 3 times the greater over 100, and $200 - 8(60 - x) =$ the number that 8 times the less falls short of 200. Hence, from the conditions, we have the equation
$$3x - 100 = 200 - 8(60 - x),$$
that is, $\qquad 3x - 100 = 200 - 480 + 8x,$
hence, $\qquad\qquad -5x = -180.$
$\therefore \ x = 36,$ the greater part.
$60 - x = 24,$ the less.

6. A line is 2 feet 4 inches long; it is required to divide it into two parts, such that one part may be three-fourths of the other part.

Let $x =$ the number of inches in the larger part,

then $\frac{3}{4}x =$ the number of inches in the other part.

Hence, from the conditions, we have the equation

$$x + \tfrac{3}{4}x = 28,$$

that is, $\qquad 4x + 3x = 112.$

$$\therefore \quad x = 16.$$

Thus one part is 16 inches long, and the other part 12 inches long.

7. Divide $47 between A, B, C, so that A may have $10 more than B, and B $8 more than C.

NOTE 2.—Here there are really three unknown quantities, but it is only necessary to represent the number of dollars the last has by a symbol.

Let $\qquad x =$ the number of dollars that C has,

then $\qquad x + 8 =$ the number of dollars that B has,

and $x + 8 + 10 =$ the number of dollars that A has.

Hence we have the equation

$$x + (x + 8) + (x + 8 + 10) = 47,$$
$$\therefore \quad 3x = 21,$$
$$\therefore \quad x = 7;$$

so that C has $7, B $15, A $25.

8. A person spent £28. 4s. in buying geese and ducks; if each goose cost 7s., and each duck cost 3s., and if the total number of birds bought was 108, how many of each did he buy?

NOTE 3.—In questions of this kind it is of essential importance to have all concrete quantities of the same kind expressed in the same denomination; in the present instance it will be convenient to express the money in shillings. In Ex. 6 it was convenient to express the length in inches.

Let $\qquad x =$ the number of geese,

then $\qquad 108 - x =$ the number of ducks.

Also $\qquad 7x =$ the number of shillings the geese cost,

and $3(108 - x) =$ the number of shillings the ducks cost.

EXAMPLES.

83

But from the conditions of the question, the whole cost of the geese and ducks is £28. 4s., i.e., 564 shillings. Hence we have the equation

$$7x + 3(108 - x) = 564,$$

that is, $7x + 324 - 3x = 564,$

$$\therefore \quad x = 60, \text{ the number of geese,}$$

and $\quad\quad\quad 108 - x = 48, \text{ the number of ducks.}$

9. A can do a piece of work in 12 hours, which B can do in 4 hours. A begins the work, but after a time B takes his place, and the whole work is finished in 6 hours from the beginning. How long did A work?

Let $\quad x = $ the number of hours that A worked,

then $\quad 6 - x = $ the number of hours that B worked.

Also $\quad \frac{1}{12} = $ the part A does in 1 hour, since he can do the whole work in 12 hours.

Therefore $\dfrac{x}{12} = $ the part done by A altogether.

Also $\quad \frac{1}{4} = $ the part B does in 1 hour, since he can do the whole work in 4 hours.

Therefore $\frac{1}{4}(6 - x) = $ the part done by B altogether.

But A and B together do the *whole* work; hence the sum of the parts of the work that they do separately must equal *unity;* and we have for the equation

$$\frac{x}{12} + \frac{1}{4}(6 - x) = 1.$$

Multiplying by 12, we have

$$x + 3(6 - x) = 12,$$
$$\therefore \quad -2x = -6.$$
$$\therefore \quad x = 3.$$

Hence A worked for 3 hours.

NOTE 4. — It should be remembered that x must always represent a *number;* what is called the unknown *quantity* is really an unknown *number.* In the above examples the unknown quantity x represents a *number* of dollars, years, inches, etc. For instance, in Ex. 6, we let x denote the number of *inches* in the longer part; beginners often say, "let $x =$ the longer part," or, "let $x =$ a part," which is not definite, because a part may be expressed in various ways, in feet, or inches, or yards. Again, in Ex. 7, we let $x =$ the number of dollars that C has; beginners often say, "let $x =$ C's money," which is not definite, because C's money may be expressed in various ways, in dollars, or in pounds, or as a fraction of the whole sum. The student must be careful to avoid beginning a solution with a vague and inexact statement.

It may seem to the student that some of the problems which are given for exercise can be readily solved by Arithmetic, and he may therefore be inclined to undervalue the power of Algebra and consider it unnecessary. We may remark, however, that by Algebra the student is enabled to solve *all* the problems given here, without any uncertainty; and also, he will find as he proceeds, that he can solve problems by Algebra, which would be extremely difficult or entirely impracticable, by Arithmetic alone.

10. The difference between two numbers is 8; if 2 be added to the greater the result will be three times the smaller : find the numbers.   *Ans.* 13, 5.

11. A man walks 10 miles, then travels a certain distance by train, and then twice as far by coach. If the whole journey is 70 miles, how far does he travel by train?

Ans. 20 miles.

12. What two numbers are those whose sum is 58, and difference 28?   *Ans.* 15, 43.

13. If 288 be added to a certain number, the result will be equal to three times the excess of the number over 12 : find the number.   *Ans.* 162.

14. Find three consecutive numbers whose sum shall equal 84.   *Ans.* 27, 28, 29.

15. Find two numbers differing by 10, whose sum is equal to twice their difference.   *Ans.* 15, 5.

16. Find a number such that if 5, 15, and 35 are added to it, the product of the first and third results may be equal to the square of the second.   *Ans.* 5.

17. A is twice as old as B, and seven years ago their united ages amounted to as many years as now represent the age of A : find the ages of A and B.   *Ans.* 28, 14.

EXAMPLLS.

Solve the following equations:

1. $3x + .5 = x + 25.$ *Ans.* 5.
2. $2x + 3 = 16 - (2x - 3).$ 4.
3. $7(25 - x) - 2x = 2(3x - 25).$ 15.
4. $5x - 17 + 3x - 5 = 6x - 7 - 8x + 115.$ 13.
5. $5(x + 2) = 3(x + 3) + 1.$ 0.
6. $2(x - 3) = 5(x + 1) + 2x - 1.$ $-2.$
7. $2(x - 1) - 3(x - 2) + 4(x - 3) + 2 = 0.$ 2.
8. $5x + 6(x + 1) - 7(x + 2) - 8(x + 3) = 0.$ $-8.$
9. $(x + 1)(2x + 1) = (x + 3)(2x + 3) - 14.$ 1.
10. $(x+1)^2 - (x^2-1) = x(2x+1) - 2(x+2)(x+1) + 20.$
     Ans. 2.
11. $6(x^2-3x+2) - 2(x^2-1) = 4(x+1)(x+2) - 24.$ 1.
12. $2x - 5\{3x - 7(4x - 9)\} = 66.$ 3.
13. $3(5 - 6x) - 5[x - 5\{1 - 3(x - 5)\}] = 23.$ 4.
14. $(x + 1)^2 + 2(x + 3)^2 = 3x(x + 2) + 35.$ 2.
15. $84 + (x+4)(x-3)(x+5) = (x+1)(x+2)(x+3).$ 1.
16. $(x + 1)(x + 2)(x + 6) = x^3 + 9x^2 + 4(7x - 1).$ 2.
17. $\dfrac{x}{5} - \dfrac{x}{4} = 1.$ $-20.$
18. $\dfrac{x - 1}{2} + \dfrac{x - 2}{3} = 3.$ 5.
19. $\frac{1}{4}(x + 1) - \frac{2}{3}(x - 1) = 3.$ $-5.$
20. $\frac{1}{2}(2 - x) - \frac{1}{5}(5x + 21) = x + 3.$ $-2\frac{13}{25}.$
21. $\dfrac{x + 1}{2} + \dfrac{x + 2}{3} + \dfrac{x + 4}{4} + 8 = 0.$ $-9\frac{5}{13}.$
22. $\dfrac{x - 5}{2} - \dfrac{x - 4}{3} = \dfrac{x - 3}{2} - (x - 2).$ $2\frac{1}{2}.$
23. $\dfrac{x + \frac{1}{3}}{2} - \dfrac{2x - \frac{1}{3}}{5} + 1\frac{1}{4} = 0.$ $-16.$
24. $\dfrac{3x + 5}{8} - \dfrac{21 + x}{2} = 5x - 15.$ 1.
25. $\dfrac{7x}{5} - \frac{1}{14}(x - 11) = \frac{3}{7}(x - 25) + 34.$ 25.

26. $\frac{1}{5}(x - 8) + \frac{4 + x}{7} + \frac{x - 1}{7} = 7 - \frac{23 - x}{5}.$

Ans. 8.

27. $x - \left(3x - \frac{2x - 5}{10}\right) = \frac{1}{6}(2x - 57) - \frac{5}{4}.$ 5.

28. $\frac{1 - 2x}{3} - \frac{4 - 5x}{6} + \frac{13}{42} = 0.$ $\frac{1}{7}$.

29. $\frac{x + 1}{3} - \frac{x - 1}{4} + 4x = 12 + \frac{2x - 1}{6}.$ 3.

30. $\frac{x + 3}{2} - \frac{x - 2}{3} = \frac{3x - 5}{12} + \frac{1}{4}.$ 28.

31. $\frac{3x - 1}{5} - \frac{13 - x}{2} = \frac{7x}{3} - \frac{11}{6}(x + 3).$ 2.

32. $\frac{2 - x}{3} + \frac{3 - x}{4} + \frac{4 - x}{5} + \frac{5 - x}{6} + \frac{3}{4} = 0.$ 4.

33. $\frac{5x - 3}{7} - \frac{9 - x}{3} = \frac{5x}{2} + \frac{19}{6}(x - 4).$ 2.

34. $\frac{5}{6}x + .25x - .\dot{3}x = x - 3.$ 12.

35. $.5x - .\dot{2}x = .\dot{3}x - 1.5.$ 27.

36. $1.5 = \frac{.36}{.2} - \frac{.09x - .18}{.9}.$ 5.

37. $(a + b)x + (a - b)x = a^2.$ $\frac{a}{2}$.

38. $(a + b)x + (b - a)x = b^2.$ $\frac{b}{2}$.

39. $\frac{1}{2}(a + x) + \frac{1}{3}(2a + x) + \frac{1}{4}(3a + x) = 3a.$ a.

40. $\frac{xa}{b} + \frac{xb}{a} = a^2 + b^2.$ ab.

41. $(a^2 + x)(b^2 + x) = (ab + x)^2.$ 0.

42. $a(x + a) + b(b - x) = 2ab.$ $b - a$.

43. $ax(x + a) + bx(x + b) = (a + b)(x + a)(x + b).$
Ans. $-\frac{1}{2}(a + b).$

44. $(x - a)^3 + (x - b)^3 + (x - c)^3 = 3(x - a)(x - b)(x - c).$
Ans. $\frac{1}{3}(a + b + c).$

45. Twenty-three times a certain number is as much above 14 as 16 is above seven times the number: find it. *Ans.* 1.

46. Divide 105 into two parts, one of which diminished by 20 shall be equal to the other diminished by 15.

Ans. 50, 55.

47. The sum of two numbers is 8, and one of them with 22 added to it is five times the other : find the numbers.

Ans. 3, 5.

48. A and B begin to play each with $60. If they play till A's money is double B's, what does A win?   *Ans.* $20.

49. The difference between the squares of two consecutive numbers is 121 : find the numbers.   *Ans.* 60, 61.

50. Divide $380 between A, B, and C, so that B may have $30 more than A, and C may have $20 more than B.

Ans. A $100, B $130, C $150.

51. A father is four times as old as his son ; in 24 years he will only be twice as old : find their ages.   *Ans.* 48, 12.

52. A is 25 years older than B, and A's age is as much above 20 as B's is below 85 : find their ages.   *Ans.* 65, 40.

53. The sum of the ages of A and B is 30 years, and five years hence A will be three times as old as B : find their present ages.   *Ans.* 25, 5.

54. The length of a room exceeds its breadth by 3 feet ; if the length had been increased by 3 feet, and the breadth diminished by 2 feet, the area would not have been altered : find the dimensions.   *Ans.* 15 ft., 12 ft.

55. There is a certain fish, the head of which is 9 inches long ; the tail is as long as the head and half the body ; and the body is as long as the head and tail together : what is the length of the fish?   *Ans.* 6 ft.

56. The sum of $76 was raised by A, B, and C together ; B contributed as much as A and $10 more, and C as much as A and B together : how much did each contribute?

Ans. $14, $24, $38.

57. After 34 gallons had been drawn out of one of two equal casks, and 80 gallons out of the other, there remained just three times as much in one cask as in the other : what did each cask contain when full?   *Ans.* 103.

58. Divide the number 20 into two parts such that the sum of three times one part, and five times the other part, may be 84. *Ans.* 8, 12.

59. A person meeting a company of beggars gave 4 cents to each, and had 16 cents left; he found that he should have required 12 cents more to enable him to give the beggars 6 cents each: how many beggars were there?
Ans. 14.

60. Divide 100 into two parts such that if a third of one part be subtracted from a fourth of the other, the remainder may be 11. *Ans.* 24, 76.

61. Divide 60 into two parts such that the difference between the greater and 64 may be equal to twice the difference between the less and 38. *Ans.* 36, 24.

62. Find a number such that the sum of its fifth and its seventh shall exceed the sum of its eighth and its twelfth by 113. *Ans.* 840.

63. An army in defeat loses one-sixth of its number in killed and wounded, and 4000 prisoners; it is re-enforced by 3000 men, but retreats, losing one-fourth of its number in doing so; there remain 18000 men: what was the original force? *Ans.* 30000.

64. One-half of a certain number of persons received 18 cents each, one-third received 24 cents each, and the rest received 30 cents each; the whole sum distributed was $5.28: how many persons were there? *Ans.* 24.

65. A father has six sons, each of whom is four years older than his next younger brother; and the eldest is three times as old as the youngest: find their respective ages.
Ans. 10, 14, 18, 22, 26, 30.

66. A man left his property to be divided between his three children in such a way that the share of the eldest was to be twice that of the second, and the share of the second twice that of the youngest; it was found that the eldest received $3000 more than the youngest: how much did each receive? *Ans.* $1000, $2000, $1000.

67. A sum of money is divided among three persons ; the first receives $10 more than a third of the whole sum ; the second receives $15 more than a half of what remains ; and the third receives what is over, which is $70 : find the original sum. *Ans.* $270.

68. In a cellar one-fifth of the wine is port and one-third claret ; besides this it contains 15 dozen of sherry and 30 bottles of spirits : how much port and claret does it contain? *Ans.* 90 port, 150 claret.

69. Two-fifths of A's money is equal to B's, and seven-ninths of B's is equal to C's : in all they have $770, what have they each? *Ans.* A $450, B $180, C $140.

70. A, B, and C have $1285 between them ; A's share is greater than five-sixths of B's by $25, and C's is four-fifteenths of B's : find the share of each. *Ans.* A $525, B $600, C $160.

71. A sum of money is to be distributed among three persons, A, B, and C ; the shares of A and B together amount to $240 ; those of A and C to $320 ; and those of B and C to $368 : find the share of each person. *Ans.* $96, $144, $224.

72. Two persons A and B are travelling together ; A has $100, and B has $48 : they are met by robbers who take twice as much from A as from B, and leave to A three times as much as to B : how much was taken from each? *Ans.* $88, $44.

73. In a mixture of wine and water the wine composed 25 gallons more than half of the mixture, and the water 5 gallons less than a third of the mixture : how many gallons were there of each? *Ans.* 85, 35.

74. A general, after having lost a battle, found that he had left fit for action 3600 men more than half of his army ; 600 men more than one-eighth of his army were wounded ; and the remainder, forming one-fifth of the army, were slain, taken prisoners, or missing : what was the number of the army? *Ans.* 24000.

CHAPTER VII.

FACTORING—GREATEST COMMON DIVISOR-. LEAST COMMON MULTIPLE.

62. Definitions. — *Factoring* is the process of resolving a quantity into its factors.

The Factors of a quantity are those quantities which multiplied together produce it. A factor of a quantity is therefore a *divisor* of the quantity, i.e., it will divide the quantity without a remainder. Thus, a is a factor or divisor of abc, and b is a factor or divisor of $ab - b^2$.

NOTE. — In Division (Chap. V.) we had given the product of two factors and one of the factors, and we showed how to find the other factor. In the present chapter we shall consider cases in which the factors of an expression can be found when none of the factors are given.

A Prime Quantity is one which has no integral factor except itself and unity. Thus, a, b, and $a + c$ are prime quantities; while ab, and $ac + bc$ are not prime.

Quantities are said to be *prime to each other* or *relatively prime*, when unity is the only integral factor common to both. Thus, ab and cd are prime to each other.

A Composite Quantity is one which is the product of two or more integral factors, neither of which is unity or the quantity itself. Thus, $ax + x^2$ is a composite quantity, the factors of which are x and $a + x$.

63. When All the Terms have one Common Factor. — When each term of a polynomial is divisible by a common factor, the polynomial may be simplified by the following

RULE.

Divide each term of the polynomial separately by the common factor, and enclose the quotient within parentheses, the common factor being placed outside as a coefficient; then the divisor will be one factor and the quotient the other.

EXAMPLES.

1. Factor the expression $3a^2 - 6ab$.

Here we see that the terms have a common factor, $3a$; therefore, dividing the polynomial by $3a$, we obtain for the quotient $a - 2b$. Hence the two factors are $3a$ and $a - 2b$.

$$\therefore \quad 3a^2 - 6ab = 3a(a - 2b).$$

Similarly

2. $5a^2bx^4 - 15ab^2x^3 - 20ab^3x^4 = 5abx^3(ax - 3b - 4b^2x)$.

Factor the following expressions:

3. $x^2 - ax$.   *Ans.* $x(x - a)$.
4. $x^3 - x^2$. $x^2(x - 1)$.
5. $a^2 - ab^2$. $a(a - b^2)$.
6. $8x - 2x^2$. $2x(4 - x)$.
7. $5ax - 5a^3x^2$. $5ax(1 - a^2x)$.
8. $x^3 - x^2y$. $x^2(x - y)$.
9. $5x - 25x^2y$. $5x(1 - 5xy)$.
10. $16x^2 + 64x^2y$. $16x^2(1 + 4y)$.
11. $54 - 81x$. $27(2 - 3x)$.
12. $3x^3 - 6x^2 + 9x$. $3x(x^2 - 2x + 3)$.
13. $6a^2bx^3 + 2ab^2x^4 + 4abx^5$. $2abx^3(3a + bx + 2x^2)$.
14. $72x^2y - 84xy^2 + 60x^2y^2$. $12xy(6x - 7y + 5xy)$.

64. Expressions containing Four Terms. — When a polynomial contains four terms which can be arranged in pairs that have a common binomial factor, the polynomial may be simplified by the following

RULE.

Divide the polynomial by the common binomial factor; then the divisor will be one factor and the quotient the other.

EXAMPLES.

1. Resolve into factors $x^2 - ax + bx - ab$.

Here we see that the first two terms contain a factor x, and the last two terms a factor b; therefore we factor the first two and last two terms by Art. 63, and obtain $x(x - a)$ and $b(x - a)$. We now see that the two pairs have the common binomial factor $x - a$. Dividing by $x - a$ we obtain the quotient $x + b$ for the other factor.

The work therefore will stand as follows:

$$x^2 - ax + bx - ab = x(x - a) + b(x - a)$$
$$= (x - a)(x + b).$$

2. Resolve into factors $6x^2 - 9ax + 4bx - 6ab$.

$$6x^2 - 9ax + 4bx - 6ab = 3x(2x - 3a) + 2b(2x - 3a)$$
$$= (2x - 3a)(3x + 2b).$$

3. Resolve into factors $12a^2 - 4ab - 3ax^2 + bx^2$.

$$12a^2 - 4ab - 3ax^2 + bx^2 = 4a(3a - b) - x^2(3a - b)$$
$$= (3a - b)(4a - x^2).$$

NOTE. — It is not necessary always to factor in the same way. In the first line of work it is usually sufficient to see that each pair contains some common factor; and any suitably chosen pairs will bring out the same result. Thus, in the last example, we may have a different arrangement, and enclose the first and third terms in one pair, and the second and fourth in another as follows:

$$12a^2 - 4ab - 3ax^2 + bx^2 = 12a^2 - 3ax^2 - (4ab - bx^2)$$
$$= 3a(4a - x^2) - b(4a - x^2)$$
$$= (4a - x^2)(3a - b),$$

which is the same result as before.

Resolve into factors

4. $a^2 + ab + ac + bc$.	*Ans.*	$(a + b)(a + c)$.
5. $a^2 - ac + ab - bc$.		$(a - c)(a + b)$.
6. $a^2c^2 + acd + abc + bd$.		$(ac + d)(ac + b)$.
7. $a^2 + 3a + ac + 3c$.		$(a + 3)(a + c)$.
8. $2ax + ay + 2bx + by$.		$(2x + y)(a + b)$.
9. $3ax - bx - 3ay + by$.		$(3a - b)(x - y)$.
10. $ax^2 + bx^2 + 2a + 2b$.		$(a + b)(x^2 + 2)$.
11. $x^2 - 3x - xy + 3y$.		$(x - 3)(x - y)$.

65. To Factor a Trinomial of the Form x^2+ax+b.—

Let $x^2 + ax + b$ be any trinomial in which the coefficient of x^2 is $+1$, and the signs of a and b either plus or minus.

Before proceeding to explain this case of resolution into factors, the student is advised to refer to Art. 40, and examine the relation that exists between two binomial factors and their product. Attention was there called to the way in which, in forming the product of two binomials, the coefficients of the different terms combined so as to give a trinomial result.

Therefore, in the converse problem, namely, the resolution of a trinomial expression into its component binomial factors, we see, by reversing the results of Art. 40, that any trinomial may be resolved into two binomial factors, when the first term is a square, and the coefficient of the second term is the sum of two quantities whose product is the third term. Hence the following

RULE

The first term of each factor is x, and the second terms are two numbers whose Algebraic sum is the coefficient of the second term, and whose product is the third term.

The application of this rule will be easily understood from the following

EXAMPLES.

1. Resolve into factors $x^2 + 11x + 24$.

Here the first term of each binomial factor is x, and the second terms of the two binomial factors must be two numbers whose sum is 11 and whose product is 24. It is clear therefore that they must be $+8$ and $+3$, since these are the only two numbers whose sum is 11 and whose product is 24.

$$\therefore \quad x^2 + 11x + 24 = (x + 8)(x + 3).$$

2. Resolve into factors $x^2 - 7x + 12$.

The first term of each factor is x, and the second terms of the factors must be such that their sum is -7, and their product is $+12$. Hence they must *both* be *negative*, and it is easy to see that they must be -4 and -3.

$$\therefore \quad x^2 - 7x + 12 = (x - 4)(x - 3).$$

3. Resolve into factors $x^2 + 5x - 24$.

The first term of each factor is x, and the second terms of the factors must be such that their Algebraic sum is $+5$, and their product is -24. Hence they must have *opposite* signs, and the greater of them must be *positive* in order to give the positive sign to their sum. It is easy to see therefore that they must be $+8$ and -3.

$$\therefore \quad x^2 + 5x - 24 = (x + 8)(x - 3).$$

4. Resolve into factors $x^2 - x - 56$.

The first term of each factor is x, and the second terms of the factors must be such that their Algebraic sum is -1, and their product is -56. Hence they must have *opposite* signs, and the greater of them must be *negative* in order to give its sign to their sum. The required terms are therefore -8 and $+7$.

$$\therefore \quad x^2 - x - 56 = (x - 8)(x + 7).$$

Note. — In examples of this kind the student should always verify his results, by forming the product, *mentally*, of the factors he has chosen, as in Art. 40.

Resolve into factors

5. $a^2 + 3a + 2$.	*Ans.* $(a + 1)(a + 2)$.
6. $x^2 - 11x + 30$.	$(x - 6)(x - 5)$.
7. $a^2 - 7a + 12$.	$(a - 4)(a - 3)$.
8. $x^2 - 15x + 56$.	$(x - 8)(x - 7)$.
9. $x^2 - 19x + 90$.	$(x - 9)(x - 10)$.
10. $x^2 + x - 2$.	$(x + 2)(x - 1)$.
11. $x^2 + x - 6$.	$(x + 3)(x - 2)$.
12. $x^2 - 2x - 3$.	$(x - 3)(x + 1)$.
13. $x^2 + 2x - 3$.	$(x + 3)(x - 1)$.
14. $x^2 + x - 56$.	$(x + 8)(x - 7)$.
15. $x^2 + 3x - 40$.	$(x + 8)(x - 5)$.

16. $x^2 - 4x - 12$. $Ans.$ $(x - 6)(x + 2)$.

17. $x^2 - x - 20$. $(x - 5)(x + 4)$.

18. $a^2 - 4a - 21$. $(a - 7)(a + 3)$.

19. $a^2 + a - 20$. $(a + 5)(a - 4)$.

20. $a^2 - 4a - 117$. $(a - 13)(a + 9)$.

21. $x^2 + 9x - 36$. $(x + 12)(x - 3)$.

NOTE. — If the term containing x^2 is negative, enclose the whole expression in a parenthesis with the minus sign prefixed. Then factor the expression within the parenthesis as in the preceding examples, and change the signs of all the terms of one of the factors. Thus

22. Factor $90 + 9x - x^2$.

$90+9x-x^2 = -(x^2-9x-90) = -(x-15)(x+6) = (15-x)(x+6)$.

23. Factor $240 + x - x^2$. $Ans.$ $(16 - x)(x + 15)$.

24. Factor $85 + 12x - x^2$. $(17 - x)(x + 5)$.

25. Factor $110 - x - x^2$. $(10 - x)(x + 11)$.

26. Factor $152 + 11x - x^2$. $(19 - x)(x + 8)$.

66. To Factor a Trinomial when the Coefficient of the Highest Power is not Unity. — Let it be required to factor $3x^2 + 14x + 8$.

The first term $3x^2$ is the product of $3x$ and x.

The third term 8 is the product of 2 and 4 or 1 and 8.

The middle term $14x$ is the result of adding together the two products $3x$ and 2 and x and 4, or $3x$ and 4 and x and 2, or $3x$ and 1 and x and 8, or $3x$ and 8 and x and 1.

Taking the first products we have $3x \times 2 + x \times 4 = 10x$; this combination therefore fails to give the correct middle term.

Next try the second products, and get $3x \times 4 + x \times 2 = 14x$, which is the correct value of the middle term.

$$\therefore \quad 3x^2 + 14x + 8 = (3x + 2)(x + 4).$$

The beginner will frequently find that it is not easy to select the proper factors at the first trial. Practice alone will enable him to detect at a glance whether any two factors are the proper ones.

EXAMPLES.

1. Resolve into factors $14x^2 + 29x - 15$.

Write down $(7x\ 5)(2x\ 3)$ for a first trial, noticing that 3 and 5 must have opposite signs. These factors give $14x^2$ and -15 for the first and third terms. But since $7 \times 3 - 2 \times 5 = 11$, this combination fails to give the correct coefficient of the middle term.

Next try $(7x\ 3)(2x\ 5)$.

Since $7 \times 5 - 3 \times 2 = 29$, these factors will be correct if we insert the signs so that the positive will predominate.

$$\therefore \quad 14x^2 + 29x - 15 = (7x - 3)(2x + 5).$$

(Verify by mental multiplication).

It is not usually necessary to put down all these steps. After a little practice the student will be able to examine the different cases rapidly, and to reject the unsuitable combinations at once.

In the factoring of such expressions as these the following hints are very useful :

(1) If the third term of the trinomial is positive, then the second terms of its factors have both the same sign as the middle term of the trinomial.

(2) If the third term of the trinomial is negative, then the second terms of its factors have opposite signs.

2. Resolve into factors $5x^2 + 17x + 6$.   . . . . (1)

$\qquad\qquad\qquad\qquad 5x^2 - 17x + 6.$ . . . . (2)

$\qquad\qquad\qquad\qquad 5x^2 + 13x - 6.$ . . . . (3)

$\qquad\qquad\qquad\qquad 5x^2 - 13x - 6.$ . . . . (4)

In (1) we notice that the factors which give 6 are both positive.

In (2) we notice that the factors which give 6 are both negative.

In (3) we notice that the factors which give 6 have opposite signs.

In (4) we notice that the factors which give 6 have opposite signs.

Therefore for (1) we write $(5x + \quad)(x + \quad)$.

(2) we write $(5x - \quad)(x - \quad)$.

(3) we write $(5x \quad 2)(x \quad 3)$, noticing that
2 and 3 have opposite signs.

(4) we write $(5x \quad 2)(x \quad 3)$, noticing that
2 and 3 have opposite signs.

Since $5 \times 3 + 1 \times 2 = 17$, we see that

$$5x^2 + 17x + 6 = (5x + 2)(x + 3).$$
$$5x^2 - 17x + 6 = (5x - 2)(x - 3).$$

And, since $5 \times 3 - 2 \times 1 = 13$, we have only to insert the proper signs in each factor.

In (3) the positive sign must predominate.

In (4) the negative sign must predominate.

Therefore $5x^2 + 13x - 6 = (5x - 2)(x + 3)$.

$$5x^2 - 13x - 6 = (5x + 2)(x - 3).$$

More generally, trinomials of the form $ax^2 + bx + c$, (a not a square) may be factored more readily as follows: Multiplying by a we get $a^2x^2 + bax + ac$. Writing z for ax, this becomes $z^2 + bz + ac$. Factor this trinomial by Art. 65, replace the value of z, and divide the result by a. Thus,

3. Resolve into factors $6x^2 - 13x + 6$.

Multiplying by 6 we get $(6x)^2 - 13(6x) + 36$.

Putting z for $6x$ we get

$$z^2 - 13z + 36,$$

which, being factored, gives

$$(z - 9)(z - 4).$$

Hence the required factors of $6x^2 - 13x + 6$ are

$$\tfrac{1}{6}(6x - 9)(6x - 4) = (2x - 3)(3x - 2).$$

Resolve into factors

4. $3x^2 + 5x + 2$.	*Ans.* $(3x + 2)(x + 1)$.
5. $2x^2 + 5x + 2$.	$(2x + 1)(x + 2)$.
6. $2x^2 + 7x + 6$.	$(2x + 3)(x + 2)$.
7. $3x^2 + 8x + 4$.	$(3x + 2)(x + 2)$.
8. $2x^2 + 11x + 5$.	$(2x + 1)(x + 5)$.

9.	$3x^2 + 10x + 3.$	*Ans.* $(3x + 1)(x + 3).$
10.	$3x^2 + 11x + 6.$	$(3x + 2)(x + 3).$
11.	$4x^2 + 11x - 3.$	$(4x - 1)(x + 3).$
12.	$3x^2 + x - 2.$	$(3x - 2)(x + 1).$
13.	$2x^2 + 3x - 2.$	$(2x - 1)(x + 2).$
14.	$2x^2 + 15x - 8.$	$(2x - 1)(x + 8).$
15.	$3x^2 - 19x - 14.$	$(3x + 2)(x - 7).$
16.	$6x^2 - 31x + 35.$	$(3x - 5)(2x - 7).$
17.	$3x^2 + 19x - 14.$	$(3x - 2)(x + 7).$
18.	$4x^2 + x - 14.$	$(4x - 7)(x + 2).$

67. To Factor an Expression which is the Difference of Two Squares. — From (3) of Art. 41, we see that the difference of two squares is equal to the product of the sum and difference of their square roots. Therefore, conversely, to find the factors we have the following

RULE.

Extract the square roots of the two terms; take the sum of the results for one factor, and the difference for the other.

EXAMPLES.

1. Resolve into factors $25x^2 - 16y^2$.

$$25x^2 - 16y^2 = (5x)^2 - (4y)^2.$$

Therefore the first factor is the sum of $5x$ and $4y$, and the second is the difference of $5x$ and $4y$.

$$\therefore \quad 25x^2 - 16y^2 = (5x + 4y)(5x - 4y).$$

The intermediate steps may usually be omitted.

Resolve into factors

2.	$x^2 - 9a^2.$	*Ans.* $(x + 3a)(x - 3a).$
3.	$121 - x^2.$	$(11 + x)(11 - x).$
4.	$y^2 - 25x^2.$	$(y + 5x)(y - 5x).$
5.	$36x^2 - 25b^2.$	$(6x + 5b)(6x - 5b).$
6.	$x^2y^2 - 36.$	$(xy + 6)(xy - 6).$
7.	$a^2b^2 - 4c^2d^2.$	$(ab + 2cd)(ab - 2cd).$

8. $81a^4 - 49x^4$.	*Ans.* $(9a^2 + 7x^2)(9a^2 - 7x^2)$.
9. $9a^4 - 121$.	$(3a^2 + 11)(3a^2 - 11)$.
10. $x^6 - 25$.	$(x^3 + 5)(x^3 - 5)$.
11. $x^4a^2 - 49$.	$(x^2a + 7)(x^2a - 7)$.
12. $a^2 - 64x^6$.	$(a + 8x^3)(a - 8x^3)$.

68. When One or Both of the Squares is a Compound Expression. — Here the same method is employed as in the last Article.

EXAMPLES.

1. Resolve into factors $(2a + b)^2 - 9x^2$.

The sum of $2a + b$ and $3x$ is $2a + b + 3x$,

and their difference is $\qquad 2a + b - 3x$.

$\therefore \quad (2a + b)^2 - 9x^2 = (2a + b + 3x)(2a + b - 3x)$.

If the factors contain like terms they should be collected so as to give the result in its simplest form.

2. Resolve into factors $(5x + 8y)^2 - (4x - 3y)^2$.

The sum of $5x + 8y$ and $4x - 3y$

is $\qquad 5x + 8y + 4x - 3y = 9x + 5y$,

and their difference is

$\qquad 5x + 8y - 4x + 3y = x + 11y$.

$\therefore \quad (5x + 8y)^2 - (4x - 3y)^2 = (9x + 5y)(x + 11y)$.

Resolve into factors

3. $(a + b)^2 - c^2$.	*Ans.* $(a + b + c)(a + b - c)$.
4. $(a - b)^2 - c^2$.	$(a - b + c)(a - b - c)$.
5. $(x + y)^2 - 4z^2$.	$(x + y + 2z)(x + y - 2z)$.
6. $(x + 2y)^2 - a^2$.	$(x + 2y + a)(x + 2y - a)$.
7. $(a - 2x)^2 - b^2$.	$(a - 2x + b)(a - 2x - b)$.
8. $(2x - 3a)^2 - 9c^2$.	$(2x - 3a + 3c)(2x - 3a - 3c)$.
9. $(x + y)^2 - x^2$.	$(2x + y)y$.
10. $x^2 - (y - x)^2$.	$y(2x - y)$.
11. $(x + 3y)^2 - 4y^2$.	$(x + 5y)(x + y)$.
12. $(7x + 3)^2 - (5x - 4)^2$.	$(12x - 1)(2x + 7)$.

69. Compound Quantities expressed as the Difference of Two Squares.

— Compound expressions, by suitably grouping the terms, can often be expressed as the difference of two squares, and so be resolved into factors.

EXAMPLES.

1. Resolve into factors $a^2 + 2ax + x^2 - 4b^2$.

From (1) of Art. 41, $a^2 + 2ax + x^2 = (a + x)^2$.

∴ $a^2 + 2ax + x^2 - 4b^2 = (a + x)^2 - 4b^2$,

which by Art. 68 $= (a + x + 2b)(a + x - 2b)$.

2. Resolve into factors $9a^2 - c^2 + 4cx - 4x^2$.

$$9a^2 - c^2 + 4cx - 4x^2 = 9a^2 - (c^2 - 4cx + 4x^2)$$
$$= 9a^2 - (c - 2x)^2 \text{ by (2) of Art. 41,}$$
$$= (3a + c - 2x)(3a - c + 2x).$$

3. Resolve into factors $24xy + 25 - 16x^2 - 9y^2$.

$$24xy + 25 - 16x^2 - 9y^2 = 25 - (16x^2 - 24xy + 9y^2)$$
$$= 25 - (4x - 3y)^2 \text{ by (2) of Art. 41,}$$
$$= (5 + 4x - 3y)(5 - 4x + 3y).$$

4. Resolve into factors $2bd - c^2 - a^2 + d^2 + b^2 + 2ac$.

Here we see that the expression is composed of two trinomials, each of which is the square of a binomial [(1) and (2) of Art. 41].

∴ $2bd - c^2 - a^2 + d^2 + b^2 + 2ac = b^2 + 2bd + d^2 - (a^2 - 2ac + c^2)$
$$= (b + d)^2 - (a - c)^2$$
$$= (b + d + a - c)(b + d - a + c).$$

Resolve into factors

5. $x^2 + 2xy + y^2 - a^2$.　　*Ans.* $(x + y + a)(x + y - a)$.

6. $x^2 - 6ax + 9a^2 - 16b^2$.　　$(x - 3a + 4b)(x - 3a - 4b)$.

7. $4a^2 + 4ab + b^2 - 9c^2$.　　$(2a + b + 3c)(2a + b - 3c)$.

8. $x^2 + a^2 + 2ax - y^2$.　　$(x + a + y)(x + a - y)$.

9. $c^2 - x^2 - y^2 + 2xy$.　　$(c + x - y)(c - x + y)$.

10. $x^2 + y^2 + 2xy - 4x^2y^2$.　　$(x + y + 2xy)(x + y - 2xy)$.

70. To Factor an Expression which can be Written as the Sum or the Difference of Two Cubes.

— Such expressions as these may be resolved into factors by Art. 51.

EXAMPLES.

1. Resolve into factors $8x^3 - 27y^3$.

By I. of Art. 51, this is divisible by $2x - 3y$.

$$\therefore \quad 8x^3 - 27y^3 = (2x)^3 - (3y)^3$$
$$= (2x - 3y)(4x^2 + 6xy + 9y^2).$$

Note. — The middle term $6xy$ is the product of $2x$ and $3y$.

2. Resolve into factors $343x^6 + 27y^3$.

$$343x^6 + 27y^3 = (7x^2 + 3y)(49x^4 - 21x^2y + 9y^2),$$
(by III. of Art. 51).

Resolve into factors

3. $8a^3 - 27y^3$.	*Ans.* $(2a - 3y)(4a^2 + 6ay + 9y^2)$.
4. $1 - 343x^3$.	$(1 - 7x)(1 + 7x + 49x^2)$.
5. $a^3b^3 + 512$.	$(ab + 8)(a^2b^2 - 8ab + 64)$.
6. $343 + 8x^3$.	$(7 + 2x)(49 - 14x + 4x^2)$.
7. $216x^3 - 343$.	$(6x - 7)(36x^2 + 42x + 49)$.
8. $27x^3 - 64y^3$.	$(3x - 4y)(9x^2 + 12xy + 16y^2)$.
9. $64x^6 + 125y^3$.	$(4x^2 + 5y)(16x^4 - 20x^2y + 25y^2)$.
10. $216x^6 - b^3$.	$(6x^2 - b)(36x^4 + 6x^2b + b^2)$.
11. $a^3 + 343b^3$.	$(a + 7b)(a^2 - 7ab + 49b^2)$.

71. Miscellaneous Cases of Resolution into Factors.

— When an expression can be arranged as the difference of two squares, it may be factored either by (II.) of Art. 51 or by (3) of Art. 41. It will be found the simplest, however, first to factor by the second method, using the rule for factoring the difference of two squares (Art. 67).

EXAMPLES.

1. Resolve into factors $16a^4 - 81b^4$.

$16a^4 - 81b^4 = (4a^2 + 9b^2)(4a^2 - 9b^2)$ (Art. 67)
$$= (4a^2 + 9b^2)(2a + 3b)(2a - 3b) \text{ (Art. 67)}.$$

2. Resolve into factors $x^6 - y^6$.

$x^6 - y^6 = (x^3 + y^3)(x^3 - y^3)$ (Art. 67)
$$= (x + y)(x^2 - xy + y^2)(x - y)(x^2 + xy + y^2)$$
(Art. 51, I. and III.).

The student should be careful in every case to remove all monomial factors that are common to each term of an expression, and place them outside a parenthesis, as explained in Art. 63.

3. Resolve into factors $28x^4y + 64x^3y - 60x^2y$.

$28x^4y + 64x^3y - 60x^2y = 4x^2y(7x^2 + 16x - 15)$
$$= 4x^2y(7x - 5)(x + 3)\text{(Art. 66)}.$$

4. Resolve into factors $x^3a^2 - 8y^3a^2 - 4x^3b^2 + 32y^3b^2$.

$x^3a^2 - 8y^3a^2 - 4x^3b^2 + 32y^3b^2 = a^2(x^3 - 8y^3) - 4b^2(x^3 - 8y^3)$
$$= (x^3 - 8y^3)(a^2 - 4b^2)$$
$$= (x - 2y)(x^2 + 2xy + 4y^2)(a + 2b)(a - 2b).$$

5. Resolve into factors $4x^2 - 25y^2 + 2x + 5y$.

$4x^2 - 25y^2 + 2x + 5y = (2x + 5y)(2x - 5y) + 2x + 5y$
$$= (2x + 5y)(2x - 5y + 1).$$

Resolve into two or more factors

6. $a^2 - y^2 - 2yz - z^2$.   *Ans.* $(a + y + z)(a - y - z)$.
7. $6x^2 - x - 77$. $(3x - 11)(2x + 7)$.
8. $x^6 - 4096$. $(x+4)(x^2-4x+16)(x-4)(x^2+4x+16)$.
9. $x^2 - a^2 + y^2 - 2xy$. $(x - y + a)(x - y - a)$.
10. $acx^2 - bcx + adx - bd$. $(cx + d)(ax - b)$.
11. $(a + b + c)^2 - (a - b - c)^2$. $4a(b + c)$.

Other expressions which, by a slight modification, can be arranged as the difference of two squares, may be factored by Art. 67.

12. Resolve into factors $x^4 + x^2y^2 + y^4$.

$$x^4 + x^2y^2 + y^4 = x^4 + 2x^2y^2 + y^4 - x^2y^2$$
$$= (x^2 + y^2)^2 - x^2y^2$$
$$= (x^2 + y^2 + xy)(x^2 + y^2 - xy).$$

13. Resolve into factors $x^4 - 15x^2y^2 + 9y^4$.

$$x^4 - 15x^2y^2 + 9y^4 = (x^2 - 3y^2)^2 - 9x^2y^2$$
$$= (x^2 - 3y^2 + 3xy)(x^2 - 3y^2 - 3xy).$$

Expressions which can be put into the form $x^3 \pm \dfrac{1}{y^3}$ may be factored by the rules for resolving the sum or the difference of two cubes (Art. 70).

14. Resolve into factors $\dfrac{8}{x^3} - 27y^6$.

$$\frac{8}{x^3} - 27y^6 = \left(\frac{2}{x}\right)^3 - (3y^2)^3$$
$$= \left(\frac{2}{x} - 3y^2\right)\left(\frac{4}{x^2} + \frac{6y^2}{x} + 9y^4\right).$$

15. Resolve $a^2x^3 - \dfrac{8a^2}{y^3} - x^3 + \dfrac{8}{y^3}$ into four factors.

$$a^2x^3 - \frac{8a^2}{y^3} - x^3 + \frac{8}{y^3} = x^3(a^2-1) - \frac{8}{y^3}(a^2-1)$$
$$= (a^2-1)\left(x^3 - \frac{8}{y^3}\right)$$
$$= (a+1)(a-1)\left(x - \frac{2}{y}\right)\left(x^2 + \frac{2x}{y} + \frac{4}{y^2}\right).$$

16. Resolve $a^9 - 64a^3 - a^6 + 64$ into six factors.

The expression
$$= a^3(a^6-64) - (a^6-64)$$
$$= (a^6-64)(a^3-1)$$
$$= (a^3+8)(a^3-8)(a^3-1)$$
$$= (a+2)(a^2-2a+4)(a-2)(a^2+2a+4)(a-1)(a^2+a+1).$$

Resolve into factors

17. $x^4 + 16x^2 + 256$.   *Ans.* $(x^2 + 4x + 16)(x^2 - 4x + 16)$.

18. $x^4 + y^4 - 7x^2y^2$. $(x^2 + 3xy + y^2)(x^2 - 3xy + y^2)$.

19. $81a^4 + 9a^2b^2 + b^4$. $(9a^2 + 3ab + b^2)(9a^2 - 3ab + b^2)$.

20. $x^4 - 19x^2y^2 + 25y^4$. $(x^2 + 3xy - 5y^2)(x^2 - 3xy - 5y^2)$.

By a skilful use of factors, the actual processes of multiplication and division can often be partially or wholly avoided.

21. Multiply $2a + 3b - c$ by $2a - 3b + c$.

The product $= [2a + (3b - c)][2a - (3b - c)]$
$= (2a)^2 - (3b - c)^2 \; [(3) \text{ of Art. } 41]$
$= 4a^2 - 9b^2 + 6bc - c^2.$

22. Divide the product of $x^2 - 5xy + 6y^2$ and $x - 4y$ by $x^2 - 7xy + 12y^2$.

We might multiply the first two expressions together and then divide the result by the third. But by factoring the first and third expressions, and denoting the division by means of a fraction (see Art. 78), the work will be much shorter.

Thus, the required quotient

$$= \frac{(x^2 - 5xy + 6y^2)(x - 4y)}{x^2 - 7xy + 12y^2}$$
$$= \frac{(x - 3y)(x - 2y)(x - 4y)}{(x - 4y)(x - 3y)}$$
$$= x - 2y.$$

23. Divide the product of $2x^2 + x - 6$ and $6x^2 - 5x + 1$ by $3x^2 + 5x - 2$. 　　　　　*Ans.* $(2x - 3)(2x - 1)$.

Find the product of

24. $2x-7y+3z$ and $2x+7y-3z$. 　$4x^2-49y^2+42yz-9z^2$.
25. $x^3+2x^2y+2xy^2+y^3$ and $x^3-2x^2y+2xy^2-y^3$. 　x^6-y^6.
26. $x^3 -4x^2+ 8x - 8$ and $x^3 + 4x^2 + 8x + 8$. 　$x^6 - 64$.

Divide

27. $2x(x^2-1)(x+2)$ by $x^2 + x - 2$. 　　　　$2x(x+1)$.
28. $(x^2 + 7x + 10)(x + 3)$ by $x^2 + 5x + 6$. 　　$x + 5$.
29. $5x(x-11)(x^2-x-156)$ by x^3+x^2-132x. 　$5(x-13)$.
30. $a^9 - b^9$ by $(a^2 + ab + b^2)(a^6 + a^3b^3 + b^6)$. 　　$a - b$.
31. $[x^2 + (a - b)x - ab][x^2 - (a - b)x - ab]$ by $x^2 + (a + b)x + ab$. 　　　　*Ans.* $(x - a)(x - b)$.

GREATEST COMMON DIVISOR.

72. Definitions.— *A Common Divisor* of two or more expressions is an expression that will divide each of them exactly.

Hence, *every factor common to two or more expressions is a common divisor of those expressions* (Art. 62).

Thus, in $4a^2b$, $6a^3b^2$, and a^4b^3, a^2 occurs as a factor of each quantity; b also occurs as a factor of each quantity; a^2 and b are therefore common divisors of these three quantities.

The Greatest Common Divisor of two or more Algebraic expressions is the expression of highest degree (Art. 18) which will divide each of them exactly.

Note. — The term *greatest common divisor*, which has been adopted from Arithmetic, does not imply in Algebra that it is numerically the greatest, but that it is the *factor of greatest degree*. The student is cautioned against being misled by the analogy between the Algebraic and the Arithmetic greatest common divisor. He should notice that no mention is made of *numerical magnitude* in the definition of the Algebraic greatest common divisor. In Arithmetic, the *greatest common divisor* of two or more whole numbers is the greatest whole number which will exactly divide each of them. But in Algebra, the terms *greater* and *less* are seldom applicable to those expressions in which definite numerical values have not been assigned to the various letters which occur. Besides, it is not always true that the Arithmetic greatest common divisor of the values of two given expressions obtained by assigning any particular values to the letters of those expressions, is the numerical value of the Algebraic greatest common divisor when those same values of the letters are substituted therein, as will be shown later (Art. 74). For this reason, some writers have used the terms, *highest common divisor*, and *highest common factor*, instead of the term *greatest common divisor*. But to avoid employing a new phrase, and in conformity with well-established usage, we shall retain the old term *greatest common divisor*.

The abbreviation G. C. D. will often be used for shortness instead of the words *greatest common divisor*.

73. The Greatest Common Divisor of Monomials, and of Polynomials which can be easily Factored. — Let it be required to find the greatest common divisor of $21a^4x^3y$, $35a^2x^4y$, $28a^3x^2y^4$, and $14a^5x^2y^2$.

By separating each expression into its prime factors, we have

$$21a^4x^3y = 7 \times 3aaaaxxxy.$$
$$35a^2x^4y = 7 \times 5aaxxxxy.$$
$$28a^3x^2y^4 = 7 \times 2 \times 2aaaxxyyyy.$$
$$14a^5x^2y^2 = 7 \times 2aaaaaxxyy.$$

By examining these expressions we find that 7, aa, xx, and y are the only factors common to all of them. Hence all the expressions can be exactly divided by either of these factors, or by their product, $7a^2x^2y$, which is therefore their greatest common divisor.

Find the G. C. D. of $4cx^3$ and $2cx^3 + 4c^2x^2$.

Resolving each expression into its factors, we have

$$4cx^3 = 2cx^2 \times 2x.$$
$$2cx^3 + 4c^2x^2 = 2cx^2(x + 2c).$$

Here it is clear that both expressions are divisible (1) by 2, which is the numerical greatest common divisor of the coefficients, (2) by c, and (3) by x^2.

$$\therefore \text{ G. C. D.} = 2cx^2.$$

Find the G. C. D. of $3a^2 + 9ab$, $a^3 - 9ab^2$, $a^3 + 6a^2b + 9ab^2$.

Resolving each expression into its factors, we have

$$3a^2 + 9ab = 3a(a + 3b).$$
$$a^3 - 9ab^2 = a(a + 3b)(a - 3b).$$
$$a^3 + 6a^2b + 9ab^2 = a(a + 3b)(a + 3b).$$
$$\therefore \text{ G. C. D.} = a(a + 3b).$$

Find the G. C. D. of $x(a - x)^2$, $ax(a - x)^3$, $2ax(a - x)^4$.

Resolving into factors, we have

$$x(a - x)^2 = x(a - x)(a - x).$$
$$ax(a - x)^3 = ax(a - x)(a - x)(a - x).$$
$$2ax(a - x)^4 = 2ax(a - x)(a - x)(a - x)(a - x).$$
$$\therefore \text{ G. C. D.} = x(a - x)^2.$$

Hence the following

RULE.

Resolve each expression into its prime factors, and take the product of all the factors common to all the expressions, giving to each factor the highest power which is common to all the given expressions.

. EXAMPLES.

Find the G. C. D. of

1. $4ab^2$, $2a^2b$, $6ab^3$.   *Ans.* $2ab$.
2. $3x^2y^2$, x^3y^2, x^2y^3. x^2y^2.
3. $6xy^2z$, $8x^2y^3z^2$, $4xyz^2$. $2xyz$.
4. $5a^3b^3$, $15abc^2$, $10a^2b^2c$. $5ab$.
5. $9x^2y^2z^2$, $12xy^3z$, $6x^3y^2z^3$. $3xy^2z$.
6. $8a^2x$, $6abxy$, $10abx^3y^2$. $2ax$.
7. $a^2 + ab$, $a^2 - b^2$. $a + b$.
8. $(x + y)^2$, $x^2 - y^2$. $x + y$.
9. $x^3 + x^2y$, $x^3 + y^3$. $x + y$.
10. $a^3 - a^2x$, $a^3 - ax^2$, $a^4 - ax^3$. $a(a - x)$.
11. $x^4 - 27a^3x$, $(x - 3a)^2$. $x - 3a$.
12. $xy - y$, $x^4y - xy$. $y(x - 1)$.
13. $ax^2 + 2a^2x + a^3$, $2ax^2 - 4a^2x - 6a^3$, $3(ax + a^2)^2$.

Ans. $a(x + a)$.

**74. The Greatest Common Divisor of Expressions
that cannot be Readily Resolved into Factors.** —
To find the G. C. D. in such cases, we adopt a method
analogous to that used in Arithmetic for finding the G. C. D.
of two or more numbers.

The method depends on two principles.

1. *If an expression contain a certain factor, any multiple
of that expression is divisible by that factor.*

Thus, if F divides A it will also divide mA. For let a
denote the quotient when A is divided by F; then $A = aF$;
therefore $mA = maF$; and therefore F divides mA.

2. *If two expressions have a common factor, it will divide
their sum and their difference; and also the sum and the
difference of any multiple of them.*

Thus, if F divides A and B, it will divide $mA \pm nB$. For
since F divides A and B, we may suppose $A = aF$, and $B = bF$;
therefore $mA \pm nB = maF \pm nbF$
                            $= F(ma \pm nb)$.

Therefore F divides $mA \pm nB$.

We can now prove the rule for finding the G. C. D. of any two compound Algebraic expressions.

Let A and B denote the two expressions. Let them be arranged in ascending or descending powers of some common letter; and let the highest power of that letter in B be either equal to or greater than the highest power in A.

Divide B by A; let p be the quotient and C the remainder. Suppose C to have a *simple* factor m. Remove this factor, and so obtain a new divisor D. Suppose further, that in order to make A divisible by D it is necessary to multiply A by a *simple* factor n. Divide nA by D; let q be the next quotient, and E the remainder. Divide D by E; let r be the quotient, and suppose that there is no remainder. Then E will be the G. C. D. required.

The operation of division will stand thus:

$$A)\,B\,(p$$
$$\underline{pA}$$
$$m)\,C$$
$$\qquad \underline{}$$
$$\quad D)\,nA\,(q$$
$$\quad\; \underline{qD}$$
$$\qquad E)\,D\,(r$$
$$\qquad\; \underline{rE}$$

First, to show that E is a common divisor of A and B. From the above division we have the following results:

$D = rE.$

$nA = qD + E = qrE + E = (qr + 1)E.$

$B = pA + C = pA + mD = \dfrac{pqrE + pE}{n} + mrE$

$$= \left(\dfrac{pqr + p}{n} + mr\right)E.$$

Therefore E is a common divisor of A and B.

Second, to show that E is the *greatest* common divisor of A and B.

By (2) of this Art. every common factor of A and B divides also $B - pA$, that is C. and therefore D (since m is a simple factor). Similarly as it divides A and D it divides $nA - qD$, that is E. But no expression of higher degree than E can divide E. Therefore E is the *greatest common divisor* of A and B.

The greatest common divisor of three expressions, A, B, C, may be obtained as follows:

First find D, the G. C. D. of any two of them, say of A and B; next find F, the G. C. D. of D and C; then F will be the G. C. D. of A, B, C. For D contains *every* factor which is common to A and B (Art. 72); and as F is the G. C. D. of D and C, it contains every factor common to D and C, and therefore every factor common to A, B, and C. Hence F is the G. C. D. of A, B, C.

EXAMPLES.

1. Find the G. C. D. of $x^2 - 4x + 3$ and $4x^3 - 9x^2 - 15x + 18$.

$$x^2 - 4x + 3) 4x^3 - 9x^2 - 15x + 18 (4x + 7$$
$$\underline{4x^3 - 16x^2 + 12x}$$
$$7x^2 - 27x + 18$$
$$\underline{7x^2 - 28x + 21}$$
$$x - 3) x^2 - 4x + 3 (x - 1$$
$$\underline{x^2 - 3x}$$
$$- x + 3$$
$$\underline{- x + 3}$$

Therefore the G. C. D. is $x - 3$.

EXPLANATION. — First arrange the given expressions according to descending powers of x. Take for dividend that expression whose first term is of the higher degree; and continue each division until the first term of the remainder is of a lower degree than the first term of the divisor. When the first remainder, $x - 3$, is made the divisor, we put the first divisor to the right of it for a dividend, and after obtaining the new quotient, $x - 1$, we have nothing for a remainder. Hence, as in Arithmetic, the last divisor, $x - 3$, is the G. C. D. required.

2. Find the G. C. D. of $8x^3 - 2x^2 - 53x - 39$ and $4x^3 - 3x^2 - 24x - 9$.

x	$4x^3 - 3x^2 - 24x - 9$	$8x^3 - 2x^2 - 53x - 39$	2
	$4x^3 - 5x^2 - 21x$	$8x^3 - 6x^2 - 48x - 18$	
$2x$	$2x^2 - 3x - 9$	$4x^2 - 5x - 21$	2
	$2x^2 - 6x$	$4x^2 - 6x - 18$	
3	$3x - 9$	$x - 3$	
	$3x - 9$		

Therefore the G. C. D. is $x - 3$.

EXPLANATION. — First arrange the given expressions according to descending powers of x. The expressions so arranged having their first terms of the same order, we take for divisor that whose highest power has the smaller coefficient, and arrange the work in parallel columns, as above. (1) At the first division we put the quotient 2 to the *right* of the dividend. (2) When the first remainder $4x^2 - 5x - 21$ is made the divisor we put the quotient x to the *left* of the dividend. (3) When the second remainder $2x^2 - 3x - 9$ is made the divisor we put the quotient 2 to the *right* of the dividend. (4) When the third remainder $x - 3$ is made the divisor we put the quotients $2x$ and 3 to the left of the dividend, and so on.

This method is used only to determine the *compound* factor of the G. C. D. *Simple* factors of the given expressions must first be separated from them, and the G. C. D. of these, if they have any, must be reserved and multiplied into the *compound* factor obtained by the rule.

3. Find the G. C. D. of $6x^4 - 26x^3 + 46x^2 - 42x$ and $18x^4 + 3x^3 - 132x^2 + 63x$.

We have

$$6x^4 - 26x^3 + 46x^2 - 42x = 2x(3x^3 - 13x^2 + 23x - 21), \quad . \ (1)$$

and

$$18x^4 + 3x^3 - 132x^2 + 63x = 3x(6x^3 + x^2 - 44x + 21). \quad . \ (2)$$

The simple factor 2 is found in the first expression and not in the second; therefore it forms no part of the G. C. D., and may be rejected. Likewise the simple factor 3, occurring in the second expression and not in the first, may be rejected as forming no part of the G. C. D. But the simple factor x is common to *both* expressions, and is therefore a factor of the G. C. D. and must be *reserved*. *Rejecting* therefore the simple factors 2 and 3 as forming no part of the G. C. D., and *reserving* the common factor x as forming a part of the G. C. D., and arranging in parallel columns, we have

$$3x^3 - 13x^2 + 23x - 21 \enspace \bigg| \begin{array}{l} 6x^3 + x^2 - 44x + 21 \\ 6x^3 - 26x^2 + 46x - 42 \\ \hline 27x^2 - 90x + 63 \end{array} \bigg| \enspace 2$$

The first division ends here, since $27x^2$ is of a lower degree than $3x^3$. If we now make $27x^2 - 90x + 63$ a divisor we find that it is not contained in $3x^3 - 13x^2 + 23x - 21$ with an *integral* quotient. But, noticing that $27x^2 - 90x + 63$ may be written in the form $9(3x^2 - 10x + 7)$, and remembering that the G. C. D. we are seeking *is contained in the remainder* $9(3x^2 - 10x + 7)$, and that, since the two expressions $3x^3 - 13x^2 + 23x - 21$ and $6x^3 + x^2 - 44x + 21$ have no *simple* factors, therefore their G. C. D. can have none, we conclude that the G. C. D. must be contained in the factor $3x^2 - 10x + 7$, and that therefore we can *reject* the simple factor 9, and go on with the divisor $3x^2 - 10x + 7$. Resuming the work, we have

$$\begin{array}{r} x \end{array} \begin{array}{l} 3x^3 - 13x^2 + 23x - 21 \\ \,3x^3 - 10x^2 + 7x \\ \hline \end{array}$$

Therefore the G. C. D is $x(3x - 7)$.

The factor 2 was removed for the same reason as the factor 9.

4. Find the G. C. D. of $2x^3 + x^2 - x - 2$ (1) and $3x^3 - 2x^2 + x - 2$ (2).

As the expressions stand neither can be divided by the other without obtaining a fractional quotient. This difficulty cannot be obviated by *removing* a simple factor, since neither expression contains a simple factor. We may however *introduce* a suitable factor into either expression, just as in Ex. 3 we removed a factor when we could no longer proceed with the division without a fractional quotient. The given expressions (1) and (2) have no common *simple* factor, therefore their G. C. D. can have no simple factor, and hence cannot be affected if we multiply either of them by any simple factor.

Multiplying (2) by 2 and taking it for dividend, we have

$$
\begin{array}{r|l|l|r}
 & 2x^3 + x^2 - x - 2 & 6x^3 - 4x^2 + 2x - 4 & 3 \\
 & 7 & 6x^3 + 3x^2 - 3x - 6 & \\ \hline
-2x & 14x^3 + 7x^2 - 7x - 14 & -7x^2 + 5x + 2 & \\
 & 14x^3 - 10x^2 - 4x & 17 & \\ \hline
17x & 17x^2 - 3x - 14 & -119x^2 + 85x + 34 & -7 \\
 & 17x^2 - 17x & -119x^2 + 21x + 98 & \\ \hline
14 & 14x - 14 & 64)\,64x - 64 & \\
 & 14x - 14 & x - 1 & \\
\end{array}
$$

Therefore the G. C. D. is $x - 1$.

In this example, after the first division the factor 7 is introduced because the first remainder $-7x^2 + 5x + 2$ will not divide the first divisor $2x^3 + x^2 - x - 2$.

After the second division the factor 17 is introduced because the second remainder $17x^2 - 3x - 14$ will not divide the second divisor $-7x^2 + 5x + 2$. Finally the factor 64 is removed as explained in Ex. 3.

NOTE. — The difference between the Algebraic G. C. D. and the Arithmetic G. C. D. can be seen by an example.

Factor (1) and (2) of last example as follows:

$$2x^3 + x^2 - x - 2 = (x - 1)(2x^2 + 3x + 2),$$
and $3x^3 - 2x^2 + x - 2 = (x - 1)(3x^2 + x + 2).$

Now since the G. C. D. of these expressions is $x - 1$, the factors $2x^2 + 3x + 2$, and $3x^2 + x + 2$, have no common factor. But if we put $x = 4$, then

$$2x^3 + x^2 - x - 2 = 138,$$
and $$3x^3 - 2x^2 + x - 2 = 162,$$

and the G. C. D. of 138 and 162 is 6, while 3 is the numerical value of the Algebraic G. C. D., $x - 1$. Thus the numerical value of the Algebraic G. C. D. does not agree with the numerical value of the Arithmetic G. C. D.

The reason may be explained as follows; the expressions $2x^2 + 3x + 2$ and $3x^2 + x + 2$ have no Algebraic common factor; but when $x = 4$ they become equal to 46 and 54 respectively, and therefore have a common Arithmetic factor 2, which, multiplied into $x - 1$ or 3, gives 6 for the numerical value of the Arithmetic G. C. D., while 3 is the numerical value of the Algebraic G. C. D. In the same way it may be shown that if we give particular numerical values to the letters in any two expressions, and in their Algebraic G. C. D., the numerical value of the G. C. D. is by no means necessarily the Arithmetic G. C. D. of the values of the expressions.

We may now enunciate the rule for finding the greatest common divisor of two compound Algebraic expressions.

Rule.

Arrange the given expressions according to the descending powers of the same letter. Divide that expression which is of the higher degree by the other; or, if both are of the same degree, divide that whose first term has the larger coefficient by the other; and if there is no remainder the first divisor will be the required greatest common divisor.

If there is a remainder divide the first divisor by it, and continue thus to divide the last divisor by the last remainder, until a divisor is obtained which leaves no remainder; the last divisor will be the greatest common divisor required.

NOTE 1. — Before beginning the division, all simple factors of the given expressions must be removed from them, and the greatest common divisor of these must be reserved as a factor of the G. C. D. required. (See Ex. 3.)

NOTE 2. — Either of the given expressions or any of the remainders may be multiplied or divided by any factor which does not divide both of the given expressions. (See Ex. 4.)

NOTE 3. — Each division must be continued until the remainder is of a lower degree than the divisor.

Find the G. C. D. of

5. $x^3 + 2x^2 - 13x + 10$, and $x^3 + x^2 - 10x + 8$.

$\quad\quad\quad\quad\quad\quad\quad\quad\quad\quad$ *Ans.* $x^2 - 3x + 2$.

6. $x^3 - 5x^2 - 99x + 40$, and $x^3 - 6x^2 - 86x + 35$.　$x^2 - 13x + 5$.

7. $x^3 - x^2 - 5x - 3$, and $x^3 - 4x^2 - 11x - 6$.　　$x^2 + 2x + 1$.

8. $x^3 + 3x^2 - 8x - 24$, and $x^3 + 3x^2 - 3x - 9$.　　$x + 3$.

9. $2x^3 + 4x^2 - 7x - 14$, and $6x^3 - 10x^2 - 21x + 35$.　　$2x^2 - 7$.

LEAST COMMON MULTIPLE.

75. Definitions. — *A Multiple* of an expression is any expression that can be divided by it exactly.

Hence, *a multiple of an expression must contain all the factors of that expression.* Thus,

$\quad\quad$ $6a^2b$ is a multiple of 3 or 2 or 6 or a or b.

A Common Multiple of two or more expressions is an expression that can be divided by each of them exactly; or, it is one of which all the given expressions are factors.

Thus, the expression ab^2c^3 is a common multiple of the expressions, a, b, c, ab, abc, ab^2, b^2c^3, etc., or of the expression itself; but it is not a multiple of a^2, nor of b^3, nor of any symbol which does not enter into it as a factor.

The Least Common Multiple * of two or more Algebraic expressions is the expression of least degree which is divisible by each of them exactly.

---

* Called also *lowest common multiple.* The term, *least common multiple,* is objected to by some, for a reason similar to the one for which they object to the term *greatest common divisor.*

Hence, *the least common multiple of two or more expressions is the product of all the factors of the expressions, each factor being taken the greatest number of times it occurs in any of the expressions.*

The abbreviation L. C. M. is often used instead of the words *least common multiple.*

Note. — Two or more expressions can have only one *least* common multiple, while they have an indefinite number of common multiples.

76. The Least Common Multiple of Monomials, and of Polynomials which can be easily Factored. — Let it be required to find the least common multiple of $21a^4x^3y$, $35a^2x^4y$, $28a^3x^2y^4$, and $14a^5x^2y^2$.

By separating each expression into its prime factors, we have

$$21a^4x^3y = 3 \times 7a^4x^3y,$$
$$35a^2x^4y = 5 \times 7a^2x^4y,$$
$$28a^3x^2y^4 = 2^2 \times 7a^3x^2y^4,$$
$$14a^5x^2y^2 = 2 \times 7a^5x^2y^2.$$

Hence, the L. C. M. $= 7 \times 3 \times 5 \times 2^2a^5x^4y^4$
$$= 420a^5x^4y^4 ;$$

for 420 is the numerical L. C. M. of the coefficients ; a^5 is the lowest power of a that is divisible by each of the quantities a^4, a^2, a^3, a^5 ; x^4 is the lowest power of x that is divisible by each of the quantities x^3, x^4, x^2 ; and y^4 is the lowest power of y that is divisible by each of the quantities y, y^4, y^2.

2. Find the L. C. M. of $6x^3(a-x)^2$, $8a^2(a-x)^3$, and $12a^2x^2(a-x)^4$.

Resolving into factors, we have

$$6x^3(a-x)^2 = 3 \times 2x^3(a-x)^2,$$
$$8a^2(a-x)^3 = 2 \times 2 \times 2a^2(a-x)^3,$$
$$12a^2x^2(a-x)^4 = 3 \times 2 \times 2a^2x^2(a-x)^4.$$

Hence, the L. C. M. $= 3 \times 2^3a^2x^3(a-x)^4$
$$= 24a^2x^3(a-x)^4.$$

For it consists of the product of (1) the numerical L. C. M. of the coefficients, and (2) the lowest power of each factor which is divisible by every power of that factor occurring in the given expressions.

3. Find the L. C. M. of $3a^2 + 9ab$, $2a^3 - 18ab^2$, $a^3 + 6a^2b + 9ab^2$, $a^3 + 5a^2b + 6ab^2$.

Resolving into factors, we have

$$3a^2 + 9ab = 3a(a + 3b),$$
$$2a^3 - 18ab^2 = 2a(a + 3b)(a - 3b),$$
$$a^3 + 6a^2b + 9ab^2 = a(a + 3b)^2,$$
$$a^3 + 5a^2b + 6ab^2 = a(a + 3b)(a + 2b).$$

Hence the L. C. M. $= 6a(a + 3b)^2(a - 3b)(a + 2b)$

Hence the following

RULE.

Resolve each expression into its prime factors, and take the product of all the factors, giving to each factor the highest exponent which it has in the given expressions.

If the expressions are prime to each other, their product is the least common multiple.

EXAMPLES.

Find the least common multiple of

1. $5a^2bc^3$, $4ab^2c$. $\qquad$ *Ans.* $20a^2b^2c^3$.
2. $12ab$, $8xy$. $\qquad$ $24abxy$.
3. $2ab$, $3bc$, $4ca$. $\qquad$ $12abc$.
4. a^2bc, b^2ca, c^2ab. $\qquad$ $a^2b^2c^2$.
5. $5a^2c$, $6cb^2$, $3bc^2$. $\qquad$ $30a^2b^2c^2$.
6. x^2, $x^2 - 3x$. $\qquad$ $x^2(x - 3)$.
7. $21x^3$, $7x^2(x + 1)$. $\qquad$ $21x^3(x + 1)$.
8. $a^2 + ab$, $ab + b^2$. $\qquad$ $ab(a + b)$.
9. $6x^2 - 2x$, $9x^2 - 3x$. $\qquad$ $6x(3x - 1)$.
10. $x^2 + 2x$, $x^2 + 3x + 2$. $\qquad$ $x(x + 2)(x + 1)$.
11. $x^2 + 4x + 4$, $x^2 + 5x + 6$. $\qquad$ $(x + 2)^2(x + 3)$.
12. $x^2 + x - 20$, $x^2 - 10x + 24$, $x^2 - x - 30$. $\qquad$ *Ans.* $(x + 5)(x - 4)(x - 6)$.

77. When the Given Expressions cannot be Resolved into Factors by Inspection. — To find the least common multiple of two compound Algebraic expressions in such cases, the expressions must be resolved by finding their G. C. D.

1. Find the L. C. M. of $2x^4 + x^3 - 20x^2 - 7x + 24$ and $2x^4 + 3x^3 - 13x^2 - 7x + 15$.

We first find their G. C. D.

$$
\begin{array}{r|l|l|l}
x & 2x^4 + x^3 - 20x^2 - 7x + 24 & 2x^4 + 3x^3 - 13x^2 - 7x + 15 & 1 \\
& 2x^4 + 7x^3 \qquad\qquad -9x & 2x^4 + x^3 - 20x^2 - 7x + 24 & \\
\cline{2-2}\cline{3-3}
-3 & -6x^3 - 20x^2 + 2x + 24 & 2x^3 + 7x^2 \qquad\quad - 9 & 2x \\
& -6x^3 - 21x^2 \qquad +27 & 2x^3 + 4x^2 - 6x & \\
\cline{2-2}\cline{3-3}
& x^2 + 2x - 3 & 3x^2 + 6x - 9 & 3 \\
& & 3x^2 + 6x - 9 &
\end{array}
$$

$$\therefore \text{ G. C. D.} = x^2 + 2x - 3.$$

Hence, by division, we obtain

$$2x^4 + x^3 - 20x^2 - 7x + 24 = (x^2 + 2x - 3)(2x^2 - 3x - 8),$$

and

$$2x^4 + 3x^3 - 13x^2 - 7x + 15 = (x^2 + 2x - 3)(2x^2 - x - 5).$$

Therefore

the L. C. M. $= (x^2 + 2x - 3)(2x^2 - 3x - 8)(2x^2 - x - 5)$.

We may now prove the rule for finding the least common multiple of any two compound Algebraic expressions.

Let A and B denote the two expressions, F their greatest common divisor, and M their least common multiple. Suppose that a and b are the respective quotients when A and B are divided by F; then

$$A = aF, \quad \text{and} \quad B = bF. \quad . \quad . \quad . \quad . \quad (1)$$

Since F contains all the factors common to A and B, the quotients a and b have no common factor, and therefore their least common multiple is ab, and hence the least common multiple of aF and bF, or of A and B, from (1) is abF, by inspection. That is

$$M = abF. \quad . \quad . \quad . \quad . \quad . \quad . \quad (2)$$

But from (1) we have

$$AB = abFF$$

$$= MF, \text{ from (2)} \quad . \quad . \quad . \quad . \quad (3)$$

or

$$M = \frac{AB}{F}.$$

RULE.

The least common multiple of two expressions may be found by dividing their product by their G. C. D., or, by dividing either of the expressions by their G. C. D. and multiplying the quotient by the other.

From (3) we see that *the product of any two expressions is equal to the product of their G. C. D. and L. C. M.*

To find the least common multiple of *three* expressions, A, B, C. First find M, the L. C. M. of A and B. Next find N, the L. C. M. of M and C; then N will be the required L. C. M. of A, B, C.

For N is the expression of least degree which is divisible by M and C, and M is the expression of least degree which is divisible by A and B. Therefore N is the expression of least degree which is divisible by all three.

In a similar manner we may find the L. C. M. of four expressions.

NOTE. — The theories of the greatest common divisor and of the least common multiple are not necessary for the subsequent chapters of the present work, and any difficulties which the student may find in them may be postponed till he has read the Theory of Equations. The examples however attached to this chapter should be carefully worked, on account of the exercise which they afford in all the fundamental processes of Algebra.

EXAMPLES.

Resolve into factors

1. $x^2 + xy$.   *Ans.* $x(x + y)$.
2. $x^3 - x^2y$. $x^2(x - y)$.
3. $10x^3 - 25x^4y$. $5x^3(2 - 5xy)$.
4. $x^3 - x^2y + xy^2$. $x(x^2 - xy + y^2)$.
5. $3a^4 - 3a^3b + 6a^2b^2$. $3a^2(a^2 - ab + 2b^2)$.
6. $38a^3x^5 + 57a^4x^2$. $19a^3x^2(2x^3 + 3a)$.
7. $ax - bx - az + bz$. $(a - b)(x - z)$.
8. $2ax + ay + 2bx + by$. $(2x + y)(a + b)$.
9. $6x^2 + 3xy - 2ax - ay$. $(2x + y)(3x - a)$.
10. $2x^4 - x^3 + 4x - 2$. $(2x - 1)(x^3 + 2)$.
11. $3x^3 + 5x^2 + 3x + 5$. $(3x + 5)(x^2 + 1)$.
12. $x^4 + x^3 + 2x + 2$. $(x + 1)(x^3 + 2)$.
13. $y^3 - y^2 + y - 1$. $(y - 1)(y^2 + 1)$.
14. $2ax^2 + 3axy - 2bxy - 3by^2$. $(2x + 3y)(ax - by)$.
15. $x^2 - 19x + 84$. $(x - 12)(x - 7)$.
16. $x^2 - 19x + 78$. $(x - 13)(x - 6)$.
17. $a^2 - 14ab + 49b^2$. $(a - 7b)(a - 7b)$.
18. $a^2 + 5ab + 6b^2$. $(a + 3b)(a + 2b)$.
19. $m^2 - 13mn + 40n^2$. $(m - 8n)(m - 5n)$.
20. $m^2 - 22mn + 105n^2$. $(m - 15n)(m - 7n)$.
21. $x^2 - 23xy + 132y^2$. $(x - 12y)(x - 11y)$.
22. $130 + 31xy + x^2y^2$. $(26 + xy)(5 + xy)$.
23. $132 - 23x + x^2$. $(12 - x)(11 - x)$.
24. $88 + 19x + x^2$. $(8 + x)(11 + x)$.
25. $65 + 8xy - x^2y^2$. $(5 + xy)(13 - xy)$.
26. $x^2 + 16x - 260$. $(x + 26)(x - 10)$.
27. $x^2 - 11x - 26$. $(x + 2)(x - 13)$.
28. $a^2b^2 - 3abc - 10c^2$. $(ab + 2c)(ab - 5c)$.
29. $x^4 - a^2x^2 - 132a^4$. $(x^2 + 11a^2)(x^2 - 12a^2)$.
30. $4x^2 + 23x + 15$. $(x + 5)(4x + 3)$.
31. $12x^2 - 23xy + 10y^2$. $(3x - 2y)(4x - 5y)$.
32. $8x^2 - 38x + 35$. $(2x - 7)(4x - 5)$.
33. $12x^2 - 31x - 15$. $(12x + 5)(x - 3)$.

34. $3 + 11x - 4x^2.$ *Ans.* $(3 - x)(1 + 4x).$

35. $6 + 5x - 6x^2.$ $(2 + 3x)(3 - 2x).$

36. $4 - 5x - 6x^2.$ $(4 + 3x)(1 - 2x).$

37. $5 + 32x - 21x^2.$ $(1 + 7x)(5 - 3x).$

38. $20 - 9x - 20x^2.$ $(5 + 4x)(4 - 5x).$

39. $(1811)^2 - (689)^2.$ $2500 \times 1122 = 2805000.$

40. $(8133)^2 - (8131)^2.$ $16264 \times 2 = 32528.$

41. $(24x + y)^2 - (23x - y)^2.$ $47x(x + 2y).$

42. $(5x + 2y)^2 - (3x - y)^2.$ $(8x + y)(2x + 3y).$

43. $9x^2 - (3x - 5y)^2.$ $5y(6x - 5y).$

44. $16x^2 - (3x + 1)^2.$ $(7x + 1)(x - 1).$

45. $a^6 + 729b^3.$ $(a^2 + 9b)(a^4 - 9a^2b + 81b^2).$

46. $x^3y^3 - 512.$ $(xy - 8)(x^2y^2 + 8xy + 64).$

47. $500x^2y - 20y^3.$ $20y(5x + y)(5x - y).$

48. $(a+b)^4 - 1.$ $[(a+b)^2 + 1](a+b+1)(a+b-1).$

Find the greatest common divisor of

49. $66a^4b^2c^3, 44a^3b^4c^2, 24a^2b^3c^4.$ $2a^2b^2c^2.$

50. $x^2 + x, (x + 1)^2, x^3 + 1.$ $x + 1.$

51. $x^3 + 8y^3, x^2 + xy - 2y^2.$ $x + 2y.$

52. $12x^2 + x - 1, 15x^2 + 8x + 1.$ $3x + 1.$

53. $2x^2 + 9x + 4, 2x^2 + 11x + 5, 2x^2 - 3x - 2.$ $2x + 1.$

54. $3x^4 - 3x^3 - 2x^2 - x - 1, 6x^4 - 3x^3 - x^2 - x - 1.$ $3x^2 + 1.$

55. $2x^3 - 9ax^2 + 9a^2x - 7a^3, 4x^3 - 20ax^2 + 20a^2x - 16a^3.$

     Ans. $x^2 - ax + a^2.$

56. $4x^5 + 14x^4 + 20x^3 + 70x^2, 8x^7 + 28x^6 - 8x^5 - 12x^4 + 56x^3.$

     Ans. $2x^2(2x + 7)$

Find the least common multiple of

57. $35ax^2, 42ay^2, 30az^2.$ $210ax^2y^2z^2.$

58. $x^2 - 3x + 2, x^2 - 1.$ $(x + 1)(x - 1)(x - 2).$

59. $x^2 - 5x + 4, x^2 - 6x + 8.$ $(x - 4)(x - 1)(x - 2).$

60. $x^2 - 1, x^3 + 1, x^3 - 1.$ $x^6 - 1.$

61. $x^2 - 1, x^2 + 1, x^4 + 1, x^8 - 1.$ $x^8 - 1.$

62. $x^2 - 1, x^3 + 1, x^3 - 1, x^6 + 1.$ $x^{12} - 1.$

63. $x^3 + 2x^2 - 3x, 2x^3 + 5x^2 - 3x.$ $x(x-1)(x+3)(2x-1).$

CHAPTER VIII.

FRACTIONS.

NOTE. — In this chapter the student will find that the definitions, rules, and demonstrations closely resemble those with which he is already familiar in Arithmetic.

78. A Fraction — Entire and Mixed Quantities. — *A Fraction* is an expression of an indicated quotient by writing the divisor under the dividend with a horizontal line between them (Art. 11). In the operation of division the divisor sometimes may be greater than the dividend, or may not be contained in it an exact number of times; in either case the quotient is expressed by means of a fraction. Thus, the expression $\frac{a}{b}$ indicates either that some unit is divided into b equal parts, and that a of these are taken, or that a times the same unit is divided into b equal parts, and one of them taken.

In any fraction the upper number, or the dividend, is called the *numerator*, and the lower number, or divisor, is called the *denominator*. Thus in the above fraction $\frac{a}{b}$, which is read a divided by b, a is called the numerator and b the denominator, and the two taken together are called the *terms* of the fraction. Thus the denominator indicates into how many equal parts the unit is to be divided, and the numerator indicates how many of those parts are to be taken.

Every integer or integral expression may be considered as a fraction whose denominator is unity; thus,

$$a = \frac{a}{1}, \quad a + b = \frac{a+b}{1}.$$

An *entire quantity* or *integral quantity* is one which has no fractional part; as ab or $a^2 - 2ab$.

A *mixed quantity* is one made up of an integer and a fraction; as $b + \dfrac{x}{a}$.

A *proper fraction* is one whose numerator is less than its denominator; as $\dfrac{a}{a + x}$.

An *improper fraction* is one whose numerator is equal to or greater than the denominator; as $\dfrac{a}{a}$ and $\dfrac{a + x}{a}$.

The *reciprocal* of a fraction is another fraction having its numerator and denominator respectively equal to the denominator and numerator of the former.

79. To Reduce a Fraction to its Lowest Terms.

— Let $\dfrac{a}{b}$ denote any fraction, and $\dfrac{ma}{mb}$ denote the same fraction with its terms multiplied by m.

Now $\dfrac{a}{b}$ means that a unit is divided into b equal parts, and that a of these are taken. . . . . (1)

And $\dfrac{ma}{mb}$ means that the same unit is divided into mb equal parts, and that ma of these are taken. . . . . . . (2)

Hence $\qquad b$ parts in (1) $= mb$ parts in (2).

$\qquad \therefore \quad$ 1 part in (1) $= m$ parts in (2),

and $\qquad \therefore \quad a$ parts in (1) $= am$ parts in (2),

that is, $\qquad\qquad \dfrac{a}{b} = \dfrac{ma}{mb}.$

Conversely, $\qquad\qquad \dfrac{ma}{mb} = \dfrac{a}{b}.$

Therefore, *the value of a fraction is not altered if the numerator and denominator are either both multiplied or both divided by the same quantity.*

When both numerator and denominator are divided by all the factors common to them, the fraction is said to be reduced to its *lowest terms.*

Hence to reduce a fraction to its lowest terms we have the following

RULE.

Divide both numerator and denominator by their greatest common divisor.

Dividing both terms of a fraction by a common factor is called *canceling* that factor.

EXAMPLES.

Reduce the following fractions to their lowest terms.

1.) $\dfrac{6a^2bc^2}{9ab^2c} = \dfrac{2ac}{3b}$.

The greatest common divisor $3abc$ of both terms is canceled.

2.) $\dfrac{7x^2yz}{28x^3yz^2} = \dfrac{1}{4xz}$.

The factor $7x^2yz$, which is the greatest common divisor of both terms, is canceled.

3.) $\dfrac{24a^3c^2x^2}{18a^3x^2 - 12a^2x^3} = \dfrac{24a^3c^2x^2}{6a^2x^2(3a - 2x)} = \dfrac{4ac^2}{3a - 2x}$.

Here $6a^2x^2$ is canceled since it is the greatest common divisor of both terms.

NOTE. — In each of these examples, the resulting fractions have the same value as the given fractions, but they are expressed in a simpler form. The student should be careful not to begin canceling until he has expressed both terms of the fraction in the most convenient form, by factoring when necessary. Thus,

4. $\dfrac{6x^2 - 8xy}{9xy - 12y^2} = \dfrac{2x(3x - 4y)}{3y(3x - 4y)} = \dfrac{2x}{3y}$.

Instead of reducing a fraction to its lowest terms by dividing the numerator and denominator by their G. C. D., we may divide by *any* common factor, and repeat the process till the fraction is reduced to its lowest terms. Thus,

5. $\dfrac{24a^3b^2c^3}{36a^2b^3c^3} = \dfrac{12a^2bc^2}{18ab^2c^2} = \dfrac{6ac}{9bc} = \dfrac{2a}{3b}.$

Reduce the following fractions to their lowest terms.

6. $\dfrac{8a^2bc^2}{12ab^2cd}.$ *Ans.* $\dfrac{2ac}{3bd}$

7. $\dfrac{3a^2 - 6ab}{2a^2b - 4ab^2}.$ $\dfrac{3}{2b}.$

8. $\dfrac{4x^2 - 9y^2}{4x^2 + 6xy}.$ $\dfrac{2x - 3y}{2x}.$

9. $\dfrac{20(x^3 - y^3)}{5x^2 + 5xy + 5y^2}.$ $4(x - y).$

10. $\dfrac{x^3 - 2xy^2}{x^4 - 4x^2y^2 + 4y^4}.$ $\dfrac{x}{x^2 - 2y^2}.$

When the factors of the numerator and denominator cannot be found by inspection, their greatest common divisor may be found by the rule (Art. 74), and the fraction then reduced to its lowest terms.

11. Reduce to its lowest terms $\dfrac{3x^3 - 13x^2 + 23x - 21}{15x^3 - 38x^2 - 2x + 21}.$

The G. C. D. of the numerator and denominator is $3x - 7$. Dividing the numerator and denominator by $3x - 7$, we obtain the respective quotients $x^2 - 2x + 3$ and $5x^2 - x - 3$. Therefore

$$\frac{3x^3 - 13x^2 + 23x - 21}{15x^3 - 38x^2 - 2x + 21} = \frac{(3x - 7)(x^2 - 2x + 3)}{(3x - 7)(5x^2 - x - 3)} = \frac{x^2 - 2x + 3}{5x^2 - x - 3}.$$

This example may also be solved without finding the G. C. D. by the rule (Art. 74) as follows:

By 2 of Art. 74, the G. C. D. of the numerator and denominator must divide their sum $18x^3 - 51x^2 + 21x$, that is, $3x(3x - 7)(2x - 1)$. If there be a common divisor it must clearly be $3x - 7$. Hence, arranging the numerator and denominator so as to show $3x - 7$ as a factor, we have the fraction

$$= \frac{x^2(3x - 7) - 2x(3x - 7) + 3(3x - 7)}{5x^2(3x - 7) - x(3x - 7) - 3(3x - 7)}$$

$$= \frac{(3x - 7)(x^2 - 2x + 3)}{(3x - 7)(5x^2 - x - 3)} = \frac{x^2 - 2x + 3}{5x^2 - x - 3}.$$

When either the numerator or denominator can readily be factored we may use the following method:

12. Reduce to its lowest terms $\dfrac{x^3 + 3x^2 - 4x}{7x^3 - 18x^2 + 6x + 5}$.

The numerator $= x(x^2 + 3x - 4) = x(x + 4)(x - 1)$.

The only one of these factors which can be a common divisor is $x - 1$, since the denominator does not contain x, and 5 the last term in the denominator does not contain 4. (See Art. 66.) Hence, arranging the denominator so as to show $x - 1$ as a factor,

the fraction $= \dfrac{x(x+4)(x-1)}{7x^2(x-1) - 11x(x-1) - 5(x-1)} = \dfrac{x(x+4)}{7x^2 - 11x - 5}$.

Reduce to lowest terms

13. $\dfrac{a^3 - a^2b - ab^2 - 2b^3}{a^3 + 3a^2b + 3ab^2 + 2b^3}$.   *Ans.* $\dfrac{a - 2b}{a + 2b}$.

14. $\dfrac{x^3 - 5x^2 + 7x - 3}{x^3 - 3x + 2}$. $\dfrac{x - 3}{x + 2}$.

15. $\dfrac{4a^3 + 12a^2b - ab^2 - 15b^3}{6a^3 + 13a^2b - 4ab^2 - 15b^3}$. $\dfrac{2a + 5b}{3a + 5b}$.

80. To Reduce a Mixed Quantity to the Form of a Fraction. — Let it be required to reduce $a + \dfrac{bc}{b + c}$ to the form of a fraction.

The entire part $a = \dfrac{a}{1}$ (Art. 78) $= \dfrac{a(b + c)}{b + c}$ (Art. 79).

Hence $a + \dfrac{bc}{b + c} = \dfrac{a(b + c)}{b + c} + \dfrac{bc}{b + c}$

$$= \dfrac{ab + ac + bc}{b + c}.$$

Hence we have the following

RULE.

Multiply the entire part by the denominator, and to the product add the numerator with its proper sign; under this sum place the denominator, and the result will be the fraction required.

EXAMPLES.

Reduce the following to fractional forms:

1. $a - x + \dfrac{x^2}{a + x}$. $\qquad Ans. \ \dfrac{a^2}{a + x}$.

2. $x + 1 + \dfrac{4}{x - 3}$. $\qquad \dfrac{x^2 - 2x + 1}{x - 3}$.

3. $x^2 - xy + y^2 - \dfrac{y^3}{x + y}$. $\qquad \dfrac{x^3}{x + y}$.

4. $a + b - \dfrac{a^2 + b^2}{a - b}$. $\qquad -\dfrac{2b^2}{a - b}$.

81. To Reduce a Fraction to an Entire or Mixed Quantity. — Let it be required to reduce $\dfrac{ax + 2x^2}{a + x}$ to a mixed quantity.

Performing the division indicated, we have

$$\frac{ax + 2x^2}{a + x} = x + \frac{x^2}{a + x}.$$

Hence we have the following

RULE.

Divide the numerator by the denominator, as far as possible, for the entire part, and annex to the quotient a fraction having the remainder for numerator, and the divisor for denominator; it will be the mixed quantity required.

EXAMPLES.

Reduce to whole or mixed quantities the following :

1. $\dfrac{24a}{7}$.

    Ans. $3a + \dfrac{3a}{7}$.

2. $\dfrac{a^2 + 3ab}{a + b}$.

    $a + \dfrac{2ab}{a + b}$.

3. $\dfrac{36ac + 4c}{9}$.

    $4ac + \dfrac{4c}{9}$.

4. $\dfrac{8a^2 + 3b}{4a}$.

    $2a + \dfrac{3b}{4a}$.

5. $\dfrac{x^2 + 3x + 2}{x + 3}$.

    $x + \dfrac{2}{x + 3}$.

6. $\dfrac{2x^2 - 6x - 1}{x - 3}$.

    $2x - \dfrac{1}{x - 3}$.

7. $\dfrac{x^4 + 1}{x - 1}$.

    $x^8 + x^2 + x + 1 + \dfrac{2}{x - 1}$.

82. To Reduce Fractions to their Least Common Denominator. — Let it be required to reduce $\dfrac{a}{yz}, \dfrac{b}{zx}, \dfrac{c}{xy}$ to equivalent fractions having the least common denominator.

The least common multiple of the denominators is xyz. Dividing this L. C. M. by the denominators, yz, zx, and xy, we have the quotients x, y, and z, respectively.

By Art. 79, both terms of a fraction may be multiplied by the same number without altering its value ; therefore we may multiply both terms of the first fraction by x, both terms of the second fraction by y, and both terms of the third fraction by z, and the resulting fractions will be equivalent to the given ones.

Hence $\dfrac{a}{yz} = \dfrac{ax}{xyz}, \quad \dfrac{b}{zx} = \dfrac{by}{xyz}, \quad \dfrac{c}{xy} = \dfrac{cz}{xyz}$.

That is, the resulting fractions $\dfrac{ax}{xyz}, \dfrac{by}{xyz}$, and $\dfrac{cz}{xyz}$ have the same values respectively as the given fractions $\dfrac{a}{yz}, \dfrac{b}{zx}$, and $\dfrac{c}{xy}$, and they have the least common denominator xyz. Hence, for reducing fractions to their least common denominator, we have the following

RULE.

Find the least common multiple of the given denominators, and take it for the common denominator; divide it by the denominator of the first fraction, and multiply the numerator of this fraction by the quotient so obtained; and do the same with all the other given fractions.

NOTE 1. — It is not absolutely necessary to take the *least* common denominator. *Any* common denominator may be used. But in practice it will be found advisable to use the *least* common denominator, as the work will thereby be shortened.

NOTE 2. — It frequently happens that the denominators of the fractions to be reduced do not contain a common factor. Thus, the denominators of the fractions $\frac{a}{b}$, $\frac{c}{d}$, and $\frac{e}{f}$ have no common factor; therefore the least common denominator of these fractions is bdf, the product of all their denominators.

EXAMPLES.

Reduce the following fractions to their L. C. D.

1. $\dfrac{3}{4x}$, $\dfrac{4}{6x^2}$, $\dfrac{5}{12x^3}$. $Ans.$ $\dfrac{9x^2}{12x^3}$, $\dfrac{8x}{12x^3}$, $\dfrac{5}{12x^3}$.

2. $\dfrac{a}{x-a}$, $\dfrac{x}{x-a}$, $\dfrac{a^2}{x^2-a^2}$. $\dfrac{a(x+a)}{x^2-a^2}$, $\dfrac{x(x+a)}{x^2-a^2}$, $\dfrac{a^2}{x^2-a^2}$.

83. Rule of Signs in Fractions. — The signs of the several terms of the numerator and denominator of a fraction relate only to those terms to which they are prefixed, while the sign prefixed to the dividing line relates to the fraction as a whole, and is the *sign of the fraction.* Thus, in the fraction $-\dfrac{a-b}{a+b}$, the sign of a, the first term of the numerator, is $+$ understood, the sign of the second term b is $-$, and the sign of each term a and b of the denominator is $+$, while the sign of the fraction itself is $-$.

The symbol $\dfrac{-a}{-b}$ means the quotient resulting from the division of $-a$ by $-b$; and this is obtained by dividing a by b, and prefixing $+$, by the rule of signs in division (Art. 46).

Therefore $\qquad \dfrac{-a}{-b} = +\dfrac{a}{b} = \dfrac{a}{b}.$ . . . . . (1)

Also, $\dfrac{-a}{b}$ is the quotient of $-a$ divided by b; and this is obtained by dividing a by b, and prefixing $-$, by the rule of signs.

Therefore $\qquad \dfrac{-a}{b} = -\dfrac{a}{b}.$ . . . . . . (2)

In like manner, $\dfrac{a}{-b}$ is the quotient of a divided by $-b$; and this is obtained by dividing a by b, and prefixing $-$, by the rule of signs.

Therefore $\qquad \dfrac{a}{-b} = -\dfrac{a}{b}.$ . . . . . . (3)

Hence, we have the following rule of signs:

(1) *If the signs of both numerator and denominator be changed, the sign of the whole fraction remains unchanged.*

(2) *If the sign of the numerator alone be changed, the sign of the whole fraction will be changed.*

(3) *If the sign of the denominator alone be changed, the sign of the whole fraction will be changed.*

Or they may be stated as follows:

(1) *We may change the sign of every term in the numerator and denominator of a fraction without altering the value of the fraction.*

(2) *We may change the sign of a fraction by changing the sign of every term either in the numerator or denominator.*

EXAMPLES.

1. $\dfrac{a-b}{m-n} = \dfrac{-a+b}{-m+n} = \dfrac{b-a}{n-m}.$

2. $\dfrac{b-a}{2x} = -\dfrac{-b+a}{2x} = -\dfrac{a-b}{2x}.$

3. $\dfrac{3x}{x-x^2} = -\dfrac{3x}{-x+x^2} = -\dfrac{3x}{x^2-x}.$

The intermediate steps may usually be omitted.

From Art. 36 we have

$$\frac{abmx}{pqr} = \frac{(-a)(-b)(-m)x}{(-p)qr} = \frac{a(-b)(-m)x}{(-p)(-q)r} = \text{etc.}$$

That is, *if the terms of a fraction are composed of any number of factors, any even number of factors may have their signs changed without altering the value of the fraction; but if any odd number of factors have their signs changed, the sign of the fraction is changed.*

Thus, $\dfrac{(a-b)(b-c)}{(x-y)(y-z)} = \dfrac{(b-a)(b-c)}{(y-x)(y-z)} = \dfrac{(b-a)(c-b)}{(y-x)(z-y)}.$

When the numerator is a product, any one or more of its factors can be removed from the numerator and made the multiplier.

Thus, $\dfrac{abcd}{a+b} = ab\dfrac{cd}{a+b} = abcd\dfrac{1}{a+b}.$

Change the signs of the following fractions so as to express them altogether in four different ways.

4. $\dfrac{a-b}{x-y}.$ *Ans.* $\dfrac{b-a}{y-x}, \; -\dfrac{b-a}{x-y}, \; -\dfrac{a-b}{y-x}.$

5. $\dfrac{x}{y-z}.$ $\dfrac{-x}{z-y}, \; -\dfrac{-x}{y-z}, \; -\dfrac{x}{z-y}.$

6. $\dfrac{b-a}{a-b+c}.$ $\dfrac{a-b}{b-c-a}, \; -\dfrac{a-b}{a-b+c}, \; -\dfrac{b-a}{b-c-a}.$

7. $\dfrac{abcd}{xyz}.$ $\dfrac{a(-b)(-c)d}{xyz}, \; \dfrac{a(-b)cd}{xy(-z)}, \; \dfrac{abcd}{x(-y)(-z)}.$

84. Addition and Subtraction of Fractions. — Let

it be required to add together $\dfrac{a}{c}$ and $\dfrac{b}{c}$.

Here the unit is divided into c equal parts, and we first take a of these parts, and then b of them; i.e., we take $a + b$ of the c parts of the unit; and this is expressed by the fraction $\dfrac{a + b}{c}$.

$$\therefore \quad \frac{a}{c} + \frac{b}{c} = \frac{a + b}{c}.$$

Similarly
$$\frac{a}{c} - \frac{b}{c} = \frac{a - b}{c}.$$

Let it be required to add together $\dfrac{a}{b}$ and $\dfrac{c}{d}$.

We have
$$\frac{a}{b} = \frac{ad}{bd}, \text{ and } \frac{c}{d} = \frac{bc}{bd}.$$

Here in each case we divide the unit into bd equal parts, and we first take ad of these parts, and then bc of them; i.e., we take $ad + bc$ of the bd parts of the unit; and this

is expressed by the fraction $\dfrac{ad + bc}{bd}$.

$$\therefore \quad \frac{a}{b} + \frac{c}{d} = \frac{ad + bc}{bd}.$$

Similarly
$$\frac{a}{b} - \frac{c}{d} = \frac{ad - bc}{bd}.$$

Here the fractions have been reduced to a common denominator bd. But if b and d have a common factor, the product bd is not the least common denominator, and the fraction $\dfrac{ad + bc}{bd}$ will not be in its lowest terms. To avoid working with fractions which are not in their lowest terms, it will be found advisable to take the *least* common denominator, which is the least common multiple of the denominators of the given fractions (Art. 82, Note 1).

Hence we have the following

To add or subtract fractions, reduce them to the least common denominator; add or subtract the numerators, and write the result over the least common denominator.

EXAMPLES.

1. Add $\dfrac{a+c}{b}$, $\dfrac{a-c}{b}$, and $\dfrac{a+d}{b}$.

Here the fractions have already a common denominator, and therefore need no reducing. Hence we have

$$\frac{a+c}{b} + \frac{a-c}{b} + \frac{a+d}{b} = \frac{a+c+a-c+a+d}{b} = \frac{3a+d}{b}.$$

2. Add $\dfrac{2x+a}{3a}$ and $\dfrac{5x-4a}{9a}$.

Here the least common denominator is $9a$. Hence we have (Art. 82)

$$\frac{2x+a}{3a} + \frac{5x-4a}{9a} = \frac{3(2x+a)}{9a} + \frac{5x-4a}{9a}$$
$$= \frac{6x+3a+5x-4a}{9a} = \frac{11x-a}{9a}.$$

3. From $\dfrac{4a-2b}{c}$ take $\dfrac{3a-3b}{c}$.

$$\frac{4a-2b}{c} - \frac{3a-3b}{c} = \frac{4a-2b-(3a-3b)}{c}$$
$$= \frac{4a-2b-3a+3b}{c} = \frac{a+b}{c}.$$

NOTE 1. — To insure accuracy, the beginner is recommended to put down the work in full; and when a fraction whose numerator is not a monomial is preceded by a — sign, he is recommended to enclose its numerator in a parenthesis as above before combining it with the other numerators.

4.) Add $\dfrac{a}{a-b}$ and $\dfrac{b}{b-a}$.

Here $\dfrac{b}{b-a} = -\dfrac{b}{a-b}$ (Art. 83).

$$\therefore \quad \frac{a}{a-b} + \frac{b}{b-a} = \frac{a}{a-b} - \frac{b}{a-b}$$

$$= \frac{a-b}{a-b} = 1.$$

5) Add $\dfrac{x-2y}{xy}$, $\dfrac{3y-a}{ay}$, and $\dfrac{2a-3x}{ax}$.

The least common denominator is axy.

$$\therefore \quad \frac{x-2y}{xy} + \frac{3y-a}{ay} + \frac{2a-3x}{ax}$$

$$= \frac{a(x-2y) + x(3y-a) + y(2a-3x)}{axy}$$

$$= \frac{ax - 2ay + 3xy - ax + 2ay - 3xy}{axy} = 0,$$

since the terms in the numerator destroy each other.

6) From $\dfrac{a+b}{a-b}$ take $\dfrac{a-b}{a+b}$.

The least common denominator is $a^2 - b^2$.

$$\therefore \quad \frac{a+b}{a-b} - \frac{a-b}{a+b} = \frac{(a+b)^2}{a^2-b^2} - \frac{(a-b)^2}{a^2-b^2}$$

$$= \frac{a^2 + 2ab + b^2 - (a^2 - 2ab + b^2)}{a^2 - b^2}$$

$$= \frac{4ab}{a^2 - b^2}.$$

7) Add $\dfrac{2x-3a}{x-2a}$ and $-\dfrac{2x-a}{x-a}$.

The least common denominator is $(x-2a)(x-a)$.

Hence we have

$$\frac{2x - 3a}{x - 2a} - \frac{2x - a}{x - a} = \frac{(2x - 3a)(x - a) - (2x - a)(x - 2a)}{(x - 2a)(x - a)}$$

$$= \frac{2x^2 - 5ax + 3a^2 - (2x^2 - 5ax + 2a^2)}{(x - 2a)(x - a)}$$

$$= \frac{2x^2 - 5ax + 3a^2 - 2x^2 + 5ax - 2a^2}{(x - 2a)(x - a)}$$

$$= \frac{a^2}{(x - 2a)(x - a)}.$$

NOTE 2. — In finding the value of an expression like
$$-(2x - a)(x - a),$$
the beginner should first express the product in parentheses, and then after multiplication, remove the parentheses, as we have done. After a little practice he will be able to take both steps together.

NOTE 3. — In practice, the foregoing general method may sometimes be modified with advantage. When the sum of several fractions is to be found, it is often best, instead of reducing at once all the fractions to their L. C. D., to take two or more of them together, and combine the results.

8. Simplify $\dfrac{a + 3}{a - 4} - \dfrac{a + 4}{a - 3} - \dfrac{8}{a^2 - 16}$.

Here, instead of reducing all the fractions to the least common denominator at once, we may take the first two fractions together, as follows:

$$\frac{a + 3}{a - 4} - \frac{a + 4}{a - 3} - \frac{8}{a^2 - 16} = \frac{a^2 - 9 - (a^2 - 16)}{(a - 4)(a - 3)} - \frac{8}{a^2 - 16}$$

$$= \frac{7}{(a - 4)(a - 3)} - \frac{8}{(a + 4)(a - 4)}$$

$$= \frac{52 - a}{(a - 4)(a + 4)(a - 3)}.$$

9. Simplify $\dfrac{1}{a - x} - \dfrac{1}{a + x} - \dfrac{2x}{a^2 + x^2} - \dfrac{4x^3}{a^4 + x^4}$.

Here we see that the first two denominators give L. C. D. $a^2 - x^2$; and this with $a^2 + x^2$ gives L. C. D. $a^4 - x^4$; and this with $a^4 + x^4$ gives L. C. D. $a^8 - x^8$. Hence, instead of reducing all the fractions to the least common denominator, we proceed as follows:

The first two terms $= \dfrac{a + x - (a - x)}{a^2 - x^2} = \dfrac{2x}{a^2 - x^2}.$

The first three terms $= \dfrac{2x}{a^2 - x^2} - \dfrac{2x}{a^2 + x^2}$

$$= \dfrac{2x(a^2 + x^2) - 2x(a^2 - x^2)}{a^4 - x^4}$$

$$= \dfrac{4x^3}{a^4 - x^4}.$$

The whole expression $= \dfrac{4x^3}{a^4 - x^4} - \dfrac{4x^3}{a^4 + x^4}$

$$= \dfrac{8x^7}{a^8 - x^8}.$$

10. Simplify $\dfrac{a}{(a-b)(a-c)} + \dfrac{b}{(b-c)(b-a)} + \dfrac{c}{(c-a)(c-b)}.$

The beginner is very liable, in this example, to take the product of the denominators for the least common denominator, and thus to render the operations very laborious. The denominator of the second fraction contains the factor $b - a$, and this factor differs from the factor $a - b$, which occurs in the denominator of the first fraction, only in the sign of each term. Also the denominator of the third fraction contains the factors $c - a$ and $c - b$, which differ from the factors $a - c$ and $b - c$ in the denominators of the first two fractions, only in the signs of each term. It is better to arrange these factors so that a precedes b or c, and that b precedes c. By Art. 83 we have

$$\frac{b}{(b - c)(b - a)} = - \frac{b}{(b - c)(a - b)},$$

and $\quad \dfrac{c}{(c - a)(c - b)} = \dfrac{c}{(a - c)(b - c)}.$

Hence the given expression may be put in the form

$$\frac{a}{(a - b)(a - c)} - \frac{b}{(b - c)(a - b)} + \frac{c}{(a - c)(b - c)},$$

whose L. C. D. we see at once is $(a - b)(a - c)(b - c)$.

Reducing the fractions to the least common denominator, the given expression becomes

$$\frac{a(b-c) - b(a-c) + c(a-b)}{(a-b)(a-c)(b-c)}$$

$$= \frac{ab - ac - ab + bc + ac - bc}{(a-b)(a-c)(b-c)} = 0.$$

The work is often made easier by completing the divisions represented by the fractions.

11. Simplify $1 + \dfrac{2x+1}{2x-2} - \dfrac{4x+5}{2x+2}$.

We have by division

$$1 + 1 + \frac{3}{2x-2} - 2 - \frac{1}{2x+2} = \frac{3}{2x-2} - \frac{1}{2x+2}$$

$$= \frac{3x+3 - x + 1}{2x^2 - 2}$$

$$= \frac{x+2}{x^2-1}.$$

12. $\dfrac{5x-1}{8} - \dfrac{3x-2}{7} + \dfrac{x-5}{4}$.   *Ans.* $\dfrac{25x-61}{56}$.

13. $\dfrac{2x+5}{x} - \dfrac{x+3}{2x} - \dfrac{27}{8x^2}$. $\dfrac{12x^2 + 28x - 27}{8x^2}$.

14. $\dfrac{x-4}{x-2} - \dfrac{x-7}{x-5}$. $\dfrac{6}{(x-2)(x-5)}$.

15. $\dfrac{4a^2 + b^2}{4a^2 - b^2} - \dfrac{2a-b}{2a+b}$. $\dfrac{4ab}{4a^2 - b^2}$.

16. $\dfrac{5}{1+2x} - \dfrac{3x}{1-2x} - \dfrac{4-13x}{1-4x^2}$. $\dfrac{1-6x^2}{1-4x^2}$.

17. $\dfrac{3}{x-2} + \dfrac{2}{3x+6} + \dfrac{5x}{x^2-4}$. $\dfrac{2(13x+7)}{3(x-2)(x+2)}$.

18. $\dfrac{3x}{1-x^2} - \dfrac{2}{x-1} - \dfrac{2}{x+1}$. $\dfrac{7x}{1-x^2}$.

19. $\dfrac{1}{(a-b)(a-c)} + \dfrac{1}{(b-c)(b-a)} + \dfrac{1}{(c-a)(c-b)}$. $0.$

85. To Multiply a Fraction by an Integer. —

RULE.

Multiply the numerator by that integer; or divide the denominator by that integer.

The rule may be proved as follows:

(1) Let $\dfrac{a}{b}$ denote any fraction, and c any integer; then

will $\dfrac{a}{b} \times c = \dfrac{ac}{b}$; for in each of the fractions $\dfrac{a}{b}$ and $\dfrac{ac}{b}$ the unit is divided into b equal parts, and the number of parts taken in $\dfrac{ac}{b}$ is c times the number taken in $\dfrac{a}{b}$.

(2) Let $\dfrac{a}{bc}$ denote any fraction, and c any integer; then

$$\frac{a}{bc} \times c = \frac{ac}{bc}, \text{ by } (1),$$

$$= \frac{a}{b} \text{ (Art. 79)}.$$

86. To Divide a Fraction by an Integer. —

RULE.

Divide the numerator by that integer; or multiply the denominator by that integer.

The rule may be proved as follows:

(1) Let $\dfrac{ac}{b}$ denote any fraction, and c any integer; then

will $\dfrac{ac}{b} \div c = \dfrac{a}{b}$; for $\dfrac{ac}{b}$ is c times $\dfrac{a}{b}$ (Art. 85); and therefore $\dfrac{a}{b}$ is the quotient of $\dfrac{ac}{b}$ divided by c.

(2) Let $\dfrac{a}{b}$ denote any fraction, and c any integer; then

$$\frac{a}{b} = \frac{ac}{bc} \text{ (Art. 79)}.$$

$$\therefore \quad \frac{a}{b} \div c = \frac{ac}{bc} \div c = \frac{a}{bc}, \text{ by } (1).$$

87. To Multiply Fractions. — Let it be required to multiply $\dfrac{a}{b}$ by $\dfrac{c}{d}$.

Put $\dfrac{a}{b} = x$, and $\dfrac{c}{d} = y$.   Then (Art. 85)

$$a = bx, \quad \text{and} \quad c = dy\,;$$

therefore $\qquad\qquad\qquad ac = bdxy\,;$

divide by bd, $\qquad\qquad\quad \dfrac{ac}{bd} = xy.$

But $\qquad\qquad\qquad\quad xy = \dfrac{a}{b} \times \dfrac{c}{d}.$

$$\therefore \quad \dfrac{a}{b} \times \dfrac{c}{d} = \dfrac{ac}{bd}\,;$$

and ac is the product of the numerators, and bd the product of the denominators.   Hence the following

RULE.

Multiply the numerators together for the numerator of the product, and the denominators for the denominator of the product.

Similarly, the rule may be demonstrated when more than two fractions are multiplied together.

NOTE. — If either of the factors is a mixed quantity, it is usually best to reduce it to a fractional form before applying the rule.   Also, it is advisable to indicate the multiplication of the numerators and denominators, and to examine if they have common factors; and, if so, to cancel them before performing the multiplication.

EXAMPLES.

1. Multiply together $\dfrac{3a}{4b}$ and $\dfrac{8c}{9a}$.

$$\dfrac{3a}{4b} \times \dfrac{8c}{9a} = \dfrac{3a \times 8c}{4b \times 9a} = \dfrac{2c}{3b} \quad \textbf{(Art. 79).}$$

2. Multiply $\dfrac{2a^2}{a^2 - b^2}$ by $\dfrac{(a + b)^2}{4a^2b}$.

$$\frac{2a^2}{a^2 - b^2} \times \frac{(a + b)^2}{4a^2b} = \frac{2a^2(a + b)(a + b)}{(a + b)(a - b)4a^2b}$$

$$= \frac{a + b}{2b(a - b)},$$

·y canceling those factors which are common to both numerator and denominator.

3. Multiply $\dfrac{2a^2 + 3a}{4a^3}$ and $\dfrac{4a^2 - 6a}{12a + 18}$ together.

$$\frac{2a^2 + 3a}{4a^3} \times \frac{4a^2 - 6a}{12a + 18} = \frac{a(2a + 3)2a(2a - 3)}{4a^3(2a + 3)6}$$

$$= \frac{2a - 3}{12a}.$$

4. Multiply $\dfrac{a}{b} + \dfrac{b}{a} + 1$ by $\dfrac{a}{b} + \dfrac{b}{a} - 1$.

$$\frac{a}{b} + \frac{b}{a} + 1 = \frac{a^2 + b^2 + ab}{ab} \quad \text{(Art. 84),}$$

and $\qquad \dfrac{a}{b} + \dfrac{b}{a} - 1 = \dfrac{a^2 + b^2 - ab}{ab}.$

$$\frac{a^2 + b^2 + ab}{ab} \times \frac{a^2 + b^2 - ab}{ab} = \frac{(a^2 + b^2)^2 - a^2b^2}{a^2b^2} \quad \text{[(3) of Art. 41]}$$

$$= \frac{a^4 + b^4 + a^2b^2}{a^2b^2}.$$

Otherwise thus:

$$\left(\frac{a}{b} + \frac{b}{a} + 1\right)\left(\frac{a}{b} + \frac{b}{a} - 1\right) = \left(\frac{a}{b} + \frac{b}{a}\right)^2 - 1$$

$$= \frac{a^2}{b^2} + \frac{b^2}{a^2} + 1 = \frac{a^4 + b^4 + a^2b^2}{a^2b^2}.$$

Simplify

5. $\dfrac{x + 1}{x - 1} \times \dfrac{x + 2}{x^2 - 1} \times \dfrac{x - 1}{(x + 2)^2}.$ *Ans.* $\dfrac{1}{(x - 1)(x + 2)}.$

6. $\dfrac{xa}{x + a} \times \left(\dfrac{x}{a} - \dfrac{a}{x}\right).$ $\qquad\qquad$ $x - a.$

7. $\left(a + \dfrac{ab}{a-b}\right)\left(b - \dfrac{ab}{a+b}\right).$ *Ans.* $\dfrac{a^2b^2}{a^2 - b^2}.$

8. $\dfrac{x(a-x)}{a^2 + 2ax + x^2} \times \dfrac{a(a+x)}{a^2 - 2ax + x^2}.$ $\dfrac{ax}{a^2 - x^2}.$

88. To Divide Fractions. — Let it be required to divide $\dfrac{a}{b}$ by $\dfrac{c}{d}.$

Denote the quotient by x. Then, since the quotient multiplied by the divisor gives the dividend (Art. 44), we have

$$x \times \frac{c}{d} = \frac{a}{b}.$$

Multiplying by $\dfrac{d}{c}$, we have

$$x \times \frac{c}{d} \times \frac{d}{c} = \frac{a}{b} \times \frac{d}{c}.$$

Therefore, Art. 87, and canceling factors common to the numerator and denominator, we have

$$x = \frac{ad}{bc}.$$

That is, $\dfrac{a}{b} \div \dfrac{c}{d} = \dfrac{ad}{bc} = \dfrac{a}{b} \times \dfrac{d}{c}.$

Hence the following

RULE.

Invert the divisor, and proceed as in multiplication.

EXAMPLES.

1. Divide a by $\dfrac{b}{c}.$

$$a = \frac{a}{1} \text{ (Art. 78)}.$$

$$\frac{a}{1} \div \frac{b}{c} = \frac{a}{1} \times \frac{c}{b} = \frac{ac}{b}.$$

2. Divide $\dfrac{ab - b^2}{(a + b)^2}$ by $\dfrac{b}{a^2 - b^2}$.

$$\dfrac{ab - b^2}{(a + b)^2} \div \dfrac{b}{a^2 - b^2} = \dfrac{ab - b^2}{(a + b)^2} \times \dfrac{a^2 - b^2}{b}$$

$$= \dfrac{b(a - b)(a + b)(a - b)}{b(a + b)^2} = \dfrac{(a - b)^2}{a + b}.$$

Simplify

3. $\dfrac{14x^2 - 7x}{12x^3 + 24x^2} \div \dfrac{2x - 1}{x^2 + 2x}$. *Ans.* $\dfrac{7}{12}$.

4. $\dfrac{a^2b^2 + 3ab}{4a^2 - 1} \div \dfrac{ab + 3}{2a + 1}$. $\dfrac{ab}{2a - 1}$.

5. $\dfrac{a^2 - 121}{a^2 - 4} \div \dfrac{a + 11}{a + 2}$. $\dfrac{a - 11}{a - 2}$.

6. $\dfrac{2x^2 + 13x + 15}{4x^2 - 9} \div \dfrac{2x^2 + 11x + 5}{4x^2 - 1}$. $\dfrac{2x - 1}{2x - 3}$.

7. $\dfrac{x^2 - 14x - 15}{x^2 - 4x - 45} \div \dfrac{x^2 - 12x - 45}{x^2 - 6x - 27}$. $\dfrac{x + 1}{x + 5}$.

89. Complex Fractions. — A fraction whose numerator and denominator are whole numbers is called a *Simple Fraction*. A fraction whose numerator or denominator is itself a fraction is called a *Complex Fraction*. Thus,

$$\dfrac{\frac{a}{b}}{c}, \quad \dfrac{a}{\frac{b}{c}}, \quad \dfrac{\frac{a}{b}}{\frac{c}{d}} \quad \text{are complex fractions.}$$

Since a fraction may be regarded as representing the quotient of the numerator by the denominator (Art. 78), a complex fraction may be regarded in the same way; therefore, to simplify a complex fraction,

Divide the numerator by the denominator, as in division of fractions (**Art. 88**).

EXAMPLES.

1. Simplify $\dfrac{\frac{1}{a}}{\frac{a}{b}}$, $\dfrac{\frac{a}{1}}{\frac{1}{b}}$, and $\dfrac{\frac{1}{a}}{\frac{1}{b}}$.

Here $\qquad \dfrac{\frac{1}{a}}{\frac{a}{b}} = 1 \div \dfrac{a}{b} = 1 \times \dfrac{b}{a} = \dfrac{b}{a}.$

$$\dfrac{\frac{a}{1}}{\frac{1}{b}} = a \div \dfrac{1}{b} = a \times b = ab.$$

$$\dfrac{\frac{1}{a}}{\frac{1}{b}} = \dfrac{1}{a} \div \dfrac{1}{b} = \dfrac{1}{a} \times \dfrac{b}{1} = \dfrac{b}{a}.$$

The student should be able to write down the above results readily without the intermediate steps.

2. Simplify $\dfrac{x + \frac{a^2}{x}}{x - \frac{a^4}{x^3}}$.

This fraction $= \left(x + \dfrac{a^2}{x} \right) \div \left(x - \dfrac{a^4}{x^3} \right)$

$$= \dfrac{x^2 + a^2}{x} \div \dfrac{x^4 - a^4}{x^3}$$

3. Simplify $\dfrac{\dfrac{a^2 + b^2}{a^2 - b^2} - \dfrac{a^2 - b^2}{a^2 + b^2}}{\dfrac{a + b}{a - b} - \dfrac{a - b}{a + b}}$.

$$\frac{a^2 + b^2}{a^2 - b^2} - \frac{a^2 - b^2}{a^2 + b^2} = \frac{(a^2 + b^2)^2 - (a^2 - b^2)^2}{(a^2 - b^2)(a^2 + b^2)}$$

$$= \frac{4a^2b^2}{(a^2 - b^2)(a^2 + b^2)}.$$

$$\frac{a + b}{a - b} - \frac{a - b}{a + b} = \frac{(a + b)^2 - (a - b)^2}{a^2 - b^2} = \frac{4ab}{a^2 - b^2}.$$

Hence the fraction

$$= \frac{4a^2b^2}{(a^2 - b^2)(a^2 + b^2)} \div \frac{4ab}{a^2 - b^2}$$

$$= \frac{4a^2b^2}{(a^2 - b^2)(a^2 + b^2)} \times \frac{a^2 - b^2}{4ab} = \frac{ab}{a^2 + b^2}.$$

Note 1. — In this example the factors $a - b$ and $a + b$ are multiplied together, and the result $a^2 - b^2$ is used instead of $(a - b)(a + b)$. In general, however, the student will find it advisable not to multiply the factors together till after he has canceled all the common factors from the numerator and denominator.

Note 2. — When the numerator and denominator are somewhat complicated, to insure accuracy and neatness, the beginner is advised to simplify each separately as in the above example.

Note 3. — The terms of the simple fractions which enter into the numerator and denominator of the complex fraction are sometimes called *Minor Terms*. Thus in Ex. 2, a^2 and a^4 are minor numerators, and x and x^3 are minor denominators.

It is often shorter to reduce a complex fraction to a simple one by multiplying both terms of the fraction by the least common multiple of all the minor denominators.

4. Simplify $\dfrac{x + 5 + \dfrac{6}{x}}{1 + \dfrac{6}{x} + \dfrac{8}{x^2}}$.

Multiplying both terms by x^2 we get

$$\frac{x^3 + 5x^2 + 6x}{x^2 + 6x + 8} = \frac{x(x + 2)(x + 3)}{(x + 2)(x + 4)} = \frac{x(x + 3)}{x + 4}.$$

5. Simplify $\dfrac{3x - 8}{x - 1 - \dfrac{1}{1 - \dfrac{x}{4 + x}}}$.

In the case of *Continued Fractions*, we begin with the lowest complex fraction, and simplify step by step. Here multiplying both terms of the fraction which follows $x - 1$ by $4 + x$, the given fraction becomes at once

$$\frac{3x - 8}{x - 1 - \dfrac{4 + x}{4 + x - x}} = \frac{3x - 8}{x - 1 - \dfrac{4 + x}{4}};$$

and now multiplying both terms by 4, we have

$$\frac{4(3x - 8)}{4(x - 1) - (4 + x)} = \frac{4(3x - 8)}{3x - 8} = 4.$$

Simplify

6. $\dfrac{x - \dfrac{1}{x}}{1 + \dfrac{1}{x}}$. *Ans. $x - 1$.*

7. $\dfrac{\dfrac{1}{x} - \dfrac{2}{x^2} - \dfrac{3}{x^3}}{\dfrac{9}{x} - x}$. $-\dfrac{x + 1}{x^2(x + 3)}$.

90. A Single Fraction Expressed as a Group of Fractions. — Let it be required to express the fraction

$$\frac{5x^2y - 10xy^2 + 15y^3 - 5x^3}{10x^2y^2}$$

as a group of four fractions.

$$\text{The fraction} = \frac{5x^2y}{10x^2y^2} - \frac{10xy^2}{10x^2y^2} + \frac{15y^3}{10x^2y^2} - \frac{5x^3}{10x^2y^2}$$

$$= \frac{1}{2y} - \frac{1}{x} + \frac{3y}{2x^2} - \frac{x}{2y^2}.$$

Express each of the following fractions as a group of simple fractions in lowest terms.

1. $\dfrac{3x^2y + xy^2 - y^3}{9xy}$. $Ans.\ \dfrac{x}{3} + \dfrac{y}{9} - \dfrac{y^2}{9x}$.

2. $\dfrac{3a^3x - 4a^2x^2 + 6ax^3}{12ax}$. $\dfrac{a^2}{4} - \dfrac{ax}{3} + \dfrac{x^2}{2}$.

3. $\dfrac{bc + ca + ab}{abc}$. $\dfrac{1}{a} + \dfrac{1}{b} + \dfrac{1}{c}$.

EXAMPLES.

Reduce to lowest terms the following examples:

1. $\dfrac{x^2 + 3x + 2}{x^2 + 6x + 5}$. $Ans.\ \dfrac{x + 2}{x + 5}$.

2. $\dfrac{x^2 + 10x + 21}{x^2 - 2x - 15}$. $\dfrac{x + 7}{x - 5}$.

3. $\dfrac{3x^2 + 23x - 36}{4x^2 + 33x - 27}$. $\dfrac{3x - 4}{4x - 3}$.

4. $\dfrac{x^2 - 10x + 21}{x^3 - 46x - 21}$. $\dfrac{x - 3}{x^2 + 7x + 3}$.

Here we see at once that the numerator $= (x - 7)(x - 3)$; and we find by trial that $x - 7$ is a factor of the denominator.

5. $\dfrac{x^2 + 9x + 20}{x^3 + 7x^2 + 14x + 8}$. $Ans.\ \dfrac{x + 5}{x^2 + 3x + 2}$.

6. $\dfrac{x^2 + x - 42}{x^3 - 10x^2 + 21x + 18}$. $\dfrac{x + 7}{x^2 - 4x - 3}$.

7. $\dfrac{6x^2 - 11x + 5}{3x^3 - 2x^2 - 1}$. $\dfrac{6x - 5}{3x^2 + x + 1}$.

8. $\dfrac{20x^2 + x - 12}{12x^3 - 5x^2 + 5x - 6}$. $\dfrac{5x + 4}{3x^2 + x + 2}$.

9. $\dfrac{x^3 - 8x - 3}{x^4 - 7x^2 + 1}$. $\dfrac{x - 3}{x^2 - 3x + 1}$.

10. $\dfrac{x^3 - x^2 - 7x + 3}{x^4 + 2x^3 + 2x - 1}$. $\dfrac{x - 3}{x^2 + 1}$.

11. $\dfrac{x^4 - 1}{x^6 - 1}$. $Ans.\ \dfrac{x^2 + 1}{x^4 + x^2 + 1}$.

12. $\dfrac{x^4 + x^3 + x^2 + x + 1}{x^5 - 1}$. $\dfrac{1}{x - 1}$.

13. $\dfrac{x^3 - 6x^2 - 37x + 210}{x^3 + 4x^2 - 47x - 210}$. $\dfrac{x - 5}{x + 5}$.

Reduce to fractional forms the following examples :

14. $a + x + \dfrac{a^2 - ax}{x}$. $Ans.\ \dfrac{a^2 + x^2}{x}$.

15. $a^2 - ax + x^2 - \dfrac{2x^3}{a + x}$. $\dfrac{a^3 - x^3}{a + x}$.

16. $x + 5 - \dfrac{2x - 15}{x - 3}$. $\dfrac{x^2}{x - 3}$.

17. $a^3 + ax^2 + \dfrac{ax^4}{a^2 - x^2}$. $\dfrac{a^5}{a^2 - x^2}$.

Reduce to whole or mixed quantities the following examples :

18. $\dfrac{ax - x^2}{a}$. $Ans.\ x - \dfrac{x^2}{a}$.

19. $\dfrac{1 - 2x^2}{1 + x}$. $1 - x - \dfrac{x^2}{1 + x}$.

20. $\dfrac{1 + 2x}{1 - 3x}$. $1 + 5x + \dfrac{15x^2}{1 - 3x}$.

21. $\dfrac{x + 7}{x + 2}$. $1 + \dfrac{5}{x + 2}$.

22. $\dfrac{3x - 2}{x + 5}$. $3 - \dfrac{17}{x + 5}$.

23. $\dfrac{2x^2 - 7x - 1}{x - 3}$. $2x - 1 - \dfrac{4}{x - 3}$.

24. $\dfrac{x^3 - 2x^2}{x^2 - x + 1}$. $x - 1 - \dfrac{2x - 1}{x^2 - x + 1}$.

25. $\dfrac{x^4 + 1}{x - 1}$. $x^3 + x^2 + x + 1 + \dfrac{2}{x - 1}$.

26. $\dfrac{x^4 - 1}{x + 1}$. $x^3 - x^2 + x - 1$.

Perform the additions and subtractions indicated in the following examples:

27. $\dfrac{1}{x + y} + \dfrac{2y}{x^2 - y^2}.$ *Ans.* $\dfrac{1}{x - y}.$

28. $\dfrac{a}{a - x} + \dfrac{3a}{a + x} - \dfrac{2ax}{a^2 - x^2}.$ $\dfrac{4a}{a + x}.$

29. $\dfrac{3}{x} - \dfrac{5}{2x - 1} - \dfrac{2x - 7}{4x^2 - 1}.$ $\dfrac{2x - 3}{x(4x^2 - 1)}.$

30. $\dfrac{a}{a - b} + \dfrac{a}{a + b} + \dfrac{4a^2b^2}{a^4 - b^4}.$ $\dfrac{2a^4 + 6a^2b^2}{a^4 - b^4}.$

31. $\dfrac{2}{x + 4} - \dfrac{x - 3}{x^2 - 4x + 16} + \dfrac{x^2}{x^3 + 64}.$ $\dfrac{2x^2 - 9x + 44}{x^3 + 64}$

32. $\dfrac{x^2 + ax + a^2}{x^3 - a^3} - \dfrac{x^2 - ax + a^2}{x^3 + a^3}.$ $\dfrac{2a}{x^2 - a^2}.$

33. $\dfrac{x^2 + y^2}{xy} - \dfrac{x^2}{xy + y^2} - \dfrac{y^2}{x^2 + xy}.$ 1.

34. $\dfrac{x^2 - 2x + 3}{x^3 + 1} + \dfrac{x - 2}{x^2 - x + 1} - \dfrac{1}{x + 1}.$ $\dfrac{x^2 - 2x}{x^3 + 1}.$

35. $\dfrac{1}{(x-3)(x-4)} - \dfrac{2}{(x-2)(x-4)} + \dfrac{1}{(x-2)(x-3)}.$ 0.

36. $\dfrac{1}{4(1 + x)} + \dfrac{1}{4(1 - x)} + \dfrac{1}{2(1 + x^2)}.$ $\dfrac{1}{1 - x^4}.$

37. $\dfrac{a-c}{(a-b)(x-a)} - \dfrac{b-c}{(b-a)(b-x)}.$ (See Art. 84, Ex. 10.)

Ans. $\dfrac{x - c}{(x - a)(x - b)}.$

38. $\dfrac{1}{(a^2-b^2)(x^2+b^2)} + \dfrac{1}{(b^2-a^2)(x^2+a^2)} - \dfrac{1}{(x^2+a^2)(x^2+b^2)}.$

Ans. 0.

39. $\dfrac{1}{a(a - b)(a - c)} + \dfrac{1}{b(b - a)(b - c)} - \dfrac{1}{abc}.$

Ans. $\dfrac{1}{c(a - c)(c - b)}.$

40. $\dfrac{a^2}{(a-b)(a-c)} + \dfrac{b^2}{(b-a)(b-c)} + \dfrac{c^2}{(c-a)(c-b)}.$

Ans. 1.

41. $\dfrac{1}{(a-b)(a-c)(x-a)} + \dfrac{1}{(b-a)(b-c)(x-b)} + \dfrac{1}{(c-a)(c-b)(x-c)}.$

Ans. $\dfrac{1}{(x-a)(x-b)(x-c)}.$

Simplify the following examples in multiplication and division:

42. $\dfrac{2x^2 + 5x + 2}{x^2 - 4} \times \dfrac{x^2 + 4x}{2x^2 + 9x + 4}.$ *Ans.* $\dfrac{x}{x-2}.$

43. $\dfrac{2x^2 - x - 1}{2x^2 + 5x + 2} \times \dfrac{4x^2 + x - 14}{16x^2 - 49}.$ $\dfrac{x-1}{4x+7}.$

44. $\dfrac{x^2+x-2}{x^2-x-20} \times \dfrac{x^2+5x+4}{x^2-x} \div \left(\dfrac{x^2+3x+2}{x^2-2x-15} \times \dfrac{x+3}{x^2} \right).$ $x.$

45. $\dfrac{x^4 - 8x}{x^2 - 4x - 5} \times \dfrac{x^2 + 2x + 1}{x^3 - x^2 - 2x} \div \dfrac{x^2 + 2x + 4}{x - 5}.$ 1.

46. $\dfrac{(a+b)^2 - c^2}{a^2 + ab - ac} \times \dfrac{a}{(a+c)^2 - b^2} \times \dfrac{(a-b)^2 - c^2}{ab - b^2 - bc}$ $\dfrac{1}{b}.$

47. $\dfrac{\dfrac{3x}{2} + \dfrac{x-1}{3}}{1\frac{3}{6}(x+1) - \dfrac{x}{3} - 2\frac{1}{2}}.$ 1.

48. $\dfrac{1 - a^2}{(1 + ax)^2 - (a + x)^2} \div \frac{1}{2}\left(\dfrac{1}{1-x} + \dfrac{1}{1+x} \right).$ 1.

49. $1 - \dfrac{1}{1 + \dfrac{1}{x}}.$ $\dfrac{1}{x+1}.$

50. $1 + \dfrac{x}{1 + x + \dfrac{2x^2}{1 - x}}.$ $\dfrac{1+x}{1+x^2}.$

51. $\left(\dfrac{2x}{x+y} + \dfrac{y}{x-y} - \dfrac{y^2}{x^2 - y^2} \right) \div \left(\dfrac{1}{x+y} + \dfrac{x}{x^2 - y^2} \right).$ $x.$

CHAPTER IX.

HARDER SIMPLE EQUATIONS OF ONE UNKNOWN QUANTITY.

91. Solution of Harder Equations.—We shall now give some simple equations, involving Algebraic fractions, which are a little more difficult than those in Chapter VI. These may be solved, by help of the preceding chapter on fractions, and by the same methods as the easier equations given in Chapter VI.

The following examples worked in full will sufficiently illustrate the most useful methods.

EXAMPLES.

1. Solve $\dfrac{6x - 3}{2x + 7} = \dfrac{3x - 2}{x + 5}$.

The L. C. M. of the denominators is $(2x + 7)(x + 5)$.

Clearing the equation of fractions by multiplying each term by $(2x + 7)(x + 5)$, we have *

$$(6x - 3)(x + 5) = (3x - 2)(2x + 7),$$

or $\qquad 6x^2 + 27x - 15 = 6x^2 + 17x - 14 ;$

$$\therefore \; 10x = 1 ; \qquad \therefore \; x = \tfrac{1}{10}.$$

We may verify this result by putting $\tfrac{1}{10}$ for x in the original equation, as in Chapter VI.; it will be found that each member then becomes $-\tfrac{1}{3}$.

NOTE 1.—When the denominators of the fractions involved contain both simple and compound factors, it is frequently best to multiply the equation by the simple factors first, and then to collect the integral terms; after this the simplification is readily completed by "multiplying across" by the compound factors.

---

* This is called "multiplying across."

2. Solve $\dfrac{8x + 23}{20} - \dfrac{5x + 2}{3x + 4} = \dfrac{2x + 3}{5} - 1.$

Multiplying by 20, the L. C. M. of the simple factors in the denominators, we have

$$8x + 23 - \frac{20(5x + 2)}{3x + 4} = 8x + 12 - 20.$$

Transposing, $31 = \dfrac{20(5x + 2)}{3x + 4}.$

Multiplying across by $3x + 4$, we have

$$93x + 124 = 20(5x + 2),$$

or $84 = 7x;$ $\therefore$ $x = 12.$

We may verify this result as before; it will be found that each side becomes $\frac{22}{5}$.

NOTE 2. — The student will see that, even when the denominators of the fractions contain all simple factors, it is sometimes advantageous to clear of fractions *partially*, and then to effect some reductions, before removing the remaining fractions.

3. Solve $\dfrac{x + 6}{11} - \dfrac{2x - 18}{3} + \dfrac{2x + 3}{4} = 5\tfrac{1}{3} + \dfrac{3x + 4}{12}.$

Multiplying by 12, the L. C. M. of 3, 4, 12,

$$\frac{12(x + 6)}{11} - 4(2x - 18) + 3(2x + 3) = 16 \times 4 + 3x + 4,$$

or $\dfrac{12(x + 6)}{11} - 8x + 72 + 6x + 9 = 64 + 3x + 4.$

Transposing and reducing, we have

$$\frac{12(x + 6)}{11} = 5x - 13.$$

Multiplying by 11, we have

$$12(x + 6) = 11(5x - 13),$$

or $12x + 72 = 55x - 143;$

$\therefore$ $43x = 215;$ $\therefore$ $x = 5.$

We may verify this result as before.

NOTE 3. — When two or more fractions have the same denominator, they should be taken together and simplified.

4. Solve $\dfrac{13 - 2x}{x + 3} + \dfrac{23x + 8\frac{1}{3}}{4x + 5} = \dfrac{16 - \frac{1}{4}x}{x + 3} + 4.$

Transposing, we have

$$\dfrac{23x + 8\frac{1}{3}}{4x + 5} - 4 = \dfrac{16 - \frac{1}{4}x - 13 + 2x}{x + 3};$$

then

$$\dfrac{7x - \frac{35}{3}}{4x + 5} = \dfrac{3 + \frac{7}{4}x}{x + 3}.$$

Multiplying across, we have

$$(x + 3)(7x - \tfrac{35}{3}) = (3 + \tfrac{7}{4}x)(4x + 5),$$

or $7x^2 - \tfrac{35}{3}x + 21x - 35 = 12x + 7x^2 + 15 + \tfrac{35}{4}x;$

$\therefore \; -\tfrac{137}{12}x = 50; \qquad \therefore \; x = -\tfrac{600}{137}.$

5. Solve $\dfrac{x - 8}{x - 10} + \dfrac{x - 4}{x - 6} = \dfrac{x - 5}{x - 7} + \dfrac{x - 7}{x - 9}.$

NOTE 4. — This equation might be solved by clearing of fractions, by multiplying by the four denominators, but the work would be very laborious. The solution will be much simplified by transposing two of the fractions as follows:

Transposing, we have

$$\dfrac{x - 8}{x - 10} - \dfrac{x - 5}{x - 7} = \dfrac{x - 7}{x - 9} - \dfrac{x - 4}{x - 6}.$$

Simplifying each side separately, we have

$$\dfrac{(x-8)(x-7)-(x-5)(x-10)}{(x-10)(x-7)} = \dfrac{(x-7)(x-6)-(x-4)(x-9)}{(x-9)(x-6)},$$

or

$$\dfrac{x^2-15x+56-(x^2-15x+50)}{(x-10)(x-7)} = \dfrac{x^2-13x+42-(x^2-13x+36)}{(x-9)(x-6)},$$

or

$$\dfrac{6}{(x-10)(x-7)} = \dfrac{6}{(x-9)(x-6)}.$$

Dividing by 6 and clearing of fractions, we have

$$(x - 9)(x - 6) = (x - 10)(x - 7),$$

or

$$x^2 - 15x + 54 = x^2 - 17x + 70;$$

$$\therefore \; x = 8.$$

NOTE 5. — This example may also be solved very neatly by writing the equation at first in the form

$$\frac{x-10+2}{x-10}+\frac{x-6+2}{x-6}=\frac{x-7+2}{x-7}+\frac{x-9+2}{x-9}.$$

Reducing each fraction to a mixed number (Art. 81), we have

$$1+\frac{2}{x-10}+1+\frac{2}{x-6}=1+\frac{2}{x-7}+1+\frac{2}{x-9},$$

which gives

$$\frac{1}{x-10}+\frac{1}{x-6}=\frac{1}{x-7}+\frac{1}{x-9}.$$

Transposing,

$$\frac{1}{x-10}-\frac{1}{x-7}=\frac{1}{x-9}-\frac{1}{x-6}.$$

$$\therefore \frac{3}{(x-10)(x-7)}=\frac{3}{(x-9)(x-6)},$$

and the solution may be completed as before.

6. Solve $\dfrac{5x-64}{x-13}-\dfrac{2x-11}{x-6}=\dfrac{4x-55}{x-14}-\dfrac{x-6}{x-7}.$

Proceeding as in the second method of Ex. 5, we have

$$5+\frac{1}{x-13}-\left(2+\frac{1}{x-6}\right)=4+\frac{1}{x-14}-\left(1+\frac{1}{x-7}\right);$$

$$\therefore \frac{1}{x-13}-\frac{1}{x-6}=\frac{1}{x-14}-\frac{1}{x-7}.$$

Simplifying each side separately, we have

$$\frac{7}{(x-13)(x-6)}=\frac{7}{(x-14)(x-7)}.$$

Clearing of fractions, or, since the numerators are equal, the denominators must be equal, we have

$$(x-13)(x-6)=(x-14)(x-7);$$
$$\therefore \quad x^2-19x+78=x^2-21x+98;$$
$$\therefore \quad x=10.$$

Solve the following equations:

7. $\dfrac{12}{x}+\dfrac{1}{12x}=2\frac{9}{24}.$ *Ans.* 10.

8. $\dfrac{45}{2x+3}=\dfrac{57}{4x-5}.$ 6.

9. $\dfrac{3x-1}{2}-\dfrac{2x-5}{3}+\dfrac{x-3}{4}-\dfrac{x}{6}=x+1.$ −7.

10. $\dfrac{6x + 8}{2x + 1} - \dfrac{2x + 38}{x + 12} = 1.$ *Ans.* 2.

11. $\frac{1}{3}(2x-10)-\frac{1}{11}(3x-40)=15-\frac{1}{5}(57-x).$ (See Note 2).
         Ans. 17.

12. $\dfrac{x-1}{4} - \dfrac{x-5}{32} + \dfrac{15-2x}{40} = \dfrac{9-x}{2} - \frac{7}{8}.$ (See Note 2).   5.

13. $\dfrac{x + 4}{3x - 8} = \dfrac{x + 5}{3x - 7}.$ 6.

14. $\dfrac{6x + 13}{15} - \dfrac{3x + 5}{5x - 25} = \dfrac{2x}{5}.$ (See Note 1).   20.

15. $\dfrac{3x - 1}{2x - 1} - \dfrac{4x - 2}{3x - 1} = \frac{1}{6}.$ $\frac{7}{17}.$

16. $\dfrac{6x + 7}{9x + 6} = 1\frac{1}{2} + \dfrac{5x - 5}{12x + 8}.$ $-6\frac{5}{6}.$

17. $\dfrac{4}{x+3} - \dfrac{2}{x+1} = \dfrac{5}{2x+6} - \dfrac{2\frac{1}{2}}{2x+2}.$ (See Note 3).   1.

18. $\dfrac{x - 1}{x - 2} + \dfrac{x - 5}{x - 6} = \dfrac{x - 4}{x - 5} + \dfrac{x - 2}{x - 3}.$ (See Ex. 5).   4.

19. $\dfrac{x - 1}{x - 2} - \dfrac{x - 2}{x - 3} = \dfrac{x - 5}{x - 6} - \dfrac{x - 6}{x - 7}.$ $4\frac{1}{2}.$

20. $\dfrac{6x + 1}{15} - \dfrac{2x - 4}{7x - 16} = \dfrac{2x - 1}{5}.$ $-2.$

92. Harder Problems Leading to Simple Equations with One Unknown Quantity. — We shall now give some examples which lead to simple equations, but which differ from those of Art. 61 in being rather more difficult. The statement of the problem is rather more difficult than in the examples of that Article, and the equations often involve more complicated expressions.

EXAMPLES.

1. A alone can do a piece of work in 9 days, and B alone can do it in 12 days : in what time will they do it if they work together?

Let $x =$ the number of days required for both to do the work ;

then $\dfrac{1}{x} =$ the part that both can do in one day.

Also $\frac{1}{9} =$ the part that A can do in one day,

and $\frac{1}{12} =$ the part that B can do in one day.

Since the sum of the parts that A and B separately can do in one day is equal to the part that both together can do in one day, we have

$$\tfrac{1}{9} + \tfrac{1}{12} = \dfrac{1}{x}.$$

Clearing of fractions by multiplying by $36x$, we have

$$4x + 3x = 36 ; \qquad \therefore \ x = 5\tfrac{1}{7},$$

which is the number of days required.

2. A workman was employed for 60 days, on condition that for every day he worked he should receive \$3, and for every day he was absent he should forfeit \$1 ; at the end of the time he had \$48 to receive : required the number of days he worked.

Let $x =$ the number of days he worked ;

then $60 - x =$ the number of days he was absent.

Also $3x =$ the number of dollars he received,

and $60 - x =$ the number of dollars he forfeited.

Hence, from the conditions of the problem, we have

$$3x - (60 - x) = 48.$$

$$\therefore \ 4x = 108. \qquad \therefore \ x = 27.$$

That is, he worked 27 days and was absent 33 days.

3. A starts from a certain place, and travels at the rate of 7 miles in 5 hours ; B starts from the same place 8 hours after A, and travels in the same direction at the rate of 5

miles in 3 hours ; how far will A travel before he is overtaken by B?

Let $x =$ the number of hours A travels before he
is overtaken ;

then $x - 8 =$ the number of hours B travels before he
overtakes A.

Also $\frac{7}{5} =$ the part of a mile which A travels in one
hour,

and $\frac{5}{3} =$ the part of a mile which B travels in one
hour,

Therefore $\frac{7}{5}x =$ the number of miles which A travels in x
hours,

and $\frac{5}{3}(x - 8) =$ the number of miles which B travels in
$x - 8$ hours.

Since, when B overtakes A, they have traveled the same
number of miles, we have for the equation

$$\tfrac{7}{5}x = \tfrac{5}{3}(x - 8).$$

$$\therefore \ 21x = 25x - 200. \qquad \therefore \ x = 50.$$

Therefore $\frac{7}{5}x = \frac{7}{5} \times 50 = 70$ miles, the distance which A
travels before he is overtaken by B.

4. A cistern could be filled with water by means of one
pipe alone in 6 hours, and by means of another pipe alone
in 8 hours ; and it could be emptied by a tap in 12 hours if
the two pipes were closed : in what time will the cistern be
filled if the pipes and the tap are all open?

Let $x =$ the required number of hours.

Then $\frac{1}{6} =$ the part of the cistern the first pipe fills in
one hour ;

therefore $\dfrac{x}{6} =$ the part of the cistern the first pipe fills in
x hours.

And $\frac{1}{8} =$ the part of the cistern the second pipe fills
in one hour ;

therefore $\dfrac{x}{8} =$ the part of the cistern the second pipe fills
in x hours.

Also $\frac{1}{12}$ = the part of the cistern the tap empties in one hour;

therefore $\frac{x}{12}$ = the part of the cistern the tap empties in x hours.

Since in x hours the *whole* cistern is filled, we have, representing the whole by *unity*,

$$\frac{x}{6} + \frac{x}{8} - \frac{x}{12} = 1.$$

Multiplying by 24, we have

$$4x + 3x - 2x = 24.$$
$$\therefore \quad x = 4\tfrac{4}{5}.$$

5. A smuggler had a quantity of brandy which he expected would bring him $198; after he had sold 10 gallons a revenue officer seized one-third of the remainder, in consequence of which the smuggler gets only $162: required the number of gallons he had at first, and the price per gallon.

Let $\quad\quad\quad\quad x$ = the number of gallons;

then $\quad\quad\quad\quad \frac{198}{x}$ = price per gallon in dollars.

$$\frac{x - 10}{3} = \text{the number of gallons seized;}$$

and $\frac{x - 10}{3} \times \frac{198}{x}$ = the value of the quantity seized in dollars.

Hence we have the equation

$$\frac{x - 10}{3} \times \frac{198}{x} = 198 - 162 = 36.$$

Clearing of fractions

$$66(x - 10) = 36x.$$
$$\therefore \quad 30x = 660.$$
$$\therefore \quad x = 22, \text{ the number of gallons;}$$

and $\quad\quad\quad\quad \frac{198}{x} = \frac{198}{22} = \$9, \text{ the price per gallon.}$

6. A colonel, on attempting to draw up his regiment in the form of a solid square, finds that he has 31 men over, and that he would require 24 men more in his regiment in order to increase the side of the square by one man : how many men were there in the regiment?

Let x = the number of men in the side of the first square;

then $x^2 + 31$ = the number of men in the regiment.

Also $(x + 1)^2 - 24$ = the number of men in the regiment.

Hence, we have the equation

$$x^2 + 31 = (x + 1)^2 - 24,$$

or $x^2 + 31 = x^2 + 2x - 23.$ $\therefore$ $x = 27.$

Hence $(27)^2 + 31 = 760$ is the number of men in the regiment.

NOTE 1. — In this example it was convenient to let x represent the number of men in the *side of the first square* instead of the number of men in the *whole regiment*.

7. At the same time that the up-train going at the rate of 33 miles an hour passes A, the down-train going at the rate of 21 miles an hour passes B : they collide 18 miles beyond the midway station from A : how far is A from B?

Let x = the distance from A to B in miles;

then $\dfrac{x}{2}$ = half the distance.

Also $\dfrac{x}{2} + 18$ = the number of miles the up-train goes,

and $\dfrac{x}{2} - 18$ = the number of miles the down-train goes.

Now $\dfrac{\text{distance in miles}}{\text{rate in miles per hour}}$ = the time in hours.

Therefore $\dfrac{\dfrac{x}{2} + 18}{33}$ = the time the up-train takes,

and $\dfrac{\dfrac{x}{2} - 18}{21}$ = the time the down-train takes.

Hence, since these times are equal, we have the equation

$$\frac{\frac{x}{2} + 18}{33} = \frac{\frac{x}{2} - 18}{21}.$$

Solving, we get $x = 162$, which is the distance from A to B in miles.

8. A cask, A, contains 12 gallons of wine and 18 gallons of water; and another cask, B, contains 9 gallons of wine and 3 gallons of water: how many gallons must be drawn from each cask so as to produce by their mixture 7 gallons of wine and 7 gallons of water?

Let x = the number of gallons to be drawn from A; then $14 - x$ = the number of gallons to be drawn from B, since the mixture is to contain 14 gallons.

Now A contains 30 gallons, of which 12 are wine; that is, $\frac{12}{30}$ of A is wine. Also B contains 12 gallons, of which 9 are wine; that is, $\frac{9}{12}$ of B is wine.

Hence $\frac{12}{30}x$ = the number of gallons of wine in the x gallons drawn from A;

and $\frac{9}{12}(14 - x)$ = the number of gallons of wine in the $14 - x$ gallons drawn from B.

Since the mixture is to contain seven gallons of wine, we have $\frac{12}{30}x + \frac{9}{12}(14 - x) = 7$;

that is, $\frac{2}{5}x + \frac{3}{4}(14 - x) = 7.$

Solving, we get $x = 10$, the number of gallons to be drawn from A,

and $14 - x = 4$, the number of gallons to be drawn from B.

9. At what time between 4 and 5 o'clock is the minute-hand of a watch 13 minutes in advance of the hour-hand?

Let x = the required number of minutes after 4 o'clock; that is, the minute-hand will move over x minute divisions of the watch face in x minutes; and as it moves 12 times as fast as the hour-hand, the hour-hand will move over $\frac{x}{12}$ minute divisions in x minutes. At 4 o'clock the minute-hand is 20 minute divisions behind the hour-hand, and finally the minute-hand is 13 minute divisions in advance; therefore, in the x minutes, the minute-hand moves $20 + 13$, or 33, divisions more than the hour-hand.

Hence $\qquad x = \dfrac{x}{12} + 33$;

therefore $\quad 11x = 12 \times 33.$ $\quad \therefore \quad x = 36,$

or the time is 36 minutes past 4.

If the question be asked, "At what *times* between 4 and 5 o'clock will there be 13 minutes between the two hands?" we must also take into consideration the case when the minute-hand is 13 divisions *behind* the hour-hand. In this case the minute-hand gains $20 - 13$, or 7 divisions.

Hence $\qquad x = \dfrac{x}{12} + 7.$

$\therefore \quad 11x = 84. \quad \therefore \quad x = 7\frac{7}{11}.$

Therefore the *times* are $7\frac{7}{11}$ minutes past 4, and 36 minutes past 4.

NOTE 2. — The student is supposed to have obtained from Arithmetic some knowledge of *ratio* and *proportion*. When two or more unknown quantities, in any example, have to each other a given ratio, it is best to assume each of them a multiple of some other unknown quantity, so that they shall have to each other the given ratio. Thus, if two unknown numbers are to each other as 2 to 3, it is best to express the numbers by $2x$ and $3x$, since these two numbers are to each other as 2 to 3. This will be illustrated in the next two examples.

10. A number consists of two digits of which the left digit is to the right digit as 2 to 3; if 18 be added to the number the digits are reversed: what is the number?

The student must remember that any number consisting of two places of figures is equal to ten times the figure in the ten's place plus the figure in the unit's place; thus, 46 is equal to $10 \times 4 + 6$; likewise 358 is equal to $100 \times 3 + 10 \times 5 + 8$.

Let $\qquad 2x =$ the left digit;

then $\qquad 3x =$ the right digit,

and $10 \times 2x + 3x =$ number.

Hence we have the equation

$$(10 \times 2x + 3x) + 18 = (10 \times 3x + 2x),$$

or

$$20x + 3x + 18 = 30x + 2x.$$

$$\therefore \quad x = 2;$$

$$\therefore \quad 2x = 4, \text{ and } 3x = 6;$$

therefore the number is 46.

11. A hare takes 4 leaps to a greyhound's 3, but 2 of the greyhound's leaps are equivalent to 3 of the hare's; the hare has a start of 50 leaps: how many leaps must the greyhound take to catch the hare?

Let $3x =$ the number of leaps taken by the greyhound;
then $4x =$ the number of leaps taken by the hare in the same time.

Also, let a denote the number of feet in one leap of the hare;
then $\frac{3}{2}a$ denotes the number of feet in one leap of the greyhound.

Therefore $3x \times \frac{3}{2}a =$ the distance in $3x$ leaps of the greyhound;
and $(4x + 50)a =$ the distance in $4x + 50$ leaps of the hare.

Hence we have the equation

$$\tfrac{9}{2}ax = (4x + 50)a.$$

Dividing by a and multiplying by 2, we have

$$9x = 8x + 100. \qquad \therefore \quad x = 100.$$

Therefore the greyhound must take 300 leaps.

NOTE 3.—It is often convenient to introduce an auxiliary symbol, as a was introduced in the above example, to enable us to form the equation easily; this can be removed by division when the equation is formed.

12. A person bought a carriage, horse, and harness for \$600; the horse cost twice as much as the harness, and the carriage half as much again as the horse and harness: what did he give for each? *Ans.* \$360, \$160, \$80.

13. In a garrison of 2744 men, there are two cavalry soldiers to twenty-five infantry, and half as many artillery as cavalry: find the numbers of each. *Ans.* 2450, 196, 98.

14. A and B play for a stake of $5; if A loses he will have as much as B, but if A wins he will have three times as much as B: how much has each?   *Ans.* $25, $15.

15. A, B, and C have a certain sum between them; A has one-half of the whole, B has one-third of the whole, and C has $50; how much have A and B?   *Ans.* $150, $100.

16. A number of troops being formed into a solid square, it was found that there were 60 over; but when formed into a column with 5 men more in front than before and 3 less in depth, there was just one man wanting to complete it: find the number of men.   *Ans.* 1504.

17. A and B began to pay their debts; A's money was at first $\frac{2}{3}$ of B's; but after A had paid $5 less than $\frac{2}{3}$ of his money, and B had paid $5 more than $\frac{2}{9}$ of his, it was found that B had only half as much as A had left: what sum had each at first?   *Ans.* $360, $540.

18. In a mixture of copper, lead, and tin, the copper was 5 lbs. less than half the whole quantity, and the lead and tin each 5 lbs. more than a third of the remainder: find the respective quantities.   *Ans.* 20, 15, 15 lbs.

19. A and B have the same income; A lays by a fifth of his; but B, by spending annually $400 more than A, at the end of four years finds himself $1100 in debt: what was their income?   *Ans.* $625.

20. There are two silver cups and one cover for both; the first weighs 12 ozs., and with the cover weighs twice as much as the other cup without it; but the second with the cover weighs a third as much again as the first without it: find the weight of the cover.   *Ans.* $6\frac{2}{3}$ oz.

21. Two casks, A and B, contain mixtures of wine and water; in A the quantity of wine is to the quantity of water as 4 to 3; in B the like proportion is that of 2 to 3. If A contain 84 gallons what must B contain, so that when the two are put together, the new mixture may be half wine and half water?   *Ans.* 60.

EXAMPLES.

Solve the following equations.

1. $\dfrac{x-2}{4} + \tfrac{1}{3} = x - \dfrac{2x-1}{3}.$ *Ans.* $-6.$

2. $\dfrac{4x+17}{x+3} + \dfrac{3x-10}{x-4} = 7.$ 2.

3. $\dfrac{x-4}{3} + (x-1)(x-2) = x^2 - 2x - 4.$ 7.

4. $\dfrac{3(7+6x)}{2+9x} = \dfrac{35+4x}{9+2x}.$ 1.

5. $\dfrac{x}{x+2} + \dfrac{4}{x+6} = 1.$ 2.

6. $\dfrac{2x-5}{5} + \dfrac{x-3}{2x-15} = \dfrac{4x-3}{10} - 1\tfrac{1}{10}.$ 5.

7. $\dfrac{4(x+3)}{9} = \dfrac{8x+37}{18} - \dfrac{7x-29}{5x-12}.$ 6.

8. $\dfrac{7}{x-4} - \dfrac{60}{5x-30} = \dfrac{10\frac{1}{2}}{3x-12} - \dfrac{8}{x-6}.$ $-10.$

9. $\dfrac{3x^2-2x-8}{5} = \dfrac{(7x-2)(3x-6)}{35}.$ 2.

10. $\dfrac{x+10}{3} - \tfrac{2}{5}(3x-4) + \dfrac{(3x-2)(2x-3)}{6} = x^2 - \tfrac{2}{15}.$ 2.

11. $\dfrac{2}{2x-3} + \dfrac{1}{x-2} = \dfrac{6}{3x+2}.$ $\tfrac{50}{29}.$

12. $\dfrac{x-4}{x-5} - \dfrac{x-5}{x-6} = \dfrac{x-7}{x-8} - \dfrac{x-8}{x-9}.$ 7.

13. $\dfrac{x}{x-2} + \dfrac{x-9}{x-7} = \dfrac{x+1}{x-1} + \dfrac{x-8}{x-6}.$ 4.

14. $\dfrac{3-2x}{1-2x} - \dfrac{2x-5}{2x-7} = 1 - \dfrac{4x^2-1}{7-16x+4x^2}.$ $-1.$

15. $\dfrac{x-5}{7} + \dfrac{x^2+6}{3} = \dfrac{x^2-2}{2} - \dfrac{x^2-x+1}{6} + 3.$ $-23.$

16. $\dfrac{3}{4-2x} + \dfrac{30}{8(1-x)} = \dfrac{3}{2-x} + \dfrac{5}{2-2x}.$ *Ans.* $-4.$

17. $\dfrac{30+6x}{x+1} + \dfrac{60+8x}{x+3} = 14 + \dfrac{48}{x+1}.$ 3.

18. $\dfrac{x}{x-2} - \dfrac{x+1}{x-1} = \dfrac{x-8}{x-6} - \dfrac{x-9}{x-7}.$ 4.

19. $\dfrac{x+5}{x+4} - \dfrac{x-6}{x-7} = \dfrac{x-4}{x-5} - \dfrac{x-15}{x-16}.$ 6.

20. $\dfrac{x-7}{x-9} - \dfrac{x-9}{x-11} = \dfrac{x-13}{x-15} - \dfrac{x-15}{x-17}.$ 13.

21. $\dfrac{x+3}{x+6} - \dfrac{x+6}{x+9} = \dfrac{x+2}{x+5} - \dfrac{x+5}{x+8}.$ $-7.$

22. $\dfrac{x+2}{x} + \dfrac{x-7}{x-5} - \dfrac{x+3}{x+1} = \dfrac{x-6}{x-4}.$ 2.

23. $\dfrac{4x-17}{x-4} + \dfrac{10x-13}{2x-3} = \dfrac{8x-30}{2x-7} + \dfrac{5x-4}{x-1}$ $2\frac{1}{2}.$

24. $\dfrac{5x-8}{x-2} + \dfrac{6x-44}{x-7} - \dfrac{10x-8}{x-1} = \dfrac{x-8}{x-6}.$ 4.

25. $(x+1)(x+2)(x+3)=(x-1)(x-2)(x-3)+3(4x-2)(x+1).$
Ans. 3.

26. $(x-9)(x-7)(x-5)(x-1)=(x-2)(x-4)(x-6)(x-10).$
Ans. $5\frac{1}{2}.$

27. $(8x-3)^2(x-1) = (4x-1)^2(4x-5).$ $\frac{4}{13}.$

28. $\dfrac{x^2-x+1}{x-1} + \dfrac{x^2+x+1}{x+1} = 2x.$ 0.

29. $\dfrac{6x+7}{15} - \dfrac{2x-2}{7x-6} = \dfrac{2x+1}{5}.$ 3.

30. $.5x - 2 = .25x + .2x - 1.$ 20.

31. $.5x + .6x - .8 = .75x + .25.$ 3.

32. $.15x + \dfrac{.135x - .225}{.6} = \dfrac{.36}{.2} - \dfrac{.09x - .18}{.9}.$ 5.

33. $\dfrac{2x-3}{.3x-.4} = \dfrac{.4x-.6}{.06x-.07}.$ $1\frac{1}{2}.$

34. $\dfrac{.3x - 1}{.5x - .4} = \dfrac{.5 + 1.2x}{2x - .1}.$ $\qquad\qquad$ $\tfrac{1}{6}.$

35. $\dfrac{(.3x - 2)(.3x - 1)}{.2x - 1} - \tfrac{1}{6}(.3x - 2) = .4x - 2.$ $\quad$ 20.

36. $a^2(x - a) + b^2(x - b) = abx.$ $\qquad\qquad$ $a + b.$

37. $\dfrac{2x + 3a}{x + a} = \dfrac{2(3x + 2a)}{3x + a}.$ $\qquad\qquad$ $a.$

38. $\tfrac{2}{3}\left(\dfrac{x}{a} + 1\right) = \tfrac{3}{4}\left(\dfrac{x}{a} - 1\right).$ $\qquad\qquad$ $17a.$

39. $\tfrac{1}{4}x(x - a) - \left(\dfrac{x + a}{2}\right)^2 = \dfrac{2a}{3}\left(x - \dfrac{a}{2}\right).$ $\qquad$ $\dfrac{a}{17}.$

40. $x^2 + a(2a - x) - \dfrac{3b^2}{4} = \left(x - \dfrac{b}{2}\right)^2 + a^2.$ $\quad$ $a + b.$

41. $(2x-a)\left(x+\dfrac{2a}{3}\right)=4x\left(\dfrac{a}{3}-x\right)-\tfrac{1}{2}(a-4x)(2a+3x).$

$$\text{Ans. } \dfrac{2a}{21}.$$

42. $\dfrac{1}{x - a} - \dfrac{1}{x - b} = \dfrac{a - b}{x^2 - ab}.$ $\qquad$ $\dfrac{2ab}{a + b}.$

43. $\dfrac{x - a}{a - b} - \dfrac{x + a}{a + b} = \dfrac{2ax}{a^2 - b^2}.$ $\qquad$ $\dfrac{a^2}{b - a}.$

44. $\dfrac{x - a}{x - a - 1} - \dfrac{x - a - 1}{x - a - 2} = \dfrac{x - b}{x - b - 1} - \dfrac{x - b - 1}{x - b - 2}.$

$$\text{Ans. } \tfrac{1}{2}(a + b + 3).$$

45. $(x-a)^3(x+a-2b)=(x-b)^3(x-2a+b).$ $\quad$ $\tfrac{1}{2}(a+b).$

46. $\dfrac{3abc}{a+b} + \dfrac{a^2b^2}{(a+b)^3} + \dfrac{(2a+b)b^2x}{a(a+b)^2} = 3cx + \dfrac{bx}{a}.$ $\quad$ $\dfrac{ab}{a+b}.$

47. A person wishing to sell a watch by lottery, charges
$1.20 each for the tickets, by which he gains $16; whereas,
if he had made a third as many tickets again and charged
$1 each, he would have gained one-fifth as many dollars as
he had sold tickets: what was the value of the watch?

$$\text{Ans. } \$128.$$

48. There is a number of two digits, whose difference is 2, and if it be diminished by half as much again as the sum of the digits, the digits will be reversed : find the number.
Ans. 75.

49. Find a number of 3 digits, each greater by 1 than that which follows it, so that its excess above a fourth of the number formed by reversing the digits shall be 36 times the sum of the digits. *Ans.* 654.

50. A can do a piece of work in 10 days, which B can do in 8 ; after A has been at work upon it 3 days, B comes to help him : in how many days will they finish it? *Ans.* $3\frac{1}{9}$ days.

51. A and B can reap a field together in 7 days, which A alone could reap in 10 days : in what time could B alone reap it? *Ans.* $23\frac{1}{3}$ days.

52. A privateer, running at the rate of 10 miles an hour, discovers a ship 18 miles off, running at the rate of 8 miles an hour : how many miles can the ship run before it is overtaken? *Ans.* 72.

53. The distance between London and Edinburgh is 360 miles ; one traveler starts from Edinburgh and travels at the rate of 30 miles an hour, while another starts at the same time from London and travels at the rate of 24 miles an hour : how far from Edinburgh will they meet? *Ans.* 200 miles.

54. Find two numbers whose difference is 4, and the difference of their squares 112. *Ans.* 12, 16.

55. Divide the number 48 into two parts so that the excess of one part over 20 may be three times the excess of 20 over the other part. *Ans.* 32, 16.

56. A cistern could be filled in 12 minutes by two pipes which run into it, and it could be filled in 20 minutes by one alone : in what time would it be filled by the other alone?
Aus. 30 minutes.

57. Divide the number 90 into four parts so that the first increased by 2, the second diminished by 2, the third multiplied by 2, and the fourth divided by 2, may all be equal.
Ans. 18, 22, 10, 40.

58. Divide the number 88 into four parts so that the first increased by 2, the second diminished by 3, the third multiplied by 4, and the fourth divided by 5, may all be equal.

Ans. 10, 15, 3, 60.

59. If 20 men, 40 women, and 50 children receive $500 among them for a week's work, and 2 men receive as much as 3 women or 5 children, what does each woman receive for a week's work?   *Ans.* $5.

60. A cistern can be filled in 15 minutes by two pipes, A and B, running together; after A has been running by itself for 5 minutes B is also turned on, and the cistern is filled in 13 minutes more: in what time would it be filled by each pipe separately?   *Ans.* $37\frac{1}{2}$, and 25 minutes.

61. A man and his wife could drink a cask of beer in 20 days, the man drinking half as much again as his wife; but $\frac{18}{25}$ of a gallon having leaked away, they found that it only lasted them together for 18 days, and the wife herself for two days longer: how much did the cask contain when full?

Ans. 12 gallons.

Let $x =$ the number of gallons the woman could drink in a day.

62. A man, woman, and child could reap a field in 30 hours, the man doing half as much again as the woman, and the woman two-thirds as much again as the child: how many hours would they each take to do it separately?

Ans. 62, 93, 155.

Let $2x =$ the man's number of hours, $3x =$ the woman's, and $5x =$ the child's.

63. A and B can reap a field together in 12 hours, A and C in 16 hours, and A by himself in 20 hours: in what time (1) could B and C together reap it, and (2) could A, B, and C together reap it?   *Ans.* $21\frac{9}{11}$ hours, $10\frac{10}{23}$ hours.

64. A can do half as much work as B, B can do half as much work as C, and together they can complete a piece of work in 24 days: in what time could each alone complete the work?   *Ans.* 168, 84, and 42 days.

65. There are two places 154 miles apart, from which two persons start at the same time with a design to meet; one travels at the rate of 3 miles in two hours, and the other at the rate of 5 miles in four hours : when will they meet?

Ans. At the end of 56 hours.

66. Three persons, A, B, and C, can together complete a piece of work in 60 days ; and it is found that A does three-fourths of what B does, and B four-fifths of what C does : in what time could each one alone complete the work?

Ans. 240, 180, 144 days.

Let x = C's time of completing the work, in days.

67. A general, on attempting to draw up his army in the form of a solid square, finds that he has 60 men over, and that he would require 41 men more in his army in order to increase the side of the square by one man : how many men were there in the army? *Ans.* 2560.

68. A person bought a certain number of eggs, half of them at 2 for a cent, and half of them at 3 for a cent; he sold them again at the rate of 5 for two cents, and lost a cent by the bargain : what was the number of eggs? *Ans.* 60.

69. A and B are at present of the same age ; if A's age be increased by 36 years, and B's by 52 years, their ages will be as 3 to 4 ; what is the present age of each? *Ans.* 12.

70. A cistern has two supply pipes which will singly fill it in $4\frac{1}{2}$ hours and 6 hours respectively ; and it has also a leak by which it would be emptied in 5 hours : in how many hours will it be filled when all are working together? *Ans.* $5\frac{5}{17}$.

71. A person hired a laborer to do a certain work on the agreement that for every day he worked he should receive $2, but that for every day he was absent he should lose $0.75 ; he worked twice as many days as he was absent, and on the whole received $39 : how many days did he work? *Ans.* 24.

72. A sum of money was divided between A and B, so that the share of A was to that of B as 5 to 3 ; also the share of A exceeded five-ninths of the whole sum by $200 : what was the share of each person? *Ans.* $1800, 1080.

73. A gentleman left his whole estate among his four sons. The share of the eldest was $4000 less than half of the estate; the share of the second was $600 more than one-fourth of the estate; the third had half as much as the eldest; and the youngest had two-thirds of what the second had: how much did each son receive?

Ans. $11000, $8100, $5500, $5400.

Let $x =$ the number of dollars in the estate.

74. A and B shoot by turns at a target; A puts 7 bullets out of 12 into the bull's eye, and B puts in 9 out of 12; between them they put in 32 bullets: how many shots did each fire? *Ans.* 24.

75. Two casks, A and B, are filled with two kinds of sherry, mixed in the cask A in the proportion of 2 to 7, and in the cask B in the proportion of 2 to 5: what quantity must be taken from each to form a mixture which shall consist of 2 gallons of the first kind and 6 of the second kind? *Ans.* $4\frac{1}{2}$, $3\frac{1}{2}$.

76. How many minutes does it want of 4 o'clock, if three-quarters of an hour ago it was twice as many minutes past 2 o'clock? *Ans.* 25.

Let $x =$ the number of minutes it wants of 4 o'clock.

77. At what time between 3 o'clock and 4 o'clock is one hand of a watch exactly in the direction of the other hand produced? *Ans.* $49\frac{1}{11}$ minutes past three.

78. The hands of a watch are at right angles to each other at 3 o'clock: when are they next at right angles?

Ans. $32\frac{8}{11}$ minutes past three.

79. At what time between 3 and 4 o'clock is the minute-hand one minute ahead of the hour-hand?

Ans. $17\frac{5}{11}$ minutes past three.

80. An officer can form his men into a hollow square 4 deep, and also into a hollow square 8 deep; the front in the latter formation contains 16 men fewer than in the former formation: find the number of men. *Ans.* 640.

CHAPTER X.

SIMULTANEOUS SIMPLE EQUATIONS OF TWO OR MORE UNKNOWN QUANTITIES.

93. Simultaneous Equations of Two Unknown Quantities. — If we have a single equation containing two unknown quantities x and y, we cannot determine any thing definite regarding the values of x and y, because whatever value we choose to give to either of them, there will be a corresponding value of the other.

Thus, from the equation,

$$2x + 3y = 24, \quad \ldots \ldots \ldots (1)$$

we may deduce the equation,

$$y = \frac{24 - 2x}{3};$$

but we cannot find the value of y from this equation unless we know the value of x. We may give to x any value we choose, and there will be one *corresponding* value of y; and thus we may find as many *pairs of values* as we please which will satisfy the given equation.

For example, if $x = 3$, then $y = (24 - 6) \div 3 = 6$.

If $x = 6$, then $y = (24 - 12) \div 3 = 4$.

If $x = 9$, then $y = (24 - 18) \div 3 = 2$.

If $x = 20$, then $y = (24 - 40) \div 3 = -5\frac{1}{3}$,

and so on.

Any one of these pairs of values $\left(\begin{matrix} x=3 \\ y=6 \end{matrix}\right), \left(\begin{matrix} x=6 \\ y=4 \end{matrix}\right), \left(\begin{matrix} x=9 \\ y=2 \end{matrix}\right)$, etc., substituted for x and y in (1) will satisfy the equation. Hence a single equation containing two unknown quantities is not sufficient to determine the *definite* value of either.

Suppose we have a second equation of the same kind, expressing a *different relation* between the unknown quantities, as for example,

$$3x + 2y = 26 ; \quad . \quad . \quad . \quad . \quad . \quad (2)$$

then we can find as many pairs of values as we please which will satisfy this equation also.

Now suppose we wish to determine values of x and y which will satisfy both equations (1) and (2) ; we shall find that there is only *one pair* of values of x and y, i.e., only one value of x and one value of y that will satisfy both equations. For, multiply equation (1) by 2, and equation (2) by 3, and the equations become

$$4x + 6y = 48, \quad . \quad . \quad . \quad . \quad . \quad (3)$$

and $\qquad\qquad\qquad 9x + 6y = 78. \quad . \quad . \quad . \quad . \quad . \quad (4)$

The coefficients of y are now the same in (3) and (4) ; hence if we *subtract* each member of (3) from the corresponding member of (4), we shall obtain an equation which does not contain y : the equation will be

$$5x = 30 ;$$

therefore $\qquad\qquad\qquad x = 6.$

Substituting this value of x in either of the two given equations, for example in equation (1), we have

$$12 + 3y = 24.$$

$$\therefore \quad 3y = 12.$$

$$\therefore \quad y = 4.$$

Thus, if both equations are to be satisfied by the *same* values of x and y, x *must* equal 6, and y *must* equal 4 ; and the pair of values $\begin{pmatrix} x = 6 \\ y = 4 \end{pmatrix}$ is the only pair of values which will satisfy both the given equations.

Simultaneous Equations are those which are satisfied by the *same values* of the unknown quantities. Thus,

Since (1) and (2) are satisfied by the same values of a and y, they are *simultaneous* equations.

Independent Equations are those which express *different relations* between the unknown quantities, and neither can be reduced to the other. Thus,

Equations (1) and (2) are independent, because they express different relations between x and y. But $2x - 3y = 4$, and $8x - 12y = 16$, are not independent equations, since the second is derived directly from the first, by multiplying both members by 4.

Hence we see that *two independent* simultaneous equations are necessary to determine the values of two unknown quantities.

94. Elimination. — In order to solve any two simultaneous equations containing two unknown quantities, it is necessary to combine them in such a way as to deduce a third equation which contains only *one* of the unknown quantities; and this equation containing only one unknown quantity can be solved by the method given in Chapter IX. When the value of one of the unknown quantities has thus been determined, we can substitute this value in either of the given equations, and then determine the value of the other unknown quantity.

The process of combining equations so as to get rid of either of the unknown quantities is called *Elimination*. The unknown quantity which disappears is said to be *eliminated*.

There are three methods of elimination in common use:[*]
(1) *by Addition or Subtraction;* (2) *by Substitution;* and (3) *by Comparison.*

95. Elimination by Addition or Subtraction. — 1. Let it be required to determine the values of x and y in the two equations

$$8x + 7y = 100 \ . \ . \ . \ . \ . \ . \ (1)$$

$$12x - 5y = 88 \ . \ . \ . \ . \ . \ . \ (2)$$

---

[*] There is also a method by *Undetermined Multipliers*, which sometimes has the advantage over either of these three methods, especially in Higher Mathematics. — See College Algebra, Art. 99, Note.

If we wish to eliminate y we multiply (1) by 5, and (2) by 7, so as to make the coefficients of y in both equations equal. This gives

$$40x + 35y = 500 \quad \ldots \ldots \quad (3)$$

$$84x - 35y = 616 \quad \ldots \ldots \quad (4)$$

Adding (3) and (4), we have

$$124x = 1116. \quad \therefore \quad x = 9.$$

To find y, substitute this value of x in *either* of the given equations. Thus in (1)

$$72 + 7y = 100.$$

$$\therefore \quad 7y = 28.$$

$$\therefore \quad y = 4$$

and $\qquad\qquad\qquad x = 9 \Big\} .$

In this solution we eliminated y by *addition*.

Otherwise thus: Suppose that in solving these equations we wish to eliminate x instead of y. Multiply (1) by 3, and (2) by 2, so as to make the coefficients of x in both equations equal. This gives

$$24x + 21y = 300 \quad \ldots \ldots \quad (5)$$

$$24x - 10y = 176 \quad \ldots \ldots \quad (6)$$

Subtracting (6) from (5), we have

$$31y = 124. \quad \therefore \quad y = 4.$$

In this solution we eliminated x by *subtraction*.

NOTE 1. — The student will observe that we might have made the coefficients of x equal by multiplying (1) and (2) by 12 and 8 respectively, instead of by 3 and 2; but it was more convenient to use the smaller multipliers, because it enabled us to work with smaller numbers.

2. Solve $\qquad 2x + 3y = 31 \quad \ldots \ldots \quad (1)$

$$12x - 17y = -59 \quad \ldots \ldots \quad (2)$$

Here it will be more convenient to eliminate x.

Multiplying (1) by 6, to make the coefficients of x in both equations equal, we have

$$12x + 18y = 186 ; \quad \dots \quad (3)$$

and from (2) $\qquad 12x - 17y = -59 \quad \dots \quad (4)$

Subtracting (4) from (3), $35y = 245.$

$$\therefore \ y = 7.$$

Substituting this value of y in (1), we have

$$2x + 21 = 31.$$

$$\therefore \ x = 5$$

and $\qquad\qquad\qquad y = 7$.

Note 2. — When one of the unknown quantities has been found, it is immaterial which of the equations we use to complete the solution, though it is sometimes more convenient to use a particular equation on account of its being less involved than the other. Thus, in this example, we substituted the value of x in (1) rather than in (2), because it rendered the process simpler.

In these two examples we have eliminated by addition and subtraction. Hence to eliminate an unknown quantity by *addition or subtraction*, we have the following

RULE.

Multiply the given equations, if necessary, by such numbers as will make the coefficients of this unknown quantity numerically equal in the resulting equations. Then, if these equal coefficients have unlike signs, add the equations together; if they have the same sign, subtract one equation from the other.

Rem. — It is generally best to eliminate that unknown quantity which has the smaller coefficients in the two equations, or which requires the smallest multipliers to make its coefficients equal. When the coefficients of the quantity to be eliminated are prime to each other, we may take each one as the multiplier of the other equation. When these coefficients are not prime to each other, find their least common multiple; and the smallest multiplier for each equation will be the quotient obtained by dividing this L. C. M. by the coefficient in that equation. Thus, in Ex. 1, first solution, 7 and 5 (the coefficients of y) are prime to each other. We multiplied (1) by 5 and (2) by 7. In the second solution of Ex. 1, the L. C. M. of 8 and 12 (the coefficients of x) is 24; and hence the smallest multipliers of (1) and (2) are 3 and 2 respectively, which we used in that solution.

3. Solve

$$171x - 213y = 642 \quad \ldots \quad (1)$$
$$114x - 326y = 244 \quad \ldots \quad (2)$$

Here we see that 171 and 114 contain a common factor 57; so we shall make the coefficients of x in (1) and (2) equal to the *least common multiple* of 171 and 114 if we multiply (1) by 2 and (2) by 3.

Thus,
$$342x - 426y = 1284$$
$$342x - 978y = 732$$

Subtracting,
$$552y = 552.$$
$$\therefore \quad y = 1,$$

therefore
$$x = 5.$$

NOTE 3. — The solution is sometimes easily effected by first adding the given equations, or by subtracting one from the other. Thus,

4. Solve

$$127x + 59y = 1928. \quad \ldots \quad (1)$$
$$59x + 127y = 1792. \quad \ldots \quad (2)$$

By addition
$$186x + 186y = 3720.$$
$$\therefore \quad x + y = 20 \quad \ldots \quad (3)$$

Subtracting (2) from (1), $68x - 68y = 136.$
$$\therefore \quad x - y = 2 \quad \ldots \quad (4)$$

Adding (3) and (4), $\qquad 2x = 22. \quad \therefore \quad x = 11.$

Subtracting (4) from (3), $\qquad 2y = 18. \quad \therefore \quad y = 9.$

NOTE 4. — The student should look carefully for opportunities to effect such reductions as are made in this example. He will find as he proceeds that in all parts of Algebra, particular examples may be treated by methods which are shorter than the general rules; but such abbreviations can only be suggested by experience and practice.

Solve the following equations by *addition* or *subtraction:*

5. $3x + 4y = 10, 4x + y = 9.$ *Ans.* $x = 2, y = 1.$

6. $x + 2y = 13, 3x + y = 14.$ $x = 3, y = 5.$

7. $4x + 7y = 29, x + 3y = 11.$ $x = 2, y = 3.$

8. $8x - y = 34, x + 8y = 53.$ $x = 5, y = 6.$

9. $14x - 3y = 39, 6x + 17y = 35.$ $x = 3, y = 1.$

10. $35x + 17y = 86, 56x - 13y = 17.$ $x = 1, y = 3.$

11. $15x + 77y = 92,\ 55x - 33y = 22.$

Ans. $x = 1,\ y = 1.$

12. $3x + 2y = 32,\ 20x - 3y = 1.$ $x = 2,\ y = 13.$

13. $7x + 5y = 60,\ 13x - 11y = 10.$ $x = 5,\ y = 5.$

14. $10x + 9y = 290,\ 12x - 11y = 130.$ $x = 20,\ y = 10.$

96. Elimination by Substitution. — Find the values of x and y in the equations

$$4x + 3y = 22 \quad . \quad . \quad . \quad . \quad . \quad . \quad (1)$$

$$5x - 7y = 6 \quad . \quad . \quad . \quad . \quad . \quad (2)$$

Transpose $3y$ in (1), $4x = 22 - 3y;$

divide by 4, $x = \dfrac{22 - 3y}{4};$

substitute this value of x in (2), and we obtain

$$5\!\left(\frac{22 - 3y}{4}\right) - 7y = 6;$$

multiply by 4, $5(22 - 3y) - 28y = 24.$

$$\therefore \quad y = 2.$$

Substitute this value of y in *either* (1) or (2), thus in (1)

$$4x + 6 = 22.$$

$$\therefore \quad x = 4 \Big\}$$

and $y = 2$.

In this solution we eliminated x by *substitution.*

Otherwise thus : from (1) we have

$$3y = 22 - 4x;$$

divide by 3, $y = \dfrac{22 - 4x}{3};$

substitute this value of y in (2),

$$5x - 7\!\left(\frac{22 - 4x}{3}\right) = 6;$$

multiply by 3, $15x - 7(22 - 4x) = 18;$

that is, $15x - 154 + 28x = 18.$

$$\therefore \quad x = 4.$$

Substitute this value of x in *either* (1) or (2), thus in (1)

$$16 + 3y = 22. \quad \therefore \quad y = 2.$$

Here we eliminated y by *substitution*.

Hence, **to eliminate an unknown** quantity by *substitution*, we have the following

RULE.

From either equation, find the value of the unknown quantity to be eliminated, in terms of the other; and substitute this value for that quantity in the other equation.

Solve by *substitution* the following equations:

2. $3x - 4y = 2, 7x - 9y = 7.$ *Ans.* $x = 10, y = 7.$
3. $11x - 7y = 37, 8x + 9y = 41.$ $x = 4, y = 1.$
4. $6x - 7y = 42, 7x - 6y = 75.$ $x = 21, y = 12.$
5. $3x - 4y = 18, 3x + 2y = 0.$ $x = 2, y = -3.$
6. $4x - \dfrac{y}{2} = 11, 2x - 3y = 0.$ $x = 3, y = 2.$
7. $2x - y = 9, 3x - 7y = 19.$ $x = 4, y = -1.$
8. $15x + 7y = 29, 9x + 15y = 39.$ $x = 1, y = 2.$
9. $2x + y = 10, 7x + 8y = 53.$ $x = 3, y = 4.$

97. Elimination by Comparison. — Find the values of x and y in the equations

$$2x + 3y = 23 \quad \ldots \ldots \quad (1)$$
$$5x - 2y = 10 \quad \ldots \ldots \quad (2)$$

Finding the value of x in terms of y from both (1) and (2), we have,

from (1), $$x = \frac{23 - 3y}{2}, \quad \ldots \ldots \quad (3)$$

and from (2), $$x = \frac{10 + 2y}{5}. \quad \ldots \ldots \quad (4)$$

Placing these two values of x equal to each other, we have

$$\frac{10 + 2y}{5} = \frac{23 - 3y}{2}.$$

Clearing of fractions, by multiplying by 10, we have

$$20 + 4y = 115 - 15y.$$
$$\therefore\ 19y = 95. \quad \therefore\ y = 5.$$

Substitute this value of y in *either* (3) or (4), thus in (4)

$$x = \frac{10 + 10}{5} = 4.$$

In this solution we eliminated x by *comparison*.

Otherwise thus: find the values of y in terms of x from (1) and (2).

$$y = \frac{23 - 2x}{3}, \quad \ldots \ldots \ldots (5)$$

$$y = \frac{5x - 10}{2}. \quad \ldots \ldots \ldots (6)$$

Therefore $\qquad \dfrac{23 - 2x}{3} = \dfrac{5x - 10}{2}.$

Clearing of fractions

$$46 - 4x = 15x - 30.$$
$$\therefore\ 19x = 76.$$
$$\therefore\ \ x = 4,\ \text{the same as before.}$$

Substituting this value of x in either (5) or (6), we deduce $y = 5$, as before.

In this solution we eliminated y by *comparison*.

Hence, to eliminate an unknown quantity by *comparison*, we have the following

Rule.

From each equation find the value of the unknown quantity to be eliminated, in terms of the other; then place these values equal to each other.

Note. — Either of these methods of elimination may be employed, according to circumstances, and we shall always obtain the same result, whichever one we use; each method has its advantages in particular cases. Generally, the last two methods introduce *fractional* expressions, while the first method does not, if the equations be first cleared of fractions. As a general rule, the method by *addition or subtraction* is the most simple and elegant. When either of the unknown quantities has 1 for its coefficient, the method by *substitution* is advantageous. When there are more than two unknown quantities, it is often convenient to use several of the methods in the same example.

Solve by *comparison* the following equations:

2. $7x - 5y = 24, \ 4x - 3y = 11.$ *Ans.* $x = 17, \ y = 19.$

3. $\dfrac{x}{3} + 3y = 7, \ \dfrac{4x - 2}{5} = 3y - 4.$ $x = 3, \ y = 2.$

4. $6x - 5y = 1, \ 7x - 4y = 8\frac{1}{2}.$ $x = 3\frac{1}{2}, y = 4.$

5. $\dfrac{x + y}{3} + x = 15, \dfrac{x - y}{5} + y = 6.$ $x = 10, \ y = 5.$

6. $\dfrac{3x}{19} + 5y = 13, 2x + \dfrac{4 - 7y}{2} = 33.$ $x = 19, \ y = 2.$

7. $2x + \dfrac{y - 2}{5} = 21, 4y + \dfrac{x - 4}{6} = 29.$ $x = 10, \ y = 7.$

98. Fractional Simultaneous Equations of the Form

$$\frac{12}{x} + \frac{8}{y} = 8, \quad \ldots \ldots (1)$$

$$\frac{27}{x} - \frac{12}{y} = 3. \quad \ldots \ldots (2)$$

If we cleared these equations of fractions they would involve the product xy of the unknown quantities; and thus they would become quite complex. But they may be solved by the methods already given, as follows:

Multiply (1) by 3, $\dfrac{36}{x} + \dfrac{24}{y} = 24.$

Multiply (2) by 2, $\dfrac{54}{x} - \dfrac{24}{y} = 6.$

Add $\dfrac{90}{x} = 30.$

Divide by 30, $\dfrac{3}{x} = 1.$ $\therefore \ x = 3.$

Substitute this value of x in (1),

$$1\tfrac{2}{3} + \frac{8}{y} = 8.$$

Transpose, $\dfrac{8}{y} = 4.$ $\therefore$ $y = 2.$

Solve the following equations :

2. $\dfrac{9}{x} - \dfrac{4}{y} = 1,\ \dfrac{18}{x} + \dfrac{20}{y} = 16.$ *Ans.* $x = 3, y = 2.$

3. $\dfrac{8}{x} - \dfrac{9}{y} = 1,\ \dfrac{10}{x} + \dfrac{6}{y} = 7.$ $x = 2, y = 3.$

4. $\dfrac{5}{x} + \dfrac{6}{y} = 3,\ \dfrac{15}{x} + \dfrac{3}{y} = 4.$ $x = 5, y = 3.$

5. $\dfrac{6}{x} - \dfrac{7}{y} = 2,\ \dfrac{2}{x} + \dfrac{14}{y} = 3.$ $x = 2, y = 7.$

6. $\dfrac{5}{x} + \dfrac{16}{y} = 79,\ \dfrac{16}{x} - \dfrac{1}{y} = 44.$ $x = \tfrac{1}{3}, y = \tfrac{1}{4}.$

99. Literal Simultaneous Equations. — Let it be required to solve

$$ax + by = c \quad \dots \quad (1)$$
$$a'x + b'y = c' \quad \dots \quad (2)$$

Multiply (1) by a', $\overline{\quad aa'x + a'by = a'c. \quad \dots \quad (3)}$

Multiply (2) by a, $aa'x + ab'y = ac'. \quad \dots \quad (4)$

Subtract (3) from (4), $(ab' - a'b)y = ac' - a'c.$

Divide by $(ab' - a'b)$, $y = \dfrac{ac' - a'c}{ab' - a'b}.$

To find the value of x, eliminate y from (1) and (2) thus :
Multiply (1) by b' and (2) by b,

$$ab'x + bb'y = b'c \quad \dots \quad (5)$$
$$a'bx + bb'y = bc' \quad \dots \quad (6)$$

Subtract (6) from (5), $(ab' - a'b)x = b'c - bc'.$

Divide by $(ab' - a'b)$, $x = \dfrac{b'c - bc'}{ab' - a'b}.$

Solve the following literal equations by either method of elimination :

2. $x + y = a + b, bx + ay = 2ab.$ *Ans.* $x = a, y = b.$

3. $(a + c)x - by = bc, x + y = a + b.$ $x = b, y = a.$

4. $x + y = c, ax - by = c(a - b).$ $x = \dfrac{ac}{a + b}, y = \dfrac{bc}{a + b}.$

5. $\dfrac{x}{a} + \dfrac{y}{b} = 1, \dfrac{x}{b} + \dfrac{y}{a} = 1.$ $x = \dfrac{ab}{a + b}, y = \dfrac{ab}{a + b}.$

6. $\dfrac{x}{a} + \dfrac{y}{b} = c, \dfrac{x}{b} - \dfrac{y}{a} = 0.$ $x = \dfrac{ab^2c}{a^2 + b^2}, y = \dfrac{a^2bc}{a^2 + b^2}.$

7. The sum of two numbers is a and their difference is b : find the numbers.   *Ans.* Greater $\dfrac{a}{2} + \dfrac{b}{2}$; less $\dfrac{a}{2} - \dfrac{b}{2}.$

When the known quantities in a problem are represented by letters, the answer furnishes a general result or *Formula* (Art. 41); and a formula expressed in ordinary language, furnishes a *Rule.* Thus, in the present example, we have the following

Rule. *The sum and difference of two numbers being given, to find the numbers: The greater number is equal to half the sum plus half the difference; the less number is equal to half the sum minus half the difference.*

100. Simultaneous Equations with Three or More Unknown Quantities. — In order to solve simultaneous equations which contain two unknown quantities, we have seen that we must have two equations (Art. 93). Similarly we find that in order to solve simultaneous equations which contain three unknown quantities, we must have three equations. And generally, when the values of several unknown quantities are to be found, it is necessary to have as many simultaneous equations as there are unknown quantities.

Simultaneous simple equations involving three or more unknown quantities, may be solved by either of the three methods of elimination explained in the preceding articles;

but the most convenient method of elimination is generally that by addition or subtraction. The unknown quantities are to be eliminated one at a time by the following

RULE.

If there be three simple equations containing three unknown quantities, eliminate one of the unknown quantities from any two of the equations, by the methods already explained (Arts. 95, 96, 97) ; *then eliminate the same unknown quantity from the third given equation and either of the former two; two equations involving two unknown quantities are thus obtained, and the values of these unknown quantities may be found by the rules given in the preceding Articles. The remaining unknown quantity may be found by substituting these values in any one of the given equations.*

If *four* equations are given involving four unknown quantities, one of the unknown quantities must be eliminated from three pairs of the equations. Three equations involving three unknown quantities will thus be obtained, which may be solved according to the rule. If five or more equations are given, they may be solved in a similar manner.

NOTE 1.—Either of the unknown quantities may be selected, as the one to be first eliminated; but it is best to begin with the quantity which has the simplest coefficients; and when an unknown quantity is not contained in all the given equations, it is generally best to eliminate that quantity first.

EXAMPLES.

1. Solve
$$6x + 2y - 5z = 13, \quad . \quad . \quad . \quad . \quad . \quad (1)$$
$$3x + 3y - 2z = 13, \quad . \quad . \quad . \quad . \quad . \quad (2)$$
$$7x + 5y - 3z = 26. \quad . \quad . \quad . \quad . \quad . \quad (3)$$

Choose y as the first quantity to be eliminated.
Multiply (1) by 3, and (2) by 2,
$$18x + 6y - 15z = 39,$$
$$6x + 6y - 4z = 26.$$

subtracting,
$$12x - 11z = 13 \quad . \quad . \quad . \quad . \quad . \quad (4)$$

Multiply (1) by 5, and (3) by 2,

$$30x + 10y - 25z = 65,$$
$$14x + 10y - 6z = 52.$$

subtracting, $16x - 19z = 13$ . . . . . (5)

We have now to find the values of x and z from (4) and (5).

Multiply (4) by 4 and (5) by 3 (Art. 95, Rem.),

$$48x - 44z = 52,$$
$$48x - 57z = 39.$$

subtracting, $13z = 13.$ $\therefore$ $z = 1.$

Substitute this value of z in (4) ; thus

$$12x - 11 = 13.   \therefore   x = 2.$$

Substitute these values of x and z in (1) ; thus

$$12 + 2y - 5 = 13.$$
$$\therefore \quad y = 3,$$

and $\left. \begin{array}{r} z = 1, \\ x = 2. \end{array} \right\}$

NOTE 2. — Although the method of elimination given by the rule is generally the best, yet in particular examples solutions may be obtained more easily and elegantly by other means, which the student must learn by experience. After a little practice he will find that the solution may often be considerably shortened by a suitable combination of the given equations. Thus, Ex. 1 may be solved as follows:

Add (1) and (2) and subtract (3),

$$2x - 4z = 0,$$

or $x = 2z$ . . . . . . (6)

Substitute this value of x in (1) and (2), and we get

$$2y + 7z = 13,$$
$$3y + 4z = 13.$$

Subtracting, $y - 3z = 0.$
$$\therefore \quad y = 3z \quad . . . . . \circ . (7)$$

Substitute these values of x and y in (1) ; thus

$$12z + 6z - 5z = 13.$$
$$\therefore \quad z = 1;$$

therefore from (6) and (7), $\left. \begin{array}{r} x = 2, \\ y = 3. \end{array} \right\}$

2. Solve

$$\frac{1}{2x} + \frac{1}{4y} - \frac{1}{3z} = \tfrac{1}{4}; \quad \cdots \cdots (1)$$

$$\frac{1}{x} = \frac{1}{3y}; \quad \cdots \cdots (2)$$

$$\frac{1}{x} - \frac{1}{5y} + \frac{4}{z} = 2\tfrac{2}{15} \quad \cdots \cdots (3)$$

Clearing of fractional coefficients, we obtain

from (1)
$$\frac{6}{x} + \frac{3}{y} - \frac{4}{z} = 3, \quad \cdots \cdots (4)$$

from (2)
$$\frac{3}{x} - \frac{1}{y} = 0, \quad \cdots \cdots (5)$$

from (3)
$$\frac{15}{x} - \frac{3}{y} + \frac{60}{z} = 32. \quad \cdots \cdots (6)$$

Choose z as the first quantity to be eliminated (Note 1).
Multiply (4) by 15 and add the result to (6) ; thus

$$\frac{105}{x} + \frac{42}{y} = 77.$$

Divide by 7,
$$\frac{15}{x} + \frac{6}{y} = 11 \quad \cdots \cdots (7)$$

Multiply (5) by 6,
$$\frac{18}{x} - \frac{6}{y} = 0.$$

$$\therefore \quad \frac{33}{x} = 11.$$

$$\therefore \quad x = 3, \\ \text{from (5)} \quad y = 1, \\ \text{from (4)} \quad z = 2.$$

3. Solve
$$5x - 3y - z = 6, \quad \cdots \cdots (1)$$
$$13x - 7y + 3z = 14, \quad \cdots \cdots (2)$$
$$7x - 4y = 8. \quad \cdots \cdots (3)$$

Multiply (1) by 3 and add the result to (2),
$$28x - 16y = 32.$$

Divide by 4,
$$7x - 4y = 8.$$

Thus we see that the combination of equations (1) and (2) leads to an equation which is identical with (3) ; and so to find x and y, we have but a single equation, $7x - 4y = 8$, with two unknown quantities, which is not sufficient to determine the definite value of either (Art. 93). The anomaly here arises from the fact that one of these three equations is deducible from the others; in other words, that the three equations are not *independent* (Art. 93).

NOTE 3. — Sometimes it is convenient to use the following rule:
Express the values of two of the unknown quantities from two of the equations in terms of the third unknown quantity, and substitute these values in the third equation. From this, the third unknown quantity can be found, and then the other two; thus

4. Solve

$$3x + 4y - 16z = 0, \quad \ldots \ldots (1)$$

$$5x - 8y + 10z = 0, \quad \ldots \ldots (2)$$

$$2x + 6y + 7z = 52 \quad \ldots \ldots (3)$$

Multiply (1) by 2 and add to (2); thus

$$11x - 22z = 0. \qquad \therefore \quad x = 2z.$$

Multiply (1) by 5, and (2) by 3, and subtract; thus

$$44y - 110z = 0. \qquad \therefore \quad y = \frac{5z}{2}.$$

Substitute these values of x and y in (3); thus

$$4z + 15z + 7z = 52.$$

and
$$\left. \begin{array}{l} \therefore \quad z = 2, \\ x = 4, \\ y = 5. \end{array} \right\}$$

NOTE 4. — The rule in Note 3 is especially convenient when all of the unknown quantities occur in only one equation; thus

5. Solve

$$x + y + z = a + b + c, \quad \ldots \ldots (1)$$

$$x - y = b - a, \quad \ldots \ldots (2)$$

$$x - z = c - a \quad \ldots \ldots (3)$$

From (2) we have
$$y = x + a - b. \quad \ldots \ldots (4)$$

From (3) we have
$$z = x + a - c. \quad \ldots \ldots (5)$$

Substitute these values of y and z in (1),

$$x + x + a - b + x + a - c = a + b + c.$$

$$\therefore \quad 3x = -a + 2b + 2c.$$

$$\therefore \quad x = \tfrac{2}{3}(a + b + c) - a, \Big\}$$

from (4) $\qquad\qquad y = \tfrac{2}{3}(a + b + c) - b, \Big\}$

from (5) $\qquad\qquad z = \tfrac{2}{3}(a + b + c) - c. \Big\}$

Solve the following equations:

6. $7x + 3y - 2z = 16,$
   $2x + 5y + 3z = 39,$ $\qquad$ *Ans.* $\begin{cases} x = & 2, \\ y = & 4, \\ z = & 5. \end{cases}$
   $5x - y + 5z = 31.$

7. $2x + 3y + 4z = 16,$
   $3x + 2y - 5z = 8,$ $\qquad\qquad\quad \begin{cases} x = & 3, \\ y = & 2, \\ z = & 1. \end{cases}$
   $5x - 6y + 3z = 6.$

8. $x + 2y + 2z = 11,$
   $2x + y + z = 7,$ $\qquad\qquad\quad \begin{cases} x = & 1, \\ y = & 2, \\ z = & 3. \end{cases}$
   $3x + 4y + z = 14.$

9. $x + 3y + 4z = 14,$
   $x + 2y + z = 7,$ $\qquad\qquad\quad \begin{cases} x = & -2, \\ y = & 4, \\ z = & 1. \end{cases}$
   $2x + y + 2z = 2.$

10. $\dfrac{1}{x} + \dfrac{2}{y} - \dfrac{3}{z} = 1,$

$\qquad \dfrac{5}{x} + \dfrac{4}{y} + \dfrac{6}{z} = 24,$ $\qquad\qquad \begin{cases} x = & \tfrac{1}{2}, \\ y = & \tfrac{2}{3}, \\ z = & \tfrac{3}{4}. \end{cases}$

$\qquad \dfrac{7}{x} - \dfrac{8}{y} + \dfrac{9}{z} = 14.$

101. Problems Leading to Simultaneous Equations. — We shall now give some examples of problems which lead to simultaneous equations of the first degree with two or more unknown quantities. Many of the problems given in Chapter IX. really contain *two* or *more* unknown quantities, but the given relations are there of so simple a nature that it is easy to express all of the unknown quantities in terms of one unknown quantity, and thus to require but a single equation.   In the problems of the present chapter the relations between the unknown quantities are not so simple, and the solution will give rise to simultaneous equations; and in all cases the conditions of the problem must be sufficient to furnish as many independent equations as there are unknown quantities to be determined (Art. 100).

EXAMPLES.

1. Find two numbers such that the greater exceeds twice the less by 3, and that twice the greater exceeds the less by 27.

Let x = the greater number,

and y = the less number.

Then from the conditions,

$$x - 2y = 3,$$

and $$2x - y = 27.$$

Solving these equations, we have $x = 17$, $y = 7$.

2. If the numerator of a fraction be increased by 2 and the denominator by 1, it becomes equal to $\frac{5}{8}$; and if the numerator and denominator are each diminished by 1, it becomes equal to $\frac{1}{2}$: find the fraction.

Let x = the numerator,

and y = the denominator.

Then from the conditions,

$$\frac{x + 2}{y + 1} = \frac{5}{8},$$

and $$\frac{x - 1}{y - 1} = \frac{1}{2}.$$

Solving, we have $x = 8$, $y = 15$.

Hence the fraction is $\frac{8}{15}$.

3. A man and a boy can do in 15 days a piece of work which would be done in 2 days by 7 men and 9 boys: how long would it take one man to do it?

Let $x =$ the number of days in which one man would do the whole,

and let $y =$ the number of days in which one boy would do the whole.

Then $\dfrac{1}{x} =$ the part that one man can do in one day,

and $\dfrac{1}{y} =$ the part that one boy can do in one day.

Then from the conditions of the question, a man and a boy together do $\frac{1}{15}$th of the work in one day; hence we have

$$\frac{1}{x} + \frac{1}{y} = \tfrac{1}{15} \quad \cdots \cdots \cdots \quad (1)$$

Also, since 7 men and 9 boys do half the work in a day, we have

$$\frac{7}{x} + \frac{9}{y} = \tfrac{1}{2} \quad \cdots \cdots \cdots \quad (2)$$

Multiplying (1) by 9, and subtracting (2) from it, we have

$$\frac{2}{x} = \tfrac{1}{10}. \quad \therefore \quad x = 20.$$

Thus one man would do the work in 20 days.

4. A railway train after traveling an hour is detained 24 minutes, after which it proceeds at six-fifths of its former rate, and arrives 15 minutes late. If the detention had taken place 5 miles further on, the train would have arrived 2 minutes later than it did. Find the original rate of the train, and the distance traveled.

Let $\quad x =$ the original rate of the train in miles per hour ;

and $\quad y =$ the number of miles in the whole distance traveled.

Then $\quad y - x =$ the number of miles to be traveled after the detention.

$\dfrac{y - x}{x} =$ the number of hours in traveling $y - x$ miles at the original rate,

and $\dfrac{5(y - x)}{6x} =$ the number of hours in traveling $y - x$ miles at the increased rate.

Since the train is detained 24 minutes, and yet arrives only 15 minutes late, it follows that the remainder of the journey is performed in nine minutes less than it would have been if the rate had not been increased; hence we have

$$\frac{y-x}{x} - \frac{5(y-x)}{6x} = \tfrac{9}{60}. \quad \cdot \quad \cdot \quad \cdot \quad \cdot \quad (1)$$

If the detention had taken place 5 miles further on, there would have been $y - x - 5$ miles left to be traveled after the detention; hence we have

$$\frac{y-x-5}{x} - \frac{5(y-x-5)}{6x} = \tfrac{7}{60}. \quad \cdot \quad \cdot \quad \cdot \quad (2)$$

Subtract (2) from (1), $\quad \dfrac{5}{x} - \dfrac{25}{6x} = \tfrac{2}{60},$

or $\qquad\qquad\qquad\qquad \dfrac{5}{6x} = \tfrac{1}{30}.$

$$\therefore \quad x = 25, \quad\big\rbrace$$
and $\qquad\qquad\qquad\qquad y = 47\tfrac{1}{2}.$

5. There is a number consisting of three digits; the middle digit is zero, and the sum of the other digits is 11; if the digits be reversed, the number so formed exceeds the original number by 495: find the number.

Let $x =$ the digit in the unit's place,
and $y =$ the digit in the hundred's place.

Then, since the digit in the ten's place is 0, the number will be represented by $100y + x$ (Art. 92, Ex. 10); hence from the conditions, we have

$$x + y = \quad 11,$$
and $\qquad 100x + y - (100y + x) = 495.$

Solving, we get $x = 8$, $y = 3$; hence the number is 308.

6. A sum of money was divided equally among a certain number of persons; had there been three more, each would have received $1 less, and had there been two less, each would have received $1 more than he did: find the number of persons, and what each received.

Let x = the number of persons,
and y = the number of dollars which each received.

Then xy = the number of dollars to be divided, and from the conditions, we have

$$(x + 3)(y - 1) = xy,$$
and
$$(x - 2)(y + 1) = xy.$$

Solving, we get $x = 12$ and $y = 5$.

7. A train traveled a certain distance at a uniform rate ; had the speed been 6 miles an hour more, the journey would have occupied 4 hours less ; and had the speed been 6 miles an hour less, the journey would have occupied 6 hours more : find the distance.

Let x = the rate of the train in miles per hour,
and y = the time of running the journey in hours.

Then xy = the distance traversed, and from the conditions,
we have
$$(x + 6)(y - 4) = xy,$$
and
$$(x - 6)(y + 6) = xy.$$

Solving, we get $x = 30$, and $y = 24$. Hence the distance is 720 miles.

8. A, B, and C can together do a piece of work in 30 days ; A and B can together do it in 32 days ; and B and C can together do it in 120 days : find the time in which each alone could do the work.

Let x = the number of days in which A could do it,
. y = the number of days in which B could do it,
and z = the number of days in which C could do it.

Then we have from the conditions,

$$\frac{1}{x} + \frac{1}{y} + \frac{1}{z} = \tfrac{1}{30},$$

$$\frac{1}{x} + \frac{1}{y} = \tfrac{1}{32},$$

$$\frac{1}{y} + \frac{1}{z} = \tfrac{1}{120}.$$

Solving, we get $x = 40$, $y = 160$, $z = 480$.

9. Find the fraction which is equal to $\frac{1}{2}$ when its numerator is increased by unity, and is equal to $\frac{1}{3}$ when its denominator is increased by unity.   *Ans.* $\frac{3}{8}$.

10. A certain number of two digits is equal to five times the sum of its digits; and if nine be added to the number the digits are reversed : find the number.   *Ans.* 45.

11. If 15 lbs. of tea and 17 lbs. of coffee together cost $7.86, and 25 lbs. of tea and 13 lbs. of coffee together cost $10.34, find the price per pound of each.

Ans. The tea cost 32 cents, and the coffee cost 18 cents, a lb.

12. If A's money were increased by $36 he would have three times as much as B ; and if B's money were diminished by $5 he would have half as much as A : find the sum possessed by each.   *Ans.* A has $42, B has $26.

13. Find two numbers such that half the first with a third of the second may make 32, and that a fourth of the first with a fifth of the second may make 18.   *Ans.* 24, 60.

14. A farmer parting with his stock, sells to one 9 horses and 7 cows for $1200 ; and to another, at the same prices, 6 horses and 13 cows for the same sum : what was the price of each ?   *Ans.* $96, $48.

15. Having $45 to give away among a certain number of persons, I find that for a distribution of $3 to each man and $1 to each woman, I shall have $1 too little ; but that, by giving $2.50 to each man and $1.50 to each woman, I may distribute the sum exactly : how many were there of men and women ?   *Ans.* 12, 10.

16. Find three numbers, A, B, C, such that A with half of B, B with a third of C, and C with a fourth of A, may each be 1000.   *Ans.* 640, 720, 840.

17. A person spent $1.82 in buying oranges at the rate of 3 for two cents, and apples at 5 cents a dozen ; if he had bought five times as many oranges and a quarter of the number of apples he would have spent $5.30 : how many of each did he buy ?   *Ans.* 153, 192.

EXAMPLES.

Solve the following equations:

1. $5x - 7y = 0, 7x + 5y = 74.$ *Ans.* $x = 7, y = 5.$

2. $5x = 7y - 21, 21x - 9y = 75.$ $x = 7, y = 8.$

3. $6y - 5x = 18, 12x - 9y = 0.$ $x = 6, y = 8.$

4. $7x + 4y = 1, 9x + 4y = 3.$ $x = 1, y = -1\frac{1}{2}.$

5. $x - 11y = 1, 111y - 9x = 99.$ $x = 100, y = 9.$

6. $8x - 21y = 5, 6x + 14y = -26.$ $x = -2, y = -1.$

7. $39x - 8y = 99, 52x - 15y = 80.$ $x = 5, y = 12.$

8. $3x = 7y, 12y = 5x - 1.$ $x = -7, y = -3.$

9. $93x + 15y = 123, 15x + 93y = 201.$ $x = 1, y = 2.$

10. $\frac{x}{2} + \frac{y}{3} = 1, \frac{x}{4} - \frac{2y}{3} = 3.$ $x = 4, y = -3.$

11. $\frac{x + y}{3} + x = 15, \frac{x - y}{5} + y = 6.$ $x = 10, y = 5.$

12. $\frac{7x}{6} + \frac{5y}{3} = 34, \frac{7x}{8} + \frac{3y}{4} = \frac{5y}{8} + 12.$ $x = 12, y = 12.$

13. $\frac{1 - 3x}{7} + \frac{3y - 1}{5} = 2, \frac{3x + y}{11} + y = 9.$ $x = 5, y = 7.$

14. $\frac{x}{2} - \frac{1}{3}(y-2) - \frac{1}{4}(x-3) = 0, x - \frac{1}{2}(y-1) - \frac{1}{3}(x-2) = 0.$
   Ans. $x = 3\frac{2}{7}, y = 6\frac{5}{7}.$

15. $\frac{x-2}{3} - \frac{y+2}{4} = 0, \frac{2x-5}{5} - \frac{11-2y}{7} = 0.$ $x = 5, y = 2.$

16. $\frac{x}{3} + \frac{y}{4} = 3x - 7y - 37, 3x - 7y = 37.$ $x = 3, y = -4.$

17. $(x+1)(y+5) = (x+5)(y+1), xy + x + y = (x+2)(y+2).$
   Ans. $x = -2, y = -2.$

18. $xy - (y-1)(x-1) = 6(y-1), x - y = 1.$
   Ans. $x = 2\frac{1}{2}, y = 1\frac{1}{2}.$

19. $\frac{7 + x}{3} = \frac{9 + y}{5} = \frac{11 + x + y}{7}.$ $x = 8, y = 16.$

20. $.3x + .125y = x - 6, 3x - .5y = 28 - .25y.$
          Ans. $x = 10, y = 8.$

21. $.08x - .21y = .33, .12x + .7y = 3.54.$ $x = 12, y = 3.$

22. $\dfrac{9}{x} - \dfrac{4}{y} = 2, \dfrac{18}{x} + \dfrac{8}{y} = 10.$ $x = 2\frac{4}{7}, y = 2\frac{2}{3}.$

23. $\dfrac{2}{x} + \dfrac{5}{y} = 7, \dfrac{3}{x} - \dfrac{2}{y} = 11.$ $x = \frac{19}{69}, y = -19.$

24. $x + \dfrac{3}{y} = \frac{7}{2}, 3x - \dfrac{2}{y} = \frac{26}{3}.$ $x = 3, y = 6.$

25. $2x - \dfrac{3}{y} = 3, 8x + \dfrac{15}{y} = -6.$ $x = \frac{1}{2}, y = -\frac{3}{2}.$

26. $\dfrac{x + \frac{y}{2} - 3}{x - 5} + 7 = 0, \dfrac{3y - 10(x - 1)}{6} + \dfrac{x - y}{4} + 1 = 0.$
          Ans. $x = 4, y = 12.$

27. $\dfrac{x}{a} + \dfrac{y}{b} = 2, bx - ay = 0.$ $x = a, y = b.$

28. $a(x + y) + b(x - y) = 1, a(x - y) + b(x + y) = 1.$
        Ans. $x = \dfrac{1}{a + b}, y = 0.$

29. $x + y = a + b, ax - by + a^2 - b^2 = 0.$
       Ans. $x = 2b - a, y = 2a - b.$

30. $(a+b)x + (a-b)y = 2ac, (b+c)x + (b-c)y = 2bc.$
        Ans. $x = y = c.$

31. $\begin{aligned} x + 2y - 3z &= 6, \\ 2x + 4y - 7z &= 9, \\ 3x - y - 5z &= 8. \end{aligned}$ $\begin{cases} x = 8\frac{5}{7}, \\ y = 3\frac{1}{7}, \\ z = 3. \end{cases}$

32. $\begin{aligned} 2x - y + z &= 4, \\ 5x + y + 3z &= 5, \\ 2x - 3y + 4z &= 20. \end{aligned}$ $\begin{cases} x = -1, \\ y = -2, \\ z = 4. \end{cases}$

33. $\begin{aligned} x + 4y + 3z &= 17, \\ 3x + 3y + z &= 16, \\ 2x + 2y + z &= 11. \end{aligned}$ $\begin{cases} x = 2, \\ y = 3, \\ z = 1. \end{cases}$

34. $2x + 3y + 4z = 20,$
    $3x + 4y + 5z = 26,$ *Ans.* $\begin{cases} x = & 1, \\ y = & 2, \\ z = & 3. \end{cases}$
    $3x + 5y + 6z = 31.$

35. $x - \dfrac{y}{5} = 6,\ y - \dfrac{z}{7} = 8,\ z - \dfrac{x}{2} = 10.$ $x = 8,\ y = 10,\ z = 14.$

36. $\dfrac{a}{x} + \dfrac{b}{y} + \dfrac{c}{z} = 3,\ \dfrac{a}{x} + \dfrac{b}{y} - \dfrac{c}{z} = 1,\ \dfrac{2a}{x} - \dfrac{b}{y} - \dfrac{c}{z} = 0.$

   Ans. $x = a,\ y = b,\ z = c.$

37. $\dfrac{1}{x} + \dfrac{1}{y} = 1,\ \dfrac{1}{x} + \dfrac{1}{z} = 2,\ \dfrac{1}{y} + \dfrac{1}{z} = \frac{3}{2}.$ $x = \frac{4}{3}, y = 4, z = \frac{4}{5}.$

38. $\dfrac{2}{x} + \dfrac{1}{y} = \dfrac{3}{z},\ \dfrac{3}{z} - \dfrac{2}{y} = 2,\ \dfrac{1}{x} + \dfrac{1}{z} = \frac{4}{3}.$ $x = \frac{7}{6}, y = -\frac{7}{2}, z = \frac{21}{10}.$

39. $\dfrac{3y - 1}{4} = \dfrac{6z}{5} - \dfrac{x}{2} + \frac{9}{5},$

   $\dfrac{5x}{4} + \dfrac{4z}{3} = y + \frac{5}{6},$ $\begin{cases} x = & 2, \\ y = & 3, \\ z = & 1. \end{cases}$

   $\dfrac{3x + 1}{7} - \dfrac{z}{14} + \frac{1}{6} = \dfrac{2z}{21} + \dfrac{y}{3}.$

40. $\begin{cases} 7x - 3y = 1,\ 11z - 7u = 1, \\ 4z - 7y = 1,\ 19x - 3u = 1. \end{cases}$ $\begin{cases} x = 4, y = & 9, \\ z = 16, u = & 25. \end{cases}$

41. $\begin{cases} 3u - 2y = 2,\ 5x - 7z = 11, \\ 2x + 3y = 39,\ 4y + 3z = 41. \end{cases}$ $\begin{cases} u = 4, x = & 12, \\ y = 5, z = & 7. \end{cases}$

42. $\begin{cases} 2x - 3y + 2z = 13,\ 4y + 2z = 14, \\ 4u - 2x = 30,\ 5y + 3u = 32. \end{cases}$ $\begin{cases} x = 3. y = & 1, \\ u = 9, z = & 5. \end{cases}$

43. $\begin{cases} 7x - 2z + 3u = 17,\ 4y - 2z + v = 11, \\ 5y - 3x - 2u = 8,\ 4y - 3u + 2v = 9, \\ 3z + 8u = 33. \end{cases}$ $\begin{cases} x = 2, y = & 4, \\ z = 3, u = & 3, \\ v = 1. \end{cases}$

44. $\begin{cases} 3x - 4y + 3z + 3v - 6u = 11, \\ 3x - 5y + 2z - 4u = 11, \\ 10y - 3z + 3u - 2v = 2, \\ 5z + 4u + 2v - 2x = 3, \\ 6u - 3v + 4x - 2y = 6, \end{cases}$ $\begin{cases} x = & 2, \\ y = & 1, \\ z = & 3, \\ u = & -1, \\ v = & -2. \end{cases}$

45. What fraction is that, to the numerator of which if 7 be added, its value is $\frac{2}{3}$; but if 7 be taken from the denominator its value is $\frac{3}{8}$?   *Ans.* $\frac{3}{15}$.

46. A rectangular bowling-green having been measured, it was observed that, if it were 5 feet broader and 4 feet longer, it would contain 116 feet more; but if it were 4 feet broader and 5 feet longer, it would contain 113 feet more: find its area.   *Ans.* 108 sq. ft.

47. A party was composed of a certain number of men and women, and, when four of the women were gone, it was observed that there were left just half as many men again as women; they came back, however, with their husbands, and now there were only a third as many men again as women: what were the original numbers of each?   *Ans.* 12, 12.

48. The sum of the two digits of a certain number is 6 times their difference, and the number itself exceeds 6 times their sum by 3: find the number.   *Ans.* 75.

49. Divide the numbers 80 and 90 each into two parts, so that the sum of one out of each pair may be 100, and the difference of the others 30.

Ans. 30, 50, and 70, 20; or 60, 20, and 40, 50.

50. Four times B's age exceeds A's age by 20 years, and one-third of A's age is less than B's age by 2 years: find their ages.   *Ans.* A 36 years, B 14 years.

51. In 8 hours A walks 12 miles more than B does in 7 hours; and in 13 hours B walks 7 miles more than A does in 9 hours: how many miles does each walk per hour?

Ans. A 5 miles, B 4 miles.

52. The sum and the difference of a number of two digits and of the number formed by reversing the digits are 110 and 54 respectively: find the numbers.   *Ans.* 28, 82.

53. In a bag containing black and white balls, half the number of white is equal to a third of the number of black; and twice the whole number of balls exceeds three times the number of black balls by four: how many balls did the bag contain?   *Ans.* 8 white, 12 black.

54. Twenty-eight tons of goods are to be carried in carts and wagons, and it is found that this will require 15 carts and 12 wagons, or else 24 carts and 8 wagons : how much can each cart and each wagon carry?   *Ans.* ⅔ tons, ⅔ tons.

55. The first edition of a book had 600 pages, and was divided into two parts ; in the second edition one quarter of the second part was omitted and 30 pages added to the first ; the change made the two parts of the same length : what were they in the first edition?   *Ans.* 240, 360.

56. If A were to receive $10 from B he would then have twice as much as B would have left ; but if B were to receive $10 from A, B would have three times as much as A would have left : how much has each?   *Ans.* $22, $26.

57. A farmer sold 30 bushels of wheat and 50 bushels of barley for $75 ; he also sold at the same prices 50 bushels of wheat and 30 bushels of barley for $77 : what was the price of the wheat per bushel?   *Ans.* $1.

58. A certain fishing rod consists of two parts ; the length of the upper part is to the length of the lower as 5 to 7 ; and 9 times the upper part together with 13 times the lower part exceeds 11 times the whole rod by 36 inches : find the lengths of the two parts.   *Ans.* 45, 63.

59. A certain company in a tavern found, when they came to pay their bill, that if there had been 3 more persons to pay the same bill, they would have paid $1 each less than they did ; and if there had been 2 fewer persons they would have paid $1 each more than they did : find the number of persons, and the number of dollars each paid.
        Ans. 12, 5.

60. There is a rectangular floor, such that if it had been 2 feet broader, and 3 feet longer, it would have been 64 square feet larger ; but if it had been 3 feet broader, and 2 feet longer, it would have been 68 square feet larger : find the length and breadth of the floor.   *Ans.* 14 ft., 10 ft.

Let $x =$ the length, and $y =$ the breadth, of the floor in feet; then $xy =$ the surface of the floor in square feet.

61. When a certain number of two digits is doubled, and increased by 36, the result is the same as if the number had been reversed, and doubled, and then diminished by 36; also the number itself exceeds 4 times the sum of its digits by 3: find the number.   *Ans.* 59.

62. Two passengers have together 560 lbs. of luggage, and are charged for the excess above the weight allowed 62 cents and $1.18 respectively; if the luggage had all belonged to one of them he would have been charged $2.30: how much luggage is each passenger allowed without charge?
                                                  Ans. 100 lbs.

63. A farmer has 28 bushels of barley at 56 cents a bushel; with these he wishes to mix rye at 72 cents a bushel, and wheat at 96 cents a bushel, so that the mixture may consist of 100 bushels, and be worth 80 cents a bushel: how many bushels of rye and wheat must he take? *Ans.* 20, 52.

64. A and B ran a race which lasted 5 minutes; B had a start of 20 yards; but A ran 3 yards while B was running 2, and won by 30 yards: find the length of the course and the rate of each per minute.
                    Ans. 150 yards, 30 yards, 20 yards.

65. A and B can together do a certain work in 30 days; at the end of 18 days however B is called off and A finishes it alone in 20 days more: find the time in which each could do the work alone.   *Ans.* 50, 75.

66. A, B, and C can together drink a cask of beer in 15 days; A and B together drink four-thirds of what C does; and C drinks twice as much as A: find the time in which each alone could drink the cask of beer. *Ans.* 70, 42, 35.

67. A and B run a mile; at the first heat A gives B a start of 20 yards, and beats him by 30 seconds; at the second heat A gives B a start of 32 seconds, and beats him by $9\frac{5}{11}$ yards: find the rate per hour at which A runs.
                                                  Ans. 12 miles.

68. A and B are two towns situated 24 miles apart, on the same bank of a river. A man goes from A to B in 7

hours, by rowing the first half of the distance, and walking the second half. In returning he walks the first half at three-fourths of his former rate, but the stream being with him he rows at double his rate in going ; and he accomplishes the whole distance in 6 hours. Find his rates of walking and rowing up stream. *Ans.* 4 miles walking, 3 miles rowing.

69. A railway train after traveling an hour is detained 15 minutes, after which it proceeds at three-fourths of its former rate, and arrives 24 minutes late. If the detention had taken place 5 miles further on, the train would have arrived 3 minutes sooner than it did. Find the original rate of the train and the distance traveled.

Ans. $33\frac{1}{3}$ miles per hour, $48\frac{1}{3}$ distance.

70. The time which an express train takes to travel a journey of 120 miles is to that taken by an ordinary train as 9 to 14. The ordinary train loses as much time in stopping as it would take to travel 20 miles without stopping. The express train loses only half as much time in stopping as the ordinary train, and it also travels 15 miles an hour faster. Find the rate of each train. *Ans.* 45, 30 miles per hour.

71. A and B can perform a piece of work together in 48 days ; A and C in 30 days ; and B and C in $26\frac{2}{3}$ days : find the time in which each could perform the work alone.

Ans. 120, 80, 40 days.

72. There is a certain number of three digits which is equal to 48 times the sum of its digits ; and if 198 be subtracted from the number the digits will be reversed ; also the sum of the extreme digits is equal to twice the middle digit : find the number. *Ans.* 432.

73. A man bought 10 horses, 120 oxen, and 46 cows. The price of 3 oxen is equal to that of 5 cows. A horse, an ox, and a cow together cost a number of dollars greater by 300 than the whole number of animals bought ; and the whole sum spent was $9366. Find the price of a horse, an ox, and a cow respectively. *Ans.* $420, $35, $21

CHAPTER XI.

INVOLUTION AND EVOLUTION.

102. Involution is the process of raising an expression to any required power. Involution is therefore only a particular case of *multiplication*, in which the factors are equal (Art. 12). It is convenient, however, to give some rules for writing down the power at once.

It is evident from the Rule of Signs (Art. 36) that,

(1) *No even power of any quantity can be negative.*

(2) *Any odd power of a quantity will have the same sign as the quantity itself.* Thus,

$$(-a)^2 = (-a)(-a) = +a^2,$$
$$(-a)^3 = (-a)(-a)(-a) = +a^2(-a) = -a^3,$$
$$(-a)^4 = (-a)(-a)(-a)(-a) = (-a^3)(-a) = +a^4;$$

and so on.

Note. — The *square* of every expression, whether positive or negative, is *positive*.

103. Involution of Powers of Monomials. — From the definition, we have, by the rules of multiplication,

$$(a^2)^3 = (a^2)(a^2)(a^2) = a^{2+2+2} = a^6.$$
$$(-a^3)^2 = (-a^3)(-a^3) = a^{3+3} = a^6.$$
$$(-3a^3)^2 = (-3a^3)(-3a^3) = (-3)^2(a^3)^2 = 9a^6.$$

Generally,

$$(a^m)^n = a^m \cdot a^m \cdot a^m \cdot a^m \ldots \text{to } n \text{ factors}$$
$$= a^{m+m+m+m} \ldots \ldots \text{to } n \text{ terms}$$
$$= a^{mn}.$$

$$(ab)^m = ab \cdot ab \cdot ab \ldots \ldots \text{to } m \text{ factors}$$
$$= (aaaa \ldots \text{to } m \text{ factors}) \times (bbbb \ldots \text{to } m \text{ factors})$$
$$= a^m b^m.$$

Hence $(ab)^m = a^m b^m$,

and so on for any number of factors. Thus, *the mth power of a product is equal to the product of the mth powers of its factors.*

Hence $(a^x b^y c^z \dots \dots)^m = (a^x)^m (b^y)^m (c^z)^m \dots \dots$

$$= a^{xm} b^{ym} c^{zm} \dots \dots$$

Hence, to raise any power of a quantity to any other power, we have the following

RULE.

Raise the numerical coefficient to the required power by Arithmetic, multiply the exponent of each factor by the exponent of the required power, and give the proper sign to the result.

$$\left(\frac{a}{b}\right)^m = \frac{a}{b} \times \frac{a}{b} \times \frac{a}{b} \dots \dots \text{ to } m \text{ factors}$$

$$= \frac{aaa \dots \dots \text{ to } m \text{ factors}}{bbb \dots \dots \text{ to } m \text{ factors}}$$

$$= \frac{a^m}{b^m}.$$

Hence, for obtaining any power of a fraction : *Raise both the numerator and denominator to that power, and give the proper sign to the result.*

EXAMPLES.

Show that

1. $(a^2 b^3)^2 = a^4 b^6.$

2. $(-a^2 b^3)^3 = -a^6 b^9.$

3. $(a^2 b^3 c^4)^4 = a^8 b^{12} c^{16}.$

4. $(-2x^2)^5 = -32x^{10}.$

5. $(-3ab^3)^6 = 729 a^6 b^{18}.$

6. $\left(\dfrac{a^2}{b^2}\right)^3 = \dfrac{a^6}{b^6}.$

7. $\left(-\dfrac{a^3}{bc^2}\right)^5 = -\dfrac{a^{15}}{b^5 c^{10}}.$

8. $\left(-\dfrac{a}{b^2 c^3}\right)^6 = \dfrac{a^6}{b^{12} c^{18}}.$

9. $\left(\dfrac{1}{3y^2}\right)^3 = \dfrac{1}{27 y^6}.$

10. $\left(-\dfrac{3x^5}{5x^3}\right)^3 = -\dfrac{27 x^{15}}{125 x^9}.$

11. $\left(-\dfrac{x^2 y^3 z^4}{2}\right)^2 = \dfrac{x^4 y^6 z^8}{4}.$

104. Involution of Binomials. — We have already proved by actual multiplication (Art. 41), the two following cases of the involution of binomial expressions:

$$(a + b)^2 = a^2 + 2ab + b^2 \quad . \quad . \quad . \quad . \quad (1)$$

$$(a - b)^2 = a^2 - 2ab + b^2 \quad . \quad . \quad . \quad . \quad (2)$$

If we multiply (1) and (2) by $a + b$ and $a - b$ respectively, we have

$$(a + b)^3 = a^3 + 3a^2b + 3ab^2 + b^3 . \quad . \quad . \quad (3)$$

$$(a - b)^3 = a^3 - 3a^2b + 3ab^2 - b^3 . \quad . \quad . \quad (4)$$

If we multiply (3) and (4) by $a + b$ and $a - b$ respectively, we shall have

$$(a + b)^4 = a^4 + 4a^3b + 6a^2b^2 + 4ab^3 + b^4.$$

$$(a - b)^4 = a^4 - 4a^3b + 6a^2b^2 - 4ab^3 + b^4.$$

By multiplying these two results by $a + b$ and $a - b$ respectively we should obtain $(a + b)^5$ and $(a - b)^5$; and by continuing the process we could obtain any required power of $(a + b)$ or $(a - b)$. Hence the following

RULE.

Multiply the binomial by itself, until it has been taken as a factor as many times as there are units in the exponent of the required power.

This rule, however, would be very laborious in finding any high power, for instance $(a + b)^{20}$. In Chapter XVII we shall prove a theorem, called the Binomial Theorem, by the aid of which any power of a binomial expression can be obtained without the labor of actual multiplication.

Since the above formulæ are true for *all* values of a and b, we can write down the squares and the cubes of any binomial expressions. Thus,

1. $(a^4 - b^4)^2 = (a^4)^2 + 2(a^4)(-b^4) + (-b^4)^2$

$\qquad = a^8 - 2a^4b^4 + b^8.$

Show that

2. $(2x + 3y)^2 = 4x^2 + 12xy + 9y^2.$

3. $(3x + 5y)^2 = 9x^2 + 30xy + 25y^2.$

4. $(x - 2y)^3 = x^3 - 6x^2y + 12xy^2 - 8y^3.$

5. $(2ab - 3c)^3 = 8a^3b^3 - 36a^2b^2c + 54abc^2 - 27c^3.$

6. $(5a^2 - 3b^2)^3 = 125a^6 - 225a^4b^2 + 135a^2b^4 - 27b^6.$

105. Involution of Polynomials. — We may now apply the formulæ of Art. 104 to obtain the powers of any trinomial or polynomial. Thus from (1)

$$(a + b + c)^2 = [(a + b) + c]^2$$
$$= (a + b)^2 + 2(a + b)c + c^2$$
$$= a^2 + b^2 + c^2 + 2ab + 2ac + 2bc \quad . \ (1)$$

In the same way we may prove

$$(a+b+c+d)^2 = a^2+b^2+c^2+d^2+2ab+2ac+2ad+2bc+2bd+2cd \quad . \ (2)$$

We observe in both (1) and (2) that the square consists of

(1) the sum of the squares of the several terms of the given expressions ;

(2) twice the sum of the products two and two of the several terms, taken with their proper signs.

The same law holds whatever be the number of terms in the expression to be squared. Hence the following

RULE.

To find the square of any polynomial, write the square of each term together with twice the product of each term by each of the terms following it.

From (3) of Art. 104 we obtain the *cube* of a trinomial as follows :

$$(a+b+c)^3 = [a+(b+c)]^3$$
$$= a^3+3a^2(b+c)+3a(b+c)^2+(b+c)^3$$
$$= a^3+b^3+c^3+3a^2(b+c)+3b^2(a+c)+3c^2(a+b)+6abc \quad . \ (3)$$

Hence to find the cube of a trinomial we have the following

RULE.

Write the cube of each term, together with three times the product of the square of each term by the sum of the other two, and six times the product of the three terms.

Formulæ (1), (2), and (3) may be used for obtaining the squares and cubes of any polynomial expressions, as explained in Art. 104. Thus, if we require $(1 - 2x + 3x^2)^2$, in formula (1) we put 1 for a, $-2x$ for b, and $3x^2$ for c, and obtain

1. $(1-2x+3x^2)^2$

$= (1)^2 + (-2x)^2 + (3x^2)^2 + 2(1)(-2x) + 2(1)(3x^2) + 2(-2x)(3x^2)$

$= 1 + 4x^2 + 9x^4 - 4x + 6x^2 - 12x^3$

$= 1 - 4x + 10x^2 - 12x^3 + 9x^4.$

Similarly by (3) we have

2. $(1-2x+3x^2)^3$

$= (1)^3 + (-2x)^3 + (3x^2)^3 + 3(1)^2(-2x+3x^2) + 3(-2x)^2(1+3x^2)$
$\qquad\qquad\qquad\qquad + 3(3x^2)^2(1-2x) + 6(1)(-2x)(3x^2)$

$= 1 - 8x^3 + 27x^6 + 3(-2x+3x^2) + 12x^2(1+3x^2)$
$\qquad\qquad\qquad\qquad\qquad\qquad + 27x^4(1-2x) - 36x^3$

$= 1 - 6x + 21x^2 - 44x^3 + 63x^4 - 54x^5 + 27x^6.$

Show that

3. $(1 - x + x^2)^2 = 1 - 2x + 3x^2 - 2x^3 + x^4.$

4. $(1 + 3x + 2x^2)^2 = 1 + 6x + 13x^2 + 12x^3 + 4x^4.$

5. $\left(\dfrac{a}{2} - 2b + \dfrac{c}{4}\right)^2 = \dfrac{a^2}{4} + 4b^2 + \dfrac{c^2}{16} - 2ab + \dfrac{ac}{4} - bc.$

6. $(\tfrac{2}{3}x^2 - x + \tfrac{3}{2})^2 = \dfrac{4x^4}{9} - \dfrac{4x^3}{3} + 3x^2 - 3x + \tfrac{9}{4}.$

7. $(1 + x + x^2)^3 = 1 + 3x + 6x^2 + 7x^3 + 6x^4 + 3x^5 + x^6.$

8. $(1 + x - x^2)^3 = 1 + 3x - 5x^3 + 3x^5 - x^6.$

EVOLUTION.

106. Evolution — Evolution of Monomials. — *Evolution* is the operation of finding any required root of a number or expression. A *root* of any quantity is a factor which being multiplied by itself a certain number of times produces the given quantity (Art. 13). Hence Evolution is the inverse of Involution (Art. 102).

The symbol which denotes that a square root is to be extracted is $\sqrt{\ }$; and for other roots the same symbol is used, but with a figure called the *index* written above to indicate the root (Art. 13).

By the Rule of Signs (Art. 36), we see that

(1) *any even root of a positive quantity may be either positive or negative;*

(2) *every odd root of a quantity has the same sign as the quantity;*

(3) *there can be no even root of a negative quantity.* Thus, $(1) a \times a = a^2$, and $(-a)(-a) = a^2$; therefore there are *two* roots of a^2, namely, $+a$ and $-a$.

(2) $(-a)(-a)(-a) = -a^3$; therefore the cube root of $-a^3$ is $-a$.

(3) There can be no square root of $-a^2$; for if any quantity be multiplied by itself, the result is a *positive* quantity.

There can be no even root of a negative quantity, because no quantity raised to an even power can produce a negative result. Even roots are called *impossible roots* or *imaginary roots*.

Since the nth power of a^m is a^{mn} (Art. 103), it follows that the nth root of a^{mn} is a^m.

Also, the mth power of a product is the product of the mth powers of its factors (Art. 103) ; hence, conversely, the mth root of a product is the product of the mth roots of its factors. Thus,

$$\sqrt{abc} = \sqrt{a}\,\sqrt{b}\,\sqrt{c}; \quad \sqrt[m]{ab} = \sqrt[m]{a}\,\sqrt[m]{b}.$$

Again, we have (Art. 103)

$$(a^x b^y c^z \ldots)^m = a^{xm} b^{ym} c^{zm} \ldots;$$

therefore, conversely, $\sqrt[m]{a^{xm} b^{ym} c^{zm} \ldots} = a^x b^y c^z \ldots$

Hence to extract any root of a monomial, we have the following

RULE.

Extract the required root of the coefficient by Arithmetic, then divide the exponent of every factor in the expression by the index of the root, and give the proper sign to the result.

Thus, for example,

$$\sqrt{a^4} = a^2; \quad \sqrt{a^6 b^4} = a^3 b^2; \quad \sqrt[3]{-x^9} = -x^3; \quad \sqrt[5]{x^{10}} = x^2;$$

$$\sqrt{16 a^2 b^4} = 4 a b^2; \quad \sqrt[3]{-8 a^6 b^9 c^{12}} = -2 a^2 b^3 c^4.$$

To obtain any root of a fraction : *Find the root of the numerator and denominator, and give the proper sign to the result.*

This is the converse of the rule in Art. 103.

For example, $\sqrt[3]{\dfrac{-27 a^6}{64 b^3}} = -\dfrac{3 a^2}{4 b}.$

NOTE 1. — Since every positive quantity has two square roots equal in magnitude but opposite in sign, it is customary to prefix the double sign $\pm$, read *plus or minus*, to a quantity when we wish to indicate that it is either $+$ or $-$. Thus

$$\sqrt[4]{256 x^4 y^8} = \sqrt[4]{4^4 x^4 y^8} = \pm 4 x y^2.$$

NOTE 2. — Any quantity whose root can be extracted is called a *perfect power*. When the square root of an expression which is not a perfect square, or the cube root of an expression which is not a perfect cube, is required, the operation cannot be performed. Thus we cannot take the cube root of a^2 since the exponent 2 is not divisible by the index 3. At present we can only express the result thus $\sqrt[3]{a^2}$. Also, $\sqrt{a}$, $\sqrt{a^3}$, $\sqrt{a^5}$, cannot at present be otherwise expressed ; and similarly in other cases. Such quantities are called *surds* or *irrational* quantities, and will be considered in Chapter XII.

EXAMPLES.

Show that

1. $\sqrt{a^6b^2c^{12}} = \pm a^4bc^6$;

2. $\sqrt{64x^6y^{18}} = \pm 8x^3y^9$;

3. $\sqrt{\dfrac{289y^4z^6}{81x^{10}}} = \pm\dfrac{17y^2z^3}{9x^5}$;

4. $\sqrt[3]{27a^6b^3c^9} = 3a^2bc^3$;

5. $\sqrt[3]{-343a^{12}b^{18}} = -7a^4b^6$;

6. $\sqrt[3]{-\dfrac{27x^{27}}{64y^{63}}} = -\dfrac{3x^9}{4y^{21}}$;

7. $\sqrt[7]{x^{14}y^{21}z^7} = x^2y^3z$;

8. $\sqrt[5]{-x^{10}y^{15}} = -x^2y^3$.

107. Square Root of a Polynomial. — Since the square of $a + b$ is $a^2 + 2ab + b^2$, the square root of $a^2 + 2ab + b^2$ is $a + b$. We may deduce the general rule for the extraction of the square root of a polynomial by observing in what manner $a + b$ may be derived from $a^2 + 2ab + b^2$.

Arrange the terms of the square according to the descending powers of a; then the first term is a^2, and its square root is a, which is the first term of the required root. Subtract its square, a^2, from the given expression, and bring down the remainder, $2ab + b^2$.

$$a^2 + 2ab + b^2\underline{\,|\,a + b}$$
$$\underline{a^2}$$
$$2a + b)2ab + b^2$$
$$\underline{2ab + b^2}$$

Thus, b, the second term of the root, will be the quotient when $2ab$, the first term of the remainder, is divided by $2a$, i.e., by double the first term of the root. This second term, b, added to $2a$, twice the first term, completes the divisor, $2a + b$; multiply this complete divisor by b, the second term, and subtract the product, i.e., $2ab + b^2$, from the remainder, and the operation is completed.

If there were more terms we should proceed with $a + b$ as we have done with a; its square, $a^2 + 2ab + b^2$, has already been subtracted from the given expression, so we should divide the remainder by twice the first term, i.e., by $2(a + b)$, for a new term of the root. Then for a new subtrahend we should multiply the sum of $2(a + b)$ and the new term by the new term. The process must be continued till the required root is found.

Hence to extract the square root of a polynomial, we have the following

RULE.

Arrange the terms according to the powers of some letter; find the square root of the first term for the first term of the square root; place this on the right, and subtract its square from the given polynomial.

Double the root already found for a trial divisor; divide the first term of the remainder by this trial divisor for the second term of the root, and annex this second term to the root and also to the trial divisor for the complete divisor.

Multiply the complete divisor by the second term of the root, and subtract the product from the remainder.

If there are other terms remaining, repeat the process until there is no remainder, or until all the terms of the root have been obtained.

EXAMPLES.

1. Find the square root of

$$4x^4 - 20x^3 + 37x^2 - 30x + 9 \;\big|\; 2x^2 - 5x + 3.$$

$$
\begin{array}{l}
 4x^4 \\[4pt]
\hline
4x^2 - 5x \;\big|\; -20x^3 + 37x^2 \\
 -20x^3 + 25x^2 \\[4pt]
\hline
 4x^2 - 10x + 3 \;\big|\; 12x^2 - 30x + 9 \\
 12x^2 - 30x + 9
\end{array}
$$

The expression is arranged according to the descending powers of x.

The square root of $4x^4$ is $2x^2$, and this is placed at the right of the given expression for the first term of the root. By doubling this we obtain $4x^2$, which is the *trial* divisor. The second term of the root, $-5x$, is obtained by dividing $-20x^3$, the first term of the remainder, by $4x^2$, and this new term has to be annexed both to the root and divisor. Next multiply the complete divisor by $-5x$ and subtract the product from the first remainder.

We then double the root already found and obtain $4x^2 - 10x$ for a new trial divisor. Dividing $12x^2$, the first term of the remainder, by $4x^2$, the first term of the divisor, we get 3, which we annex both to the root and divisor. We now multiply the complete divisor by 3 and subtract. There is no remainder, and the root is found.

2. Find the square root of

$$15a^2x^4 - 6ax^5 + x^6 - 20a^3x^3 + a^6 + 15a^4x^2 - 6a^5x.$$

Arrange in descending powers of x.

$$x^6-6ax^5+15a^2x^4-20a^3x^3+15a^4x^2-6a^5x+a^6 \,\underline{\big|\, x^3-3ax^2+3a^2x-a^3}$$
$$x^6$$

$2x^3-3ax^2$	$-6ax^5+15a^2x^4$
	$-6ax^5+\ 9a^2x^4$
$2x^3-6ax^2+3a^2x$	$6a^2x^4-20a^3x^3+15a^4x^2$
	$6a^2x^4-18a^3x^3+\ 9a^4x^2$
$2x^3-6ax^2+6a^2x-a^3$	$-\ 2a^3x^3+\ 6a^4x^2-6a^5x+a^6$
	$-\ 2a^3x^3+\ 6a^4x^2-6a^5x+a^6$

NOTE. — All *even* roots admit of a double sign (Art. 106). Thus the square root of $a^2 + 2ab + b^2$ is either $a + b$ or $- a - b$, as may be verified. In the process of extracting the square root of $a^2 + 2ab + b^2$, we begin by extracting the square root of a^2, and this may be either a or $-a$. If we take the latter, and continue the operation by the *rule* as before, we shall obtain $- a - b$. A similar remark holds in every other case. Thus, in Ex. 2 the square root of the first term x^6 is either x^3 or $-x^3$. If we take the latter, and continue the operation as before, we shall obtain $-x^3 + 3ax^2 - 3a^2x + a^3$.

Since the *fourth* power is the square of the square, the *fourth root* of an expression may be found by extracting the square root of the square root. Similarly the *eighth* root may be found by three successive extractions of the square root; and so on.

For example, required the fourth root of

$$16x^4 - 96x^3y + 216x^2y^2 - 216xy^3 + 81y^4.$$

By the rule we find that the square root is $4x^2 - 12xy + 9y^2$; and the square root of this is $2x - 3y$, which is therefore the fourth root of the given expression.

3. Find the square root of

$$24 + \frac{16y^2}{x^2} - \frac{8x}{y} + \frac{x^2}{y^2} - \frac{32y}{x}.$$

Arranging in descending powers of y, we have *

$$\frac{16y^2}{x^2} - \frac{32y}{x} + 24 - \frac{8x}{y} + \frac{x^2}{y^2} \left| \frac{4y}{x} - 4 + \frac{x}{y} \right.$$

$$\frac{16y^2}{x^2}$$

$$\frac{8y}{x} - 4 \left| -\frac{32y}{x} + 24 \right.$$

$$-\frac{32y}{x} + 16$$

$$\frac{8y}{x} - 8 + \frac{x}{y} \left| 8 - \frac{8x}{y} + \frac{x^2}{y^2} \right.$$

$$8 - \frac{8x}{y} + \frac{x^2}{y^2}$$

Here the second term of the root, -4, is found by the rule as usual, i.e., by dividing $-\frac{32y}{x}$ by $\frac{8y}{x}$, and the third term, $\frac{x}{y}$, is found by dividing 8 by $\frac{8y}{x}$.

EXAMPLES.

Find the square root of each of the following:

4. $25x^2 - 30xy + 9y^2$.   *Ans.* $5x - 3y$.

5. $4x^4 - 12x^3 + 29x^2 - 30x + 25$. $2x^2 - 3x + 5$.

6. $1 - 10x + 27x^2 - 10x^3 + x^4$. $1 - 5x + x^2$.

7. $4x^2 + 9y^2 + 25z^2 + 12xy - 30yz - 20xz$. $2x + 3y - 5z$.

8. $34x^3 - 22x^4 + x^6 + 121x^2 - 374x + 289$. $x^3 - 11x + 17$.

---

* The reason for this arrangement will appear in Chap. XII.

9. $\dfrac{64x^2}{9y^2} + \dfrac{32x}{3y} + 4.$ *Ans.* $\dfrac{8x}{3y} + 2.$

10. $\dfrac{a^4}{4} + \dfrac{x^2}{a^2} + \dfrac{a^3}{x} - ax + \dfrac{a^2}{x^2} - 2.$ $\dfrac{a^2}{2} + \dfrac{a}{x} - \dfrac{x}{a}.$

108. Square Root of Arithmetic Numbers. — The rule which is given in Arithmetic for extracting the square root of a number is based upon the method explained in Art. 107.

Since $1 = 1^2$, $100 = 10^2$, $10000 = 100^2$, $1000000 = 1000^2$, and so on, it follows that the square root of a number between 1 and 100 is between 1 and 10; the square root of a number between 100 and 10000 is between 10 and 100; the square root of a number between 10000 and 1000000 is between 100 and 1000, and so on. That is, the square root of a number of *one* or *two* figures consists of only *one* figure : the square root of a number of *three* or *four* figures consists of *two* figures ; the square root of a number of *five* or *six* figures consists of *three* figures ; and so on. Hence the

RULE.

If a point is placed over every second figure in any number, beginning with the units' place, the number of points will show the number of figures in the square root.

Find the square root of 5329.

Point the number according to the rule. Thus, it appears that the root consists of two places of figures, i.e., of tens and units. Let a denote the value of the figure in the tens' place of the root, and b that in the units' place. Then a must be the greatest multiple of 10 whose square is less than 5300 ; this we find to be 70. Subtract a^2, i.e., the square of 70, from the given number, and the remainder is 429, which must equal $(2a + b)b$. Divide this remainder by the

$$5\overset{.}{3}2\overset{.}{9}(70 + 3 = 73.$$
$$4900$$
$$140 + 3\overline{)\begin{array}{l} 429 \\ 429 \end{array}}$$

trial divisor, $2a$, i.e., by 140, and the quotient is 3, which
is the value of b. Then the complete divisor, $2a + b$, is
$140 + 3 = 143$, and $(2a + b)b$, that is, 143×3 or 429
is the number to be subtracted; and as there is now no
remainder, we conclude that $70 + 3$ or 73 is the required
square root.

In squaring the tens, and also in doubling them, the
ciphers are omitted for the sake of brevity, though they are
understood. Also the units' figure is added to the double of
the tens by merely writing it in the units' place. The actual
operation is usually performed as follows:

If the root consists of three places of figures, let a represent the hundreds and b the tens; then hav-
ing obtained a and b as before, let a represent
the hundreds and tens as a new value; and
find a new value of b for the units; and in
general, let a represent the part of the root
already found.

$$5329(73$$
$$\underline{49}$$
$$143 \overline{)429}$$
$$\underline{429}$$

Hence for the extraction of the square root of a number,
we have the following

RULE.

*Separate the given number into periods of two figures each,
by pointing every second figure, beginning at the units' place.*

*Find the greatest number whose square is contained in the
left period, and place it on the right; this is the first figure
of the root ; subtract its square from the first period, and to
the remainder bring down the next period for a dividend.*

*Double the root already found for a trial divisor, and see
how many times it is contained in the dividend, omitting
the last figure, and annex the result to the root and also to the
trial divisor.*

*Multiply the divisor thus completed by the figure of the root
last obtained, and subtract the product from the dividend.*

*If there are more periods to be brought down, continue the
operation as before, regarding the root already found as one
term.*

Extract the square root of 132496, and 10246401.

1. 132496(364
   9
   ___
66)424
   396
   ___
724)2896
   2896
   ___

2. 10246401(3201
   9
   ___
62)124
   124
   ___
6401)6401
     6401
     ___

As the trial divisor is an *incomplete* divisor, it is sometimes found that the product of the complete divisor by the corresponding figure of the root exceeds the dividend. In such a case the last root figure must be diminished. Thus, in Ex. 1, after finding the first figure of the root, we are required by the rule to divide 42 by 6 for the next figure of the root, so that *apparently* 7 is the next figure. On multiplying however 67 by 7 we obtain a product which is greater than the dividend 424, which shows that 7 is too large, and we accordingly try 6, which is found to be correct.

The student will observe in Ex. 2 that, in consequence of the dividend, exclusive of the right hand figure, not containing the trial divisor, 64, we place a cipher in the root and also at the right of the trial divisor 64, making it 640; we then bring down the next period and proceed as before.

109. Square Root of a Decimal. — The rule for extracting the square root of a *decimal* follows from the rule of Art. 108. If any decimal be squared there will be an *even* number of decimal places in the result; thus $(.25)^2 = .0625$, and $(.111)^2 = .012321$. Therefore there cannot be an exact square root of any decimal which has an *odd* number of decimal places.

The square root of 32.49 is one-tenth of the square root of 3249. Also the square root of .0361 is one-hundredth of that of 361. Hence, for the extraction of the square root of a decimal, we have the following

Rule.

Separate the given number into periods of two figures each, by putting a point over every second figure, beginning at the units' place and continuing both to the right and to the left of it; then proceed as in the extraction of the square root of integers, and point off as many decimal places in the result as there are periods in the decimal part of the proposed number.

If there be a final remainder in extracting the square root of an integer, it indicates that the given number has not an exact square root. We may in this case place a decimal point at the end of the given number, and annex any even number of ciphers, and continue the operation to any desired extent. We thus obtain a decimal part to be added to the integral part already found.

Also, if a decimal number has no exact square root, we may annex ciphers and obtain decimal figures in the root to any desired extent.

Find the square root of 12 ; and also of .4 to three decimal places.

```
 12.000000(3.464        .400000(.632
  9                     36
64)300                123)400
   256                   369
686)4400              1262)3100
    4116                  2524
6924)28400
     27696
```

NOTE. — We see here in what sense we can be said to approximate to the square root of a number. The square of 3.464 is less than 12, and the square of 3.465 is greater than 12. Also the square of .632 is less than .4, and the square of .633 is greater than .4.

No fraction can have a square root unless the numerator and denominator are both square numbers when the fraction is in its lowest terms. But we may approximate to the square root of a fraction to any desired extent. Thus,

Let it be required to find the square root of $\frac{5}{7}$.

Here (Art. 111) $\qquad \sqrt{\frac{5}{7}} = \frac{\sqrt{5}}{\sqrt{7}}.$

Therefore we find the square root of 5 and also of 7, approximately, and divide the former by the latter.

Or, we may reduce the fraction $\frac{5}{7}$ to a decimal to any required degree of approximation, and obtain the square root of this decimal.

Otherwise thus:

$$\sqrt{\frac{5}{7}} = \sqrt{\frac{5 \times 7}{7 \times 7}} = \frac{\sqrt{5 \times 7}}{\sqrt{7 \times 7}} = \frac{\sqrt{35}}{7};$$

then find the square root of 35 approximately, and divide the result by 7. Either of these last methods is preferable to the first.

If the square root of a number consists of $2n + 1$ *figures, when the first* $n + 1$ *of these have been obtained by the ordinary method, the remaining* n *may be obtained by division.*

Let N represent the given number; a the part of the square root already found, i.e., the first $n + 1$ figures found by the rule, with n ciphers annexed; and x the part of the root which remains to be found.

Then $\qquad\qquad \sqrt{N} = a + x;$

$\therefore \qquad\qquad N = a^2 + 2ax + x^2;$

$\therefore \qquad \dfrac{N - a^2}{2a} = x + \dfrac{x^2}{2a} \quad . \; . \; . \; . \; . \; . \; . \; . \; (1)$

Now $N - a^2$ is the remainder after $n + 1$ figures of the root, represented by a, have been found; and $2a$ is the corresponding trial divisor. We see from (1) that $N - a^2$ divided by $2a$ gives x, the rest of the square root required, increased by $\dfrac{x^2}{2a}.$

Now $\dfrac{x^2}{2a}$ is a *proper fraction*, so that by neglecting the remainder arising from the division, we obtain x, the rest of the root. For, x contains n figures by supposition, so that x^2 cannot contain more than $2n$ figures; but a contains $2n + 1$ figures (the last n of which are ciphers) and thus $2a$ contains $2n + 1$ figures at *least;* therefore $\dfrac{x^2}{2a}$ is a proper fraction.

From this investigation, by putting $n = 1$, we see that at least *two* of the figures of a square root must have been obtained in order that the method of division may give the next figure of the square root correctly.

We will apply this method to finding the square root of 290 to five places of decimals. We must obtain the first four figures in the square root by the ordinary method; and then the remaining three may be found by division.

$$\dot{2}9\dot{0}\ (17.02$$
$$\underline{1}$$
$$27)\ 190$$
$$\underline{189}$$
$$3402)\ 10000$$
$$\underline{6804}$$
$$3196$$

We now divide the remainder 3196, which is $N - a^2$, by twice the root already found, 3404, which is $2a$, and obtain the next three figures. Thus,

$$3404)\ 31960\ (938$$
$$\underline{30636}$$
$$13240$$
$$\underline{10212}$$
$$30280$$
$$\underline{27232}$$
$$3048$$

Therefore to five places of decimals, $\sqrt{290} = 17.02938$.

In extracting the square root, the student will observe that each remainder brought down is the given expression minus the square of the root already found, and is therefore in the form $N - a^2$.

EXAMPLES.

Find the square roots of the following numbers:

1.	15129.	*Ans.* 123.	5.	.835396.	*Ans.* .914.
2.	103041.	321.	6.	1522756.	1234.
3.	3080.25.	55.5.	7.	29376400.	5420.
4.	41.2164.	6.42.	8.	384524.01.	620.1.

110. Cube Root of a Polynomial. — Since the cube of $a + b$ is $a^3 + 3a^2b + 3ab^2 + b^3$, the cube root of $a^3 + 3a^2b + 3ab^2 + b^3$ is $a + b$. We may deduce a general rule for the extraction of the cube root of a polynomial by observing in what manner $a + b$ may be derived from $a^3 + 3a^2b + 3ab^2 + b^3$.

Arrange the terms of the cube according to the descending powers of a, then the first term is a^3, and its cube root is a, which is the first term of the required root. Subtract its cube, a^3, from the given expression, and bring down the remainder $3a^2b + 3ab^2 + b^3$. Thus, b, the second term of the root, will be the quotient when $3a^2b$, the first term of the remainder, is divided by $3a^2$, i.e., by three times the square of the first term of the root.

Also, since $3a^2b + 3ab^2 + b^3 = (3a^2 + 3ab + b^2)b$, we add to the trial divisor $3ab + b^2$, i.e., three times the product of the first term of the root by the second, plus the square of the second, and we have the complete divisor $3a^2 + 3ab + b^2$; multiply this complete divisor by b, and subtract the product, $3a^2b + 3ab^2 + b^3$, from the remainder, and the operation is completed.

The work may be arranged as follows:

$$a^3 + 3a^2b + 3ab^2 + b^3 \underline{\lfloor a + b}$$
$$a^3$$
$$\begin{array}{r|l} 3a^2 + 3ab + b^2 & 3a^2b + 3ab^2 + b^3 \\ & 3a^2b + 3ab^2 + b^3 \end{array}$$

If there were more terms, we should proceed with $a + b$ as we have done with a; its cube, $a^3 + 3a^2b + 3ab^2 + b^3$, has already been subtracted from the given expression, so we should divide the remainder by three times the square of the first term, i.e., by $3(a + b)^2$, for a new term of the root, c say. Then for a new complete divisor we would have $3(a + b)^2 + 3(a + b)c + c^2$; and this multiplied by c would give us a new subtrahend; and so on. Hence the following

RULE.

Arrange the terms according to the powers of some letter; find the cube root of the first term for the first term of the cube root; and subtract its cube from the given polynomial.

Take three times the square of the root already found for a trial divisor; divide the first term of the remainder by this trial divisor for the second term of the root; annex this second term to the root, and complete the divisor by adding to the trial divisor three times the product of the first and second terms of the root and the square of the second term.

Multiply the complete divisor by the second term of the root, and subtract the product from the remainder.

If there are other terms remaining, take three times the square of the part of the root already found for a new trial divisor; and continue the operation until there is no remainder, or until all the terms of the root have been obtained.

EXAMPLES.

1. Find the cube root of $8x^3 - 36x^2y + 54xy^2 - 27y^3$.
The work may be arranged as follows:

$$8x^3-36x^2y+54xy^2-27y^3 \, \lfloor \, 2x - 3y$$
$$8x^3$$

$$
\begin{array}{ll}
3(2x)^2 & =12x^2 \\
3(2x)(-3y)= & -18xy \\
(-3y)^2= & +9y^2 \\
\hline
& 12x^2-18xy+9y^2
\end{array}
\quad
\begin{array}{l}
-36x^2y+54xy^2-27y^3 \\[2ex]
\hline
-36x^2y+54xy^2-27y^3
\end{array}
$$

2. Find the cube root of $\qquad\qquad \lfloor \, 3+4x-2x^2$

$$27+108x+\ 90x^2-\ 80x^3-60x^4+48x^5-8x^6$$
$$27$$

$$27+36x+16x^2 \, \lfloor \, 108x+\ 90x^2-\ 80x^3$$
$$\qquad\qquad\qquad\ \ \, 108x+144x^2+\ 64x^3$$

$$27+72x+48x^2 \qquad\ \ -54x^2-144x^3-60x^4+48x^5-8x^6$$
$$\qquad -18x^2-24x^3$$

$$\qquad\qquad\qquad +4x^4$$
$$\overline{27+72x+30x^2-24x^3+4x^4}\ \ \lfloor\ -54x^2-144x^3-60x^4+48x^5-8x^6$$

EXPLANATION. — The root is placed above the given expression for convenience. When we have obtained two terms in the root, $3 + 4x$, we form the second divisor as follows: take 3 times the square of the root already found for the trial divisor, $27 + 72x + 48x^2$; divide $-54x^2$, the first term of the remainder, by 27, the first term of the trial divisor; this gives the third term of the root, $-2x^2$. To complete the divisor we add to the trial divisor 3 times the product of $(3 + 4x)$ and $-2x^2$, and also the square of $-2x^2$. Now multiply the complete divisor by $-2x^2$ and subtract; there is no remainder and the root is found.

Find the cube root of each of the following:

3. $a^3 + 3a^2 + 3a + 1$.   *Ans.* $a + 1$.
4. $64a^3 - 144a^2b + 108ab^2 - 27b^3$. $4a - 3b$.
5. $x^6 + 3x^5 + 6x^4 + 7x^3 + 6x^2 + 3x + 1$. $x^2 + x + 1$.
6. $1 - 6x + 21x^2 - 44x^3 + 63x^4 - 54x^5 + 27x^6$. $1 - 2x + 3x^2$.

111. Cube Root of Arithmetic Numbers.—The

rule which is given in Arithmetic for extracting the cube root of a number is based upon the method explained in Art. 110.

Since $1 = 1^3$, $1000 = 10^3$, $1000000 = 100^3$, and so on, it follows that the cube root of a number between 1 and 1000 is between 1 and 10; the cube root of a number between 1000 and 1000000 is between 10 and 100; and so on. That is, the cube root of a number of *one* or *two* or *three* figures consists of only *one* figure; the cube root of a number of *four* or *five* or *six* figures consists of *two* figures; and so on. Hence the

RULE. *If a point is placed over every third figure in any number, beginning with the units' place, the number of points will show the number of figures in the cube root.*

Find the cube root of 614125.

Point the number according to the rule. Thus it appears that the root consists of two places of figures, i.e., of tens and units. Let a denote the value of the figure in the tens' place of the root, and b that in the units' place. Then a must be the greatest multiple of 10 whose cube is less than 614000; this we find to be 80. Subtract a^3, i.e., the cube of 80, from the given number, and the remainder is 102125, which must equal $(3a^2 + 3ab + b^2)b$. Divide this remainder by the trial divisor, $3a^2$, i.e., by 19200, and the quotient is 5, which is the value of b. Then adding $3ab$, or 1200, and b^2, or 25, to the trial divisor $3a^2$, or 19200, we obtain the complete divisor 20425; and multiplying the complete divisor by 5 and subtracting the product 102125, there is no remainder. Therefore 85 is the required cube root.

$$
\begin{array}{rl}
& 614125 \,\big|\, 80 + 5 \\
& 512000 \\
3a^2 = 3(80)^2 = 19200 & \big|\,102125 \\
3ab = 3(80)(5) = 1200 & \\
b^2 = (5)^2 = 25 & \\
\hline
20425 & \big|\,102125
\end{array}
$$

In cubing the tens the ciphers are omitted for the sake of brevity, though they are understood.

If the root consists of three places of figures, let a represent the hundreds and b the tens, and proceed as before. See Art. 108.

Hence for the extraction of the cube root of a number, we have the following

RULE.

Separate the given number into periods of three figures each by pointing every third figure, beginning at the units' place.

Find the greatest number whose cube is contained in the left period, and place it on the right; this is the first figure of the root; subtract its cube from the first period, and to the remainder bring down the next period for a dividend.

Take three times the square of the root already found for a trial divisor, and see how many times it is contained in the dividend, omitting the last two figures, and annex the result to the root. Add together, the trial divisor with two ciphers annexed; three times the product of the last figure of the root by the rest, with one cipher annexed; and the square of the last figure of the root.

Multiply the divisor thus completed by the figure of the root last obtained, and subtract the product from the dividend.

If there are more periods to be brought down, the operation must be repeated, regarding the root already found as one term.

Extract the cube root of 109215352.

$$109215352(478$$
$$\underline{64}$$

$$
\begin{array}{rl}
3(4)^2 = 4800 & |45215 \\
3(4)(7) = 840 & \\
(7)^2 = \underline{49} & \\
5689 & |39823 \\
\hline
3(47)^2 = 662700 & |5392352 \\
3(47)(8) = 11280 & \\
(8)^2 = \underline{64} & \\
674044 & |5392352
\end{array}
$$

After finding the first figure of the root we are required by the rule to divide 452 by 48 for the next figure of the root, so that *apparently* 8 or 9 is the next figure. On trial we find that these numbers are too large; so we try 7, which is found to be correct. As in the case of the square root (Art. 113), we are liable occasionally to try too large a figure, especially at the early stages of the operation.

112. Cube Root of a Decimal. — If the cube root have any number of decimal places, the cube will have three times as many. Hence for the extraction of the cube root of a decimal, we have the following

RULE.

Separate the given number into periods of three figures each, by putting a point over every third figure, beginning at the units' place and continuing both to the right and to the left of it; then proceed as in the extraction of the cube root of integers, and point off as many decimal places in the result as there are periods in the decimal part of the proposed number.

If there be a final remainder in extracting the cube root of any number, integral or decimal, it indicates that the number has no exact cube root. We may in this case, as in the extraction of the square root (Art. 109), annex any number of ciphers, and continue the operation to any desired extent.

Extract the cube root of 1481.544

$$\begin{array}{r|l} \overset{\text{.}}{1}48\overset{\text{.}}{1}.54\overset{\text{.}}{4} & \underline{11.4} \\ 1 & \\ \hline \end{array}$$

$$\begin{array}{r|l} 300 & 481 \\ 30 & \\ 1 & \\ \hline 331 & 331 \\ \hline 36300 & 150544 \\ 1320 & \\ 16 & \\ \hline 37636 & 150544 \\ \hline \end{array}$$

The cube root is 11.4.

EXAMPLES.

Show that

1. $(-2a^5)^4 = 16a^{20}$.
2. $(-a^4)^5 = -a^{20}$.
3. $(-3a^7b^5c)^3 = -27a^{21}b^{15}c^3$.
4. $(-5a^2b^3c^4)^3 = -125a^6b^9c^{12}$.
5. $(-3a^7b^2c)^4 = 81a^{28}b^8c^4$.

Find the value of

6. $(-ax^4 + by^3)^2$.   *Ans.* $a^2x^8 - 2abx^4y^3 + b^2y^6$.
7. $(2a^4 + 3b^3)^2$. $4a^8 + 12a^4b^3 + 9b^6$.
8. $(2a^2 - 3b^2)^3$. $8a^6 - 36a^4b^2 + 54a^2b^4 - 27b^6$.
9. $(2x + 3)^4$. $16x^4 + 96x^3 + 216x^2 + 216x + 81$.
10. $(1 + x)^5 - (1 - x)^5$. $2(5x + 10x^3 + x^5)$.

11. $(1 - 3x + 3x^2)^2$. $1 - 6x + 15x^2 - 18x^3 + 9x^4$.

12. $(2 + 3x + 4x^2)^2 + (2 - 3x + 4x^2)^2$. $2(4 + 25x^2 + 16x^4)$.

13. $(1 + 3x + 2x^2)^3$.

Ans. $1 + 9x + 33x^2 + 63x^3 + 66x^4 + 36x^5 + 8x^6$.

14. $(2 + 3x + 4x^2)^3 - (2 - 3x + 4x^2)^3$.

Ans. $2(36x + 171x^3 + 144x^5)$.

15. $(1 + 4x + 6x^2 + 4x^3 + x^4)^2$.

Ans. $1 + 8x + 28x^2 + 56x^3 + 70x^4 + 56x^5 + 28x^6 + 8x^7 + x^8$.

Show that

16. $\sqrt[5]{32x^5 y^{10}} = 2xy^2$

17. $\sqrt[8]{256a^8 x^{64}} = 2ax^8$.

18. $\sqrt[5]{-32x^{10}y^{15}} = -2x^2 y^3$.

19. $\sqrt[9]{\dfrac{a^{18}}{b^{27}c^{36}}} = \dfrac{a^2}{b^3 c^4}$.

Find the square roots of the following expressions:

20. $9x^4 - 12x^3 - 2x^2 + 4x + 1$.   Ans. $3x^2 - 2x - 1$.

21. $16x^6 + 16x^7 - 4x^8 - 4x^9 + x^{10}$. $4x^3 + 2x^4 - x^5$.

22. $25x^4 - 30ax^3 + 49a^2 x^2 - 24a^3 x + 16a^4$. $5x^2 - 3ax + 4a^2$.

23. $x^4 - 4x^3 + 8x + 4$. $x^2 - 2x - 2$.

24. $x^6 + 4ax^5 - 10a^3 x^3 + 4a^5 x + a^6$. $x^3 + 2ax^2 - 2a^2 x - a^3$.

25. $x^4 - 2ax^3 + (a^2 + 2b^2)x^2 - 2ab^2 x + b^4$. $x^2 - ax + b^2$.

26. $16 - 96x + 216x^2 - 216x^3 + 81x^4$. $4 - 12x + 9x^2$.

27. $9x^6 - 12x^5 + 22x^4 + x^2 + 12x + 4$. $3x^3 - 2x^2 + 3x + 2$.

28. $4x^8 - 4x^6 - 7x^4 + 4x^2 + 4$. $2x^4 - x^2 - 2$.

29. $1 - xy - \frac{15}{4}x^2 y^2 + 2x^3 y^3 + 4x^4 y^4$. $1 - \frac{1}{2}xy - 2x^2 y^2$.

30. $\dfrac{x^4}{4} + 4x^2 + \dfrac{ax^2}{3} + \dfrac{a^2}{9} - 2x^3 - \dfrac{4ax}{3}$. $\dfrac{x^2}{2} - 2x + \dfrac{a}{3}$.

Find the square roots of

31. 165649.   Ans. 407.

32. 384524.01.   621.1.

33. 4981.5364.   70.58.

34. .24373969.   .4937.

35. 144168049.   12007.

36. .5687573056. Ans. .75416.

37. 3.25513764.   1.8042.

38. 4.54499761.   2.1319.

39. 196540602241.   443329.

Find the square roots of the following to five decimals:

40. .9.   *Ans.* .94868.
41. 6.21.   2.49198.
42. .43.   .65574.
43. .00852.   *Ans.* .09230.
44. 17.   4.12310.
45. 129.   11.35781.

Find the cube roots of the following expressions:

46. $1728x^6 + 1728x^4y^3 + 576x^2y^6 + 64y^9$.   *Ans.* $12x^2 + 4y^3$.
47. $x^6 - 3ax^5 + 5a^3x^3 - 3a^5x - a^6$. $x^2 - ax - a^2$.
48. $8x^6 + 48cx^5 + 60c^2x^4 - 80c^3x^3 - 90c^4x^2 + 108c^5x - 27c^6$.
   Ans. $2x^2 + 4cx - 3c^2$.
49. $1 - 9x + 39x^2 - 99x^3 + 156x^4 - 144x^5 + 64x^6$.
   Ans. $1 - 3x + 4x^2$.
50. $27x^6 - 27x^5 - 18x^4 + 17x^3 + 6x^2 - 3x - 1$.
   Ans. $3x^2 - x - 1$.
51. $\dfrac{x^3}{y^3} + \dfrac{6x^2}{y^2} + \dfrac{9x}{y} - \dfrac{y^3}{x^3} + \dfrac{6y^2}{x^2} - \dfrac{9y}{x} - 4$. $\dfrac{x}{y} + 2 - \dfrac{y}{x}$.

Find the fourth roots of

52. $1 + 4x + 6x^2 + 4x^3 + x^4$.   *Ans.* $1 + x$.
53. $1 - 4x + 10x^2 - 16x^3 + 19x^4 - 16x^5 + 10x^6 - 4x^7 + x^8$.
   Ans. $1 - x + x^2$.

Find the sixth root of

54. $1 + 12x + 60x^2 + 160x^3 + 240x^4 + 192x^5 + 64x^6$.
   Ans. $1 + 2x$.

Find the cube roots of

55. 2628072.   *Ans.* 138.
56. 3241792.   148.
57. 60236.288.   39.2.
58. .220348864.   *Ans.* .604.
59. 1371330631.   1111.
60. 20910518875.   2755.

CHAPTER XII.

THE THEORY OF EXPONENTS — SURDS.

113. Exponents that are Positive Integers. — Hitherto we have supposed that an *exponent* was always a *positive integer.* Thus, in Art. 12, we defined a^m as the product of m factors each equal to a, which would have no meaning unless the exponent was a positive integer.

When m and n are positive integers, we have

$$a^m = a \cdot a \cdot a \ldots \text{ to } m \text{ factors ;}$$

and $\quad a^n = a \cdot a \cdot a \ldots \text{ to } n \text{ factors.}$

$$\therefore \; a^m \times a^n = (a \cdot a \cdot a \ldots \text{ to } m \text{ factors}) \times (a \cdot a \cdot a \ldots \text{ to } n \text{ factors})$$

$$= a \cdot a \cdot a \ldots \text{ to } m+n \text{ factors}$$

$$= a^{m+n} \text{ by definition (Art. 12)} \quad \ldots \ldots (1)$$

$$\text{Also } a^m \div a^n = \frac{a^m}{a^n} = \frac{a \cdot a \cdot a \ldots \text{ to } m \text{ factors}}{a \cdot a \cdot a \ldots \text{ to } n \text{ factors}}$$

$$= a \cdot a \cdot a \ldots \text{ to } m - n \text{ factors}$$

$$= a^{m-n}. \quad \ldots \ldots \ldots \ldots \ldots (2)$$

From Art. 103 we have

$$(a^m)^n = a^{mn}, \quad \ldots \ldots \ldots (3)$$

and $\qquad\qquad a^m \times b^m = (ab)^m \quad \ldots \ldots \ldots (4)$

These four fundamental laws of combining exponents are proved directly from a definition which has meaning only when the exponents are *positive* and *integral.*

114. Fractional Exponents. — It is often found convenient to use fractional and negative exponents, such as $a^{\frac{3}{2}}$, a^{-5}, which at present have no intelligible meaning, because we cannot write a $1\frac{1}{2}$ times or -5 times as a factor. It is very important that Algebraic symbols should always obey

the same laws; and to secure this result in the case of
exponents, the definition should be extended so as to include
fractional and negative values. Now it is found convenient
to give such definitions to fractional and negative exponents
as will make the relation

$$a^m \times a^n = a^{m+n} \quad . \quad . \quad . \quad . \quad . \quad (1)$$

always true, whatever m and n may be.

To find the meaning of $a^{\frac{1}{2}}$.

Since (1) is to be true for all values of m and n, we must
have

$$a^{\frac{1}{2}} \times a^{\frac{1}{2}} = a^{\frac{1}{2}+\frac{1}{2}} = a^1 = a.$$

Thus $a^{\frac{1}{2}}$ must be such a number that its square is a. But
the *square root* of a is such a number (Art. 13). Therefore

$$a^{\frac{1}{2}} = \sqrt{a} \quad . \quad . \quad . \quad . \quad . \quad . \quad (2)$$

To find the meaning of $a^{\frac{1}{3}}$.

By (1) $a^{\frac{1}{3}} \times a^{\frac{1}{3}} \times a^{\frac{1}{3}} = a^{\frac{1}{3}+\frac{1}{3}+\frac{1}{3}} = a^1 = a.$

Hence $a^{\frac{1}{3}}$ must be such a number that when taken three
times as a factor it produces a; that is, $a^{\frac{1}{3}}$ must be equivalent
to the cube root of a.

$$\therefore \quad a^{\frac{1}{3}} = \sqrt[3]{a} \quad . \quad . \quad . \quad . \quad . \quad (3)$$

To find the meaning of $a^{\frac{3}{4}}$.

By (1) $a^{\frac{3}{4}} \times a^{\frac{3}{4}} \times a^{\frac{3}{4}} \times a^{\frac{3}{4}} = a^3.$

$$\therefore \quad a^{\frac{3}{4}} = \sqrt[4]{a^3} \quad . \quad . \quad . \quad (4)$$

To find the meaning of $a^{\frac{1}{n}}$, where n is any positive integer.
By (1)

$$a^{\frac{1}{n}} \times a^{\frac{1}{n}} \times a^{\frac{1}{n}} \times \ldots \text{to } n \text{ factors} = a^{\frac{1}{n}+\frac{1}{n}+\frac{1}{n}+\ldots \text{to } n \text{ terms}} = a^1 = a;$$

therefore $a^{\frac{1}{n}}$ must be such that its nth power is a.

$$\therefore \quad a^{\frac{1}{n}} = \sqrt[n]{a} \quad . \quad . \quad . \quad . \quad . \quad (5)$$

*To find the meaning of $a^{\frac{m}{n}}$, where m and n are any positive
integers.*

By (1) $a^{\frac{m}{n}} \times a^{\frac{m}{n}} \times a^{\frac{m}{n}} \times \ldots \ldots$ to n factors

$= a^{\frac{m}{n} + \frac{m}{n} + \frac{m}{n} + \ldots \ldots \text{to } n \text{ terms}} = a^m$

therefore $a^{\frac{m}{n}}$ must be equal to the nth root of a^m; that is,

$$a^{\frac{m}{n}} = \sqrt[n]{a^m} \quad \ldots \ldots \ldots (6)$$

Also, $a^{\frac{1}{n}} \times a^{\frac{1}{n}} \times a^{\frac{1}{n}} \times \ldots \ldots$ to m factors $= a^{\frac{m}{n}}$;

therefore $a^{\frac{m}{n}}$ means also the mth power of $a^{\frac{1}{n}}$; that is, from (5)

$$a^{\frac{m}{n}} = (\sqrt[n]{a})^m \quad \ldots \ldots \ldots (7)$$

Therefore from (6) and (7), $a^{\frac{m}{n}} = \sqrt[n]{a^m} = (\sqrt[n]{a})^m \quad \ldots \; (8)$

Hence $a^{\frac{m}{n}}$ *means the* n^{th} *root of the* m^{th} *power of* a, *or the* m^{th} *power of the* n^{th} *root of* a; *that is, in a fractional exponent the numerator denotes a power and the denominator a root.*

EXAMPLES. $x^{\frac{5}{7}} = \sqrt[7]{x^5}$, $a^{\frac{5}{2}} = \sqrt{a^5}$, $4^{\frac{3}{2}} = \sqrt{4^3} = \sqrt{64} = 8$.

115. Negative Exponents. — (1) *To find the meaning of* a^0.

By (1) of Art. 114, $a^0 \times a^n = a^{0+n} = a^n$;

$$\therefore \quad a^0 = a^n \div a^n = 1 \quad \ldots \; (1)$$

Hence, *any number whose exponent is zero is equal to 1.* (See Art. 45).

(2) *To find the meaning of* a^{-n}, *where* n *is any positive number.*

By (1) of Art. 114, $a^n \times a^{-n} = a^{n-n} = a^0 = 1$ [from (1)].

Hence $\quad a^n = \dfrac{1}{a^{-n}}, \quad \text{and} \quad a^{-n} = \dfrac{1}{a^n} \quad \ldots \; (2)$

Thus we see that any quantity may be changed from the numerator to the denominator, or from the denominator to the numerator, of a fraction, if the sign of its exponent be changed.

EXAMPLES. $x^{-2} = \dfrac{1}{x^2}$; $\dfrac{1}{x^{-\frac{1}{2}}} = x^{\frac{1}{2}} = \sqrt{x}$; $\dfrac{1}{x^{-\frac{2}{3}}} = x^{\frac{2}{3}} = \sqrt[3]{x^2}$;

$$\dfrac{a^2b^3}{xy^3} = a^2b^3x^{-1}y^{-3} = \dfrac{1}{a^{-2}b^{-3}xy^3}.$$

$27^{-\frac{2}{3}} = \dfrac{1}{27^{\frac{2}{3}}} = \dfrac{1}{\sqrt[3]{27^2}} = \dfrac{1}{\sqrt[3]{3^6}} = \dfrac{1}{3^2} = \frac{1}{9}$ (by (8) of Art. 114).

Otherwise thus: $\dfrac{1}{27^{\frac{2}{3}}} = \dfrac{1}{(\sqrt[3]{27})^2} = \dfrac{1}{3^2} = \frac{1}{9}.$

(3) *To prove that* $a^m \div a^n = a^{m-n}$ *for all values of* m *and* n.

$a^m \div a^n = \dfrac{a^m}{a^n} = a^m \times a^{-n} = a^{m-n}$, by the fundamental law.

EXAMPLES. $a^8 \div a^5 = a^{8-5} = a^{-2} = \dfrac{1}{a^2}.$

$$a \div a^{-\frac{2}{5}} = a^{1+\frac{2}{5}} = a^{1\frac{2}{5}}.$$

$$\dfrac{2a^{\frac{1}{2}} \times a^{\frac{2}{3}} \times 6a^{-\frac{2}{3}}}{9a^{-\frac{5}{3}} \times a^{\frac{3}{2}}} = \tfrac{4}{3}a^{\frac{1}{2}+\frac{2}{3}-\frac{2}{3}+\frac{5}{3}-\frac{3}{2}} = \tfrac{4}{3}a^{-1} = \dfrac{4}{3a}.$$

$$\dfrac{\sqrt{x^3} \times \sqrt[3]{y^2}}{\sqrt[6]{y^{-2}} \times \sqrt[4]{x^6}} = \dfrac{x^{\frac{3}{2}} \times y^{\frac{2}{3}}}{y^{-\frac{1}{3}} \times x^{\frac{3}{2}}} = x^{\frac{3}{2}-\frac{3}{2}}y^{\frac{2}{3}+\frac{1}{3}} = x^0y = y.$$

NOTE. — It appears that it is not *absolutely necessary* to introduce fractional and negative exponents into Algebra, since they merely supply us with a new notation in addition to one we already had. It is simply a *convenient* notation, which the student will learn to appreciate as he proceeds.

EXAMPLES.

Express with positive exponents:

1. $2x^{-\frac{1}{4}}a^{-\frac{2}{3}}.$ *Ans.* $\dfrac{2}{x^{\frac{1}{4}}a^{\frac{2}{3}}}.$

2. $\dfrac{2x^{\frac{1}{2}} \times 3x^{-1}}{\sqrt{x^3}}.$ $\dfrac{6}{x^2}.$

3. $\dfrac{xy^2 \times x^{-1} \times \sqrt[4]{x^3}}{\sqrt{y^{-8}}}.$ *Ans.* $x^{\frac{3}{4}}y^{\frac{1}{2}}.$

4. $\sqrt[6]{a^{-8}} \div \sqrt[6]{a^7}.$ $\dfrac{1}{a^2}.$

Express with radical signs:

5. $5a^{\frac{1}{2}}x^{-\frac{1}{2}}b^{-\frac{3}{4}}$. *Ans.* $\dfrac{5\sqrt{a}}{\sqrt{xb^3}}$.

6. $a^{-\frac{1}{3}} \times 2a^{-\frac{1}{2}}$. $\dfrac{2}{\sqrt[6]{a^5}}$.

7. $x^{-\frac{2}{3}} \div 2a^{-\frac{1}{2}}$. *Ans.* $\dfrac{\sqrt{a}}{2\sqrt[3]{x^2}}$.

8. $7a^{-\frac{1}{2}} \times 3a^{-1}$. $\dfrac{21}{\sqrt{a^3}}$.

116. To Prove that $(a^m)^n = a^{mn}$ is Universally True for All Values of m and n.

1. *Let n be a positive integer, and m have any value.* Then from the definition of a positive integral exponent

$$(a^m)^n = a^m \times a^m \times a^m \times \ldots\ldots \text{ to } n \text{ factors}$$

$$= a^{m+m+m+\ldots\ldots\text{to } n \text{ terms}}$$

$$= a^{mn} \quad . \quad . \quad . \quad . \quad . \quad . \quad . \quad . \quad . \quad (1)$$

2. *Let n be a positive fraction $\dfrac{p}{q}$, where p and q are positive integers, and m unrestricted as before.* Then

$$(a^m)^n = (a^m)^{\frac{p}{q}} = \sqrt[q]{(a^m)^p} \text{ (Art. 114)}$$

$$= \sqrt[q]{a^{mp}} \text{ by (1)}$$

$$= a^{\frac{mp}{q}} \text{ (Art. 114)}$$

$$= a^{mn}. \quad . \quad . \quad . \quad . \quad . \quad (2)$$

3. *Let n be negative, and equal to $-p$, where p is a positive integer, and m unrestricted as before.* Then

$$(a^m)^n = (a^m)^{-p} = \dfrac{1}{(a^m)^p} \text{ (Art. 115)}$$

$$= \dfrac{1}{a^{mp}} \text{ by (1) and (2)}$$

$$= a^{-mp} = a^{mn}. \quad . \quad . \quad . \quad . \quad . \quad (3)$$

Hence $(a^m)^n = (a^n)^m = a^{mn}$ for all values of m and n.

4. Let $n = \dfrac{1}{n}$.

Then we have

228 TO PROVE THAT $(ab)^n = a^n b^n$ FOR ANY VALUE OF n.

That is, *the nth root of the mth power of a is equal to the mth power of the nth root of a.*

5. Let $m = \dfrac{1}{m}$ and $n = \dfrac{1}{n}$.

$$(a^{\frac{1}{m}})^{\frac{1}{n}} = (a^{\frac{1}{n}})^{\frac{1}{m}} = a^{\frac{1}{mn}} \quad \ldots \ldots (5)$$

That is, *the nth root of the mth root, or the mth root of the nth root of a is equal to the mn^{th} root of a.*

EXAMPLES. $(b^{\frac{2}{3}})^{\frac{6}{7}} = b^{\frac{2}{3} \times \frac{6}{7}} = b^{\frac{4}{7}}.$

$[(x^{-2})^3]^{-\frac{1}{3}} = (x^{-6})^{-\frac{1}{3}} = x^{-6(-\frac{1}{3})} = x^8.$

$(\sqrt[6]{3})^3 = (3^{\frac{1}{6}})^3 = 3^{\frac{1}{2}} = \sqrt{3}.$

$\sqrt[3]{\sqrt{27x^3}} = [(27x^3)^{\frac{1}{2}}]^{\frac{1}{3}} = [(27x^3)^{\frac{1}{3}}]^{\frac{1}{2}} = \sqrt{3x}.$

117. To Prove that $(ab)^n = a^n b^n$ for Any Value of n. — This has already been shown to be true when n is a positive integer (Art. 103).

1. *Let n be a positive fraction $\dfrac{p}{q}$, where p and q are positive integers.* Then

$$(ab)^n = (ab)^{\frac{p}{q}}.$$

Now $\qquad [(ab)^{\frac{p}{q}}]^q = (ab)^p$ (Art. 116)

$$= a^p b^p \text{ (Art. 103)}$$

$$= (a^{\frac{p}{q}} b^{\frac{p}{q}})^q.$$

$$\therefore \ (ab)^{\frac{p}{q}} = a^{\frac{p}{q}} b^{\frac{p}{q}} \quad \ldots \ldots \ldots (1)$$

2. *Let n have any negative value, say $-r$, where r is a positive integer.* Then

$$(ab)^n = (ab)^{-r} = \frac{1}{(ab)^r}$$

$$= \frac{1}{a^r b^r} = a^{-r} b^{-r}, \quad \ldots \ldots (2)$$

which proves the proposition generally.

In this proof the quantities a and b are *wholly unrestricted*, and may themselves have exponents.

Let $\dfrac{p}{q} = \dfrac{1}{n}$. Then from (1) we have

$$(ab)^{\frac{1}{n}} = a^{\frac{1}{n}} b^{\frac{1}{n}}.$$

$$\therefore \ \sqrt[n]{ab} = \sqrt[n]{a} \cdot \sqrt[n]{b} \quad \ldots \ldots (3)$$

That is, *the nth root of the product is equal to the product of the nth roots.*

EXAMPLES.

1. $(x^{\frac{1}{2}} y^{-\frac{1}{2}})^{\frac{1}{3}} \div (x^2 y^{-1})^{-\frac{1}{3}} = x^{\frac{1}{6}} y^{-\frac{1}{6}} \div x^{-\frac{2}{3}} y^{\frac{1}{3}} = x^{\frac{5}{6}} y^{-1}.$

Express with positive exponents

2. $\left(\dfrac{16x^2}{y^{-2}}\right)^{-\frac{1}{4}}$. *Ans.* $\dfrac{1}{2x^{\frac{1}{2}} y^{\frac{1}{2}}}$.

3. $\left(\dfrac{27x^3}{8a^{-3}}\right)^{-\frac{2}{3}}$. $\dfrac{4}{9a^2 x^2}$.

4. $\left(\dfrac{a^{-\frac{1}{2}}}{4c^2}\right)^{-2}$. $16ac^4$.

5. $(x^a y^{-b})^3 (x^3 y^2)^{-a}$. *Ans.* $\dfrac{1}{y^{2a+3b}}$.

6. $\sqrt[4]{x \sqrt[3]{x^{-1}}}$. $x^{\frac{1}{6}}$.

7. $(4a^{-2} \div 9x^2)^{-\frac{1}{2}}$. $\dfrac{3ax}{2}$.

8. $\sqrt[3]{ab^{-1}c^{-2}} \times (a^{-1}b^{-2}c^{-4})^{-\frac{1}{6}}$. $a^{\frac{1}{2}}$.

9. $\sqrt[6]{a^{4b} x^6} \times (a^{\frac{2}{3}} x^{-1})^{-b}$. x^{b+1}.

10. $(a^{-\frac{1}{2}} \sqrt[3]{x})^{-3} \times \sqrt{x^{-2} \sqrt[2]{a^{-6}}}$. $\dfrac{1}{x^2}$.

11. $\sqrt[3]{(a+b)^5} \times (a+b)^{-\frac{2}{3}}$. $a+b$.

REM. — Since the laws of the exponents * just proved are universally true, all the ordinary operations of multiplication, division, involution, and evolution are applicable to any expressions which contain fractional and negative exponents.

---

* Called the *index laws*.

The reason for the arrangement in Ex. 3, Art. 107, may now be seen. Thus the descending powers of x are

$$x^3, \ x^2, \ x, \ 1, \frac{1}{x}, \ \frac{1}{x^2}, \ \frac{1}{x^3} \ \cdots \cdots,$$

as may be seen (Art. 115) by writing the terms as follows:

$$x^3, \ x^2, \ x^1, \ x^0, \ x^{-1}, \ x^{-2}, \ x^{-3} \ \cdots \cdots$$

12. Multiply $3x^{-\frac{1}{3}} + x + 2x^{\frac{2}{3}}$ by $x^{\frac{1}{3}} - 2$.

Arrange in descending powers of x.

$$\begin{array}{l} x \ + 2x^{\frac{2}{3}} + 3x^{-\frac{1}{3}} \\ x^{\frac{1}{3}} - 2 \\ \hline x^{\frac{4}{3}} + 2x \ + 3 \\ \quad \ - 2x \ - 4x^{\frac{2}{3}} - 6x^{-\frac{1}{3}} \\ \hline x^{\frac{4}{3}} - 4x^{\frac{2}{3}} + 3 \quad - 6x^{-\frac{1}{3}}. \end{array}$$

13. Divide $3x^{\frac{2}{3}}y^{-\frac{1}{3}} + x^{\frac{1}{2}} - 3x^{\frac{1}{3}}y^{-\frac{1}{6}} - y^{-\frac{1}{2}}$ by $x^{\frac{1}{3}} + y^{-\frac{1}{3}} - 2x^{\frac{1}{6}}y^{-\frac{1}{6}}$.

Arrange in descending powers of x.

$$x^{\frac{1}{3}} - 2x^{\frac{1}{6}}y^{-\frac{1}{6}} + y^{-\frac{1}{3}} \ \big) \ x^{\frac{1}{2}} - 3x^{\frac{1}{3}}y^{-\frac{1}{6}} + 3x^{\frac{1}{6}}y^{-\frac{1}{3}} - y^{-\frac{1}{2}} \ \big(\ x^{\frac{1}{6}} - y^{-\frac{1}{6}}$$

$$\begin{array}{l} \quad \quad x^{\frac{1}{2}} - 2x^{\frac{1}{3}}y^{-\frac{1}{6}} + \ x^{\frac{1}{6}}y^{-\frac{1}{3}} \\ \hline \quad \quad - \ x^{\frac{1}{3}}y^{-\frac{1}{6}} + 2x^{\frac{1}{6}}y^{-\frac{1}{3}} - y^{-\frac{1}{2}} \\ \quad \quad - \ x^{\frac{1}{3}}y^{-\frac{1}{6}} + 2x^{\frac{1}{6}}y^{-\frac{1}{3}} - y^{-\frac{1}{2}} \\ \hline \end{array}$$

Multiply

14. $x^{\frac{3}{4}} + y^{\frac{3}{4}}$ by $x^{\frac{3}{4}} - y^{\frac{3}{4}}$.   *Ans.* $x^{\frac{3}{2}} - y^{\frac{3}{2}}$.

15. $x^4 + x^2 + 1$ by $x^{-4} - x^{-2} + 1$. $x^4 + 1 + x^{-4}$.

16. $a^{-\frac{2}{3}} + a^{-\frac{1}{3}} + 1$ by $a^{-\frac{1}{3}} - 1$. $a^{-1} - 1$.

Divide

17. $21x + x^{\frac{2}{3}} + x^{\frac{1}{3}} + 1$ by $3x^{\frac{1}{3}} + 1$. $7x^{\frac{2}{3}} - 2x^{\frac{1}{3}} + 1$.

18. $15a - 3a^{\frac{1}{3}} - 2a^{-\frac{1}{3}} + 8a^{-1}$ by $5a^{\frac{2}{3}} + 4$. $3a^{\frac{1}{3}} - 3a^{-\frac{1}{3}} + 2a^{-1}$.

Find the square root of

19. $9x - 12x^{\frac{1}{2}} + 10 - 4x^{-\frac{1}{2}} + x^{-1}$. $3x^{\frac{1}{2}} - 2 + x^{-\frac{1}{2}}$.

20. $4x^n + 9x^{-n} + 28 - 24x^{-\frac{n}{2}} - 16x^{\frac{n}{2}}$. $2x^{\frac{n}{2}} - 4 + 3x^{-\frac{n}{2}}$.

SURDS (RADICALS).

118. Surds. Definitions. — When the indicated root of a quantity cannot be exactly obtained, it is called an *irrational quantity* or a *Surd.*

Thus, $\sqrt{2}$, $\sqrt[3]{4}$, $\sqrt[5]{a^3}$, $\sqrt{a^2 + b^2}$, $a^{\frac{2}{3}}$, are surds.

When the indicated root *can* be exactly obtained, it is called a *rational quantity.* Thus $\sqrt[3]{x^6}$, $\sqrt{9}$, $\sqrt[4]{a^4}$, are rational quantities, though in the *form* of surds.

The *order* * of a surd is indicated by the index of the root. Thus $\sqrt[3]{x}$, $\sqrt[5]{a}$ are respectively surds of the *third* and *fifth* orders.

The surds of the most common occurrence are those of the *second* order; they are sometimes called *quadratic surds.* Thus $\sqrt{3}$, $\sqrt{a}$, $\sqrt{x + y}$ are quadratic surds. Surds of the *third* and *fourth* orders are called *cubic* and *biquadratic* surds respectively.

When the same root is required to be taken, the surds are said to be of the *same order.* Thus, $\sqrt[3]{a}$, $\sqrt[3]{a + b}$, and $5^{\frac{2}{3}}$ are all surds of the *third* order or *cubic* surds.

Surds are said to be *like* or *similar* when they are of the same order, or can be reduced to the same order, with the same quantity under the radical sign.

Thus, $4\sqrt{7}$ and $5\sqrt{7}$ are like or similar surds; also $5\sqrt[3]{2}$ and $3\sqrt[3]{16}$ are like surds; $2\sqrt{3}$ and $3\sqrt{2}$ are unlike surds.

A *mixed* surd is the product of a rational factor and a surd factor. Thus $4\sqrt{5}$, and $3\sqrt{7}$ are mixed surds.

When there is no rational factor outside of the radical sign, the surd is said to be *entire.* Thus $\sqrt{2}$ and $\sqrt{3}$ are entire surds.

The rules for operating with surds follow from the propositions of the preceding Articles of this Chapter. ·

---

* Sometimes called *degree.*

119. To Reduce a Rational Quantity to a Surd Form. — It is often desirable to write a rational quantity in the form of a surd. Thus

$$a = \sqrt{a^2} = \sqrt[3]{a^3} = \sqrt[n]{a^n}; \quad 3 = \sqrt{9} = \sqrt[3]{27}.$$

In the same way the form of any surd may be altered. Hence the

RULE.

Any rational quantity may be expressed in the form of a surd of any required order by raising it to the power corresponding to the root indicated by the surd, and prefixing the radical sign.

EXAMPLES. $5 = \sqrt{25} = \sqrt[3]{125} = \sqrt[n]{5^n}.$

$$(a + b) = \sqrt{(a + b)^2} = \sqrt[3]{(a + b)^3}.$$

120. To Introduce the Coefficient of a Surd under the Radical Sign. — We have

$$3\sqrt{2} = \sqrt{9} \times \sqrt{2} \text{ (Art. 119)}$$
$$= \sqrt{9 \times 2} \text{ [(3) of Art. 117]} = \sqrt{18}.$$
$$2\sqrt[3]{5} = \sqrt[3]{2^3} \times \sqrt[3]{5} = \sqrt[3]{2^3 \times 5} = \sqrt[3]{40}.$$
$$x\sqrt{2a - x} = \sqrt{2ax^2 - x^3}.$$

RULE. *Reduce the coefficient to the form of the surd and then multiply the surds together.*

EXAMPLES.

Express as entire surds.

1. $11\sqrt{2}$.	*Ans.* $\sqrt{242}$.	3. $14\sqrt{5}$.	*Ans.* $\sqrt{980}$.
2. $5\sqrt[3]{6}$.	$\sqrt[3]{750}$.	4. $6\sqrt[3]{4}$.	$\sqrt[3]{864}$.

121. To Reduce an Entire to a Mixed Surd. — We have

$$\sqrt{32} = \sqrt{16 \times 2} = \sqrt{16} \times \sqrt{2} = 4\sqrt{2};$$

also $\sqrt[3]{a^6b} = \sqrt[3]{a^6} \times \sqrt[3]{b} = a^2\sqrt[3]{b}.$

RULE. *Resolve the quantity under the radical sign into two factors, one of which is the greatest perfect power corresponding to the root indicated; extract the required root of this factor, and prefix the result as a coefficient to the indicated root of the other.*

When a surd is so reduced that the *smallest possible integer* is left under the radical sign, it is said to be in its *simplest form*. Thus,

The simplest form of $\sqrt{128} = \sqrt{64 \times 2} = 8\sqrt{2}$.

EXAMPLES.

Express in the simplest form:

1. $\sqrt{288}$.   *Ans.* $12\sqrt{2}$. $\quad$ | $\quad$ 3. $\sqrt{36a^3}$.   *Ans.* $6a\sqrt{a}$.

2. $\sqrt[3]{1029}$. $7\sqrt[3]{3}$. $\quad$ | $\quad$ 4. $\sqrt{27a^3b^5}$. $3ab^2\sqrt{3ab}$.

122. Reduction of Surds to Equivalent Surds.—
To reduce surds of different orders to equivalent surds of the same order.

For example, take $\sqrt{5}$ and $\sqrt[3]{11}$.

$$\sqrt{5} = 5^{\frac{1}{2}} = 5^{\frac{3}{6}} = \sqrt[6]{5^3} = \sqrt[6]{125}.$$

$$\sqrt[3]{11} = 11^{\frac{1}{3}} = 11^{\frac{2}{6}} = \sqrt[6]{11^2} = \sqrt[6]{121}.$$

In general, let $\sqrt[n]{a^m}$ and $\sqrt[q]{b^p}$ be two surds of different orders. Then we have to change these into equivalent surds whose fractional exponents have the same denominator.

We can reduce both surds to the order nq as follows:

$$\sqrt[n]{a^m} = a^{\frac{m}{n}} = a^{\frac{mq}{nq}} = \sqrt[nq]{a^{mq}},$$

and $\qquad \sqrt[q]{b^p} = b^{\frac{p}{q}} = b^{\frac{pn}{qn}} = \sqrt[nq]{b^{pn}}.$

Thus the equivalent surds of the same order are $\sqrt[nq]{a^{mq}}$ and $\sqrt[nq]{b^{pn}}$.

RULE. *Represent the surds with fractional exponents; reduce these fractions to their least common denominator; then express the resulting fractional exponents with radical signs, and reduce the expressions under the radical signs to their simplest forms.*

Note. — In this way, surds of *different orders* may be compared. Thus, if we wish to know which is the greater, $\sqrt{5}$ or $\sqrt[3]{11}$, we have only to reduce them to the same order, as above; we see that the former is greater because 125 is greater than 121.

EXAMPLES.

Express as surds of the twelfth order, with positive exponents :

1. $x^{\frac{1}{3}}$.   *Ans.* $\sqrt[12]{x^4}$.

2. $a^{-1} \div a^{-\frac{1}{2}}$. $\dfrac{1}{\sqrt[12]{a^6}}$.

3. $\dfrac{1}{a^{-\frac{3}{4}}}$. $\sqrt[12]{a^9}$.

4. $x^{\frac{3}{4}}$.   *Ans.* $\sqrt[12]{x^9}$.

5. $\sqrt[3]{x^6 y^{\frac{1}{4}}}$. $\sqrt[12]{x^{24} y}$.

6. $\sqrt[4]{ax^3} \times \sqrt[3]{a^{-1} x^{-2}}$. $\sqrt[12]{\dfrac{x}{a}}$.

Express as surds of the same lowest order :

7. $\sqrt{a}, \sqrt[9]{a^5}$. *Ans.* $\sqrt[18]{a^9}, \sqrt[18]{a^{10}}$.

8. $\sqrt[5]{a^3}, \sqrt{a}$. $\sqrt[10]{a^6}, \sqrt[10]{a^5}$.

9. $\sqrt[16]{x^4}, \sqrt[12]{x^{10}}$. *Ans.* $\sqrt[12]{x^3}, \sqrt[12]{x^{10}}$.

10. $\sqrt[21]{a^3 b^4}, \sqrt[7]{ab}$. $\sqrt[21]{a^3 b^4}, \sqrt[21]{a^3 b^3}$.

11. Which is the greater $\sqrt[3]{14}$ or $\sqrt{6}$? $\sqrt{6}$.

123. Addition and Subtraction of Surds. — Let it be required to find the sum of $\sqrt{12}, \sqrt{75}, -\sqrt{48}$, and $\sqrt{50}$.

Here we have (Art. 121)

$$\sqrt{12} + \sqrt{75} - \sqrt{48} + \sqrt{50} = 2\sqrt{3} + 5\sqrt{3} - 4\sqrt{3} + 5\sqrt{2}$$
$$= (2 + 5 - 4)\sqrt{3} + 5\sqrt{2}$$
$$= 3\sqrt{3} + 5\sqrt{2}.$$

Rule. *Reduce the surds to their simplest form; then add or subtract the coefficients of similar surds and prefix the result to the common surd, and indicate the addition or subtraction of unlike surds.*

Thus, $3\sqrt{20} + 4\sqrt{5} + \sqrt{\frac{1}{5}} + \sqrt{75}$

$$= 6\sqrt{5} + 4\sqrt{5} + \tfrac{1}{5}\sqrt{5} + 5\sqrt{3}$$
$$= 10\tfrac{1}{5}\sqrt{5} + 5\sqrt{3}.$$

EXAMPLES.

Find the value of the following:

1. $3\sqrt{45} + 7\sqrt{5} - \sqrt{20}$.   *Ans.* $14\sqrt{5}$.

2. $4\sqrt{63} + 5\sqrt{7} - 8\sqrt{28}$. $\sqrt{7}$.

3. $\sqrt{44} - 5\sqrt{176} + 2\sqrt{99}$. $-12\sqrt{11}$.

4. $2\sqrt{363} - 5\sqrt{243} + \sqrt{192}$. $-15\sqrt{3}$.

5. $2\sqrt[3]{\frac{1}{4}} + 8\sqrt[3]{\frac{1}{32}}$. $3\sqrt[3]{2}$.

6. $\sqrt[3]{40} - \frac{1}{2}\sqrt[3]{320} + \sqrt[3]{135}$. $3\sqrt[3]{5}$.

124. Multiplication of Surds. — (1) *When the surds are of the same order.*

To multiply $a\sqrt[n]{x}$ by $b\sqrt[n]{y}$.

Here $a\sqrt[n]{x} \times b\sqrt[n]{y} = ax^{\frac{1}{n}} \times by^{\frac{1}{n}}$ [Art. 114, (5)]

$$= abx^{\frac{1}{n}}y^{\frac{1}{n}} = ab(xy)^{\frac{1}{n}} \text{ (Art. 117)}$$

$$= ab\sqrt[n]{xy}.$$

(2) *When the surds are of different orders.*

To multiply $a\sqrt[n]{x}$ by $b\sqrt[m]{y}$.

Here $a\sqrt[n]{x} \times b\sqrt[m]{y} = ax^{\frac{1}{n}} \times by^{\frac{1}{m}}$

$$= abx^{\frac{m}{mn}}y^{\frac{n}{mn}} \text{ (Art. 122)}$$

$$= ab(x^m y^n)^{\frac{1}{mn}} \text{ (Art. 117)}$$

$$= ab\sqrt[mn]{x^m y^n}.$$

RULE. *When the surds are of the same order, multiply separately the rational factors and the irrational factors. When the surds are of different orders, reduce them to equivalent surds of the same order, and proceed as before.*

Thus, $3\sqrt{2} \times 7\sqrt{6} = 21\sqrt{12} = 42\sqrt{3}$.

$$5\sqrt[3]{2} \times 2\sqrt{5} = 5\sqrt[6]{2^2} \times 2\sqrt[6]{5^3}$$

$$= 10\sqrt[6]{500}.$$

A *compound surd* is an expression involving two or more

simple surds.   Thus $2\sqrt{a} - 3\sqrt{b}$, and $\sqrt[3]{a} + \sqrt[4]{b}$ are compound surds.

The multiplication of compound surds is performed like the multiplication of compound Algebraic expressions.

Multiply $2\sqrt{x} - 5$ by $3\sqrt{x}$.

The product $= 3\sqrt{x}(2\sqrt{x} - 5) = 6x - 15\sqrt{x}$.

NOTE. — To multiply a surd of the second order by itself is simply to remove the radical sign; therefore $\sqrt{x} \times \sqrt{x} = x$.

Multiply $6\sqrt{3} - 5\sqrt{2}$ by $2\sqrt{3} + 3\sqrt{2}$.

The product $= (6\sqrt{3} - 5\sqrt{2})(2\sqrt{3} + 3\sqrt{2})$
$$= 36 + 18\sqrt{6} - 10\sqrt{6} - 30 = 6 + 8\sqrt{6}.$$

The following case of the multiplication of compound surds deserves careful attention.   *The product of the sum and difference of any two quadratic surds is a rational quantity.*   Thus

$$(3\sqrt{5} + 4\sqrt{3})(3\sqrt{5} - 4\sqrt{3}) = (3\sqrt{5})^2 - (4\sqrt{3})^2$$
$$= 45 - 48 = -3.$$

Also $(\sqrt{a} + \sqrt{b})(\sqrt{a} - \sqrt{b}) = (\sqrt{a})^2 - (\sqrt{b})^2 = a - b$.

A binomial in which one or both of the terms are irrational, is called a *binomial surd.*

When two binomial quadratic surds differ only in the sign which connects their terms, they are said to be *conjugate.* Thus

$$3\sqrt{5} + 4\sqrt{3} \text{ is conjugate to } 3\sqrt{5} - 4\sqrt{3}.$$

Similarly, $a - \sqrt{a^2 - b^2}$ is conjugate to $a + \sqrt{a^2 - b^2}$.

The product of two conjugate surds is always rational.

EXAMPLES.

Find the value of

1. $2\sqrt{14} \times \sqrt{21}$.   *Ans.* $14\sqrt{6}$.
2. $3\sqrt{8} \times \sqrt{6}$. $12\sqrt{3}$.
3. $2\sqrt{15} \times 3\sqrt{5}$. $30\sqrt{3}$.
4. $\sqrt[3]{168} \times \sqrt[3]{147}$.   *Ans.* $14\sqrt[3]{9}$.
5. $\sqrt{3} \times \sqrt[3]{2}$. $\sqrt[6]{108}$.
6. $\sqrt{2} \times \sqrt[3]{3} \times \sqrt[4]{\tfrac{1}{2}} \times \sqrt[3]{\tfrac{1}{3}}$. $\sqrt[4]{2}$.

7. $(3\sqrt{x} - 5) \times 2\sqrt{x}.$ *Ans.* $6x - 10\sqrt{x}.$

8. $(\sqrt{x} - \sqrt{a}) \times 2\sqrt{x}.$ $2x - 2\sqrt{ax}.$

9. $(\sqrt{7} + 5\sqrt{3})(2\sqrt{7} - 4\sqrt{3}).$ $6\sqrt{21} - 46.$

10. $(3\sqrt{5} - 4\sqrt{2})(2\sqrt{5} + 3\sqrt{2}).$ $6 + \sqrt{10}.$

11. $(5 + 3\sqrt{2})(5 - 3\sqrt{2}).$ $7.$

12. $(3\sqrt{a} + \sqrt{x - 9a})(3\sqrt{a} - \sqrt{x - 9a}).$ $18a - x.$

125. To Rationalize the Denominator of a Fraction. — The process by which surds are removed from the denominator of any fraction is known as *rationalizing the denominator.*

(1) *When the denominator is a monomial.*

$$\frac{2}{\sqrt{3}} = \frac{2\sqrt{3}}{\sqrt{3} \times \sqrt{3}} = \frac{2\sqrt{3}}{3}.$$

$$\sqrt[3]{\frac{2}{3}} = \sqrt[3]{\frac{2 \times 9}{3 \times 9}} = \sqrt[3]{\frac{18}{27}} = \frac{\sqrt[3]{18}}{3}.$$

RULE. *Multiply both terms of the fraction by any factor which will render the denominator rational.*

(2) *When the denominator is a binomial quadratic surd.*

Rationalize the denominator of $\dfrac{b^2}{\sqrt{a^2 + b^2} + a}.$

The expression $= \dfrac{b^2}{\sqrt{a^2 + b^2} + a} \times \dfrac{\sqrt{a^2 + b^2} - a}{\sqrt{a^2 + b^2} - a}$

$$= \frac{b^2[\sqrt{a^2 + b^2} - a]}{(a^2 + b^2) - a^2} = \sqrt{a^2 + b^2} - a.$$

RULE. *Multiply both numerator and denominator of the fraction by the surd which is conjugate to the denominator.*

When the denominator of the fraction is rationalized, its numerical value can be more easily found. Thus, the numerical value of $\frac{2}{3}\sqrt{3}$ can be found more easily than that of $\dfrac{2}{\sqrt{3}}.$

Given $\sqrt{5} = 2.236068$, find the value of $\dfrac{29}{7 - 2\sqrt{5}}$.

It might seem at first sight that we must subtract twice the square root of 5 from 7, and divide 29 by the remainder — a troublesome process, as the divisor would have 7 figures. We may avoid much of this labor by rationalizing the denominator. Thus,

$$\frac{29}{7 - 2\sqrt{5}} = \frac{29(7 + 2\sqrt{5})}{49 - 20} = 7 + 2\sqrt{5}$$
$$= 11.472136.$$

EXAMPLES.

Rationalize the denominators of

1. $\dfrac{\sqrt{3}}{\sqrt{5}}$.   *Ans.* $\frac{1}{5}\sqrt{15}$.

2. $\sqrt{\frac{2}{3}}$. $\frac{1}{3}\sqrt{6}$.

3. $\sqrt{\frac{3}{8}}$. $\frac{1}{4}\sqrt{6}$.

4. $\sqrt[3]{\frac{2}{3}}$. $\frac{1}{3}\sqrt[3]{18}$.

5. $\dfrac{2 + \sqrt{5}}{\sqrt{5} - 1}$. $\dfrac{7 + 3\sqrt{5}}{4}$.

6. $\dfrac{25\sqrt{3} - 4\sqrt{2}}{7\sqrt{3} - 5\sqrt{2}}$.   *Ans.* $5 + \sqrt{6}$.

7. $\dfrac{10\sqrt{6} - 2\sqrt{7}}{3\sqrt{6} + 2\sqrt{7}}$. $8 - \sqrt{42}$.

8. $\dfrac{\sqrt{7} + \sqrt{2}}{9 + 2\sqrt{14}}$. $\dfrac{\sqrt{7} - \sqrt{2}}{5}$.

126. Division of Surds. — Since $a\sqrt[n]{x} \times b\sqrt[n]{y} = ab\sqrt[n]{xy}$ (Art. 124), therefore

$$ab\sqrt[n]{xy} \div a\sqrt[n]{x} = b\sqrt[n]{y}.$$

RULE. *When the surds are of the same order, divide separately the rational factors and the irrational factors. When the surds are of different orders, reduce them to equivalent surds of the same order, and proceed as before. Then the denominator may be rationalized (Art. 125).*

Thus, $4\sqrt{75} \div 25\sqrt{56} = \dfrac{4\sqrt{75}}{25\sqrt{56}} = \dfrac{4 \times 5\sqrt{3}}{25 \times 2\sqrt{14}}$

$$= \tfrac{2}{5}\sqrt{\frac{3}{14}} = \tfrac{2}{5}\sqrt{\frac{3 \times 14}{14 \times 14}} = \frac{\sqrt{42}}{35}.$$

The only case of division of a compound surd which we shall consider is that in which the divisor is a binomial quadratic surd. We may express the division by means of a fraction, and then practically effect the division by rationalizing the denominator. Thus,

Divide $\sqrt{3} + \sqrt{2}$ by $2\sqrt{3} - \sqrt{2}$.

The quotient $= \dfrac{\sqrt{3} + \sqrt{2}}{2\sqrt{3} - \sqrt{2}} = \dfrac{(\sqrt{3} + \sqrt{2})(2\sqrt{3} + \sqrt{2})}{(2\sqrt{3} - \sqrt{2})(2\sqrt{3} + \sqrt{2})}$

$= \dfrac{8 + 3\sqrt{6}}{12 - 2} = \dfrac{8 + 3\sqrt{6}}{10}$.

EXAMPLES.

Find the value of

1. $21\sqrt{384} \div 8\sqrt{98}$. $\quad$ *Ans.* $3\sqrt{3}$.

2. $5\sqrt{27} \div 3\sqrt{24}$. $\quad \dfrac{5\sqrt{2}}{4}$.

3. $-13\sqrt{125} \div 5\sqrt{65}$. $\quad$ *Ans.* $-\sqrt{13}$.

4. $6\sqrt{14} \div 2\sqrt{21}$. $\quad \sqrt{6}$.

5. $29 \div (11 + 3\sqrt{7})$. $\quad \dfrac{11 - 3\sqrt{7}}{2}$.

6. $(3\sqrt{2} - 1) \div (3\sqrt{2} + 1)$. $\quad \dfrac{19 - 6\sqrt{2}}{17}$.

7. $(2\sqrt{3} + 7\sqrt{2}) \div (5\sqrt{3} - 4\sqrt{2})$. $\quad 2 + \sqrt{6}$.

8. $(2x - \sqrt{xy}) \div (2\sqrt{xy} - y)$. $\quad \dfrac{\sqrt{xy}}{y}$.

127. Binomial Surds. Important Propositions.

(1) *The square root of a rational quantity cannot be partly rational and partly a quadratic surd.*

If possible, let $\sqrt{a} = b + \sqrt{c}$.

Squaring, we have $a = b^2 + 2b\sqrt{c} + c$.

$$\therefore \quad \sqrt{c} = \frac{a - b^2 - c}{2b};$$

that is, a surd is equal to a rational quantity, which is impossible.

(2) *In any equation consisting of rational quantities and quadratic surds, the rational parts on each side are equal, and also the irrational parts.*

If $x + \sqrt{y} = a + \sqrt{b}$, then will $x = a$, and $y = b$.

For if x is not equal to a, suppose $x = a + m$;

then $\qquad a + m + \sqrt{y} = a + \sqrt{b}$;

that is, $\qquad m + \sqrt{y} = \sqrt{b}$,

which is impossible by (1). Hence $x = a$, and therefore $\sqrt{y} = \sqrt{b}$.

NOTE. — When $x + \sqrt{y} = a + \sqrt{b}$, we can conclude that $x = a$ and $\sqrt{y} = \sqrt{b}$ only when $\sqrt{y}$ and $\sqrt{b}$ are really irrational. We cannot, for example, from the relation $6 + \sqrt{4} = 5 + \sqrt{9}$, conclude that $6 = 5$ and $\sqrt{4} = \sqrt{9}$.

(3) If $\sqrt{a + \sqrt{b}} = \sqrt{x} + \sqrt{y}$, then $\sqrt{a - \sqrt{b}} = \sqrt{x} - \sqrt{y}$.

For by squaring the first equation we have

$$a + \sqrt{b} = x + y + 2\sqrt{xy}.$$

$$\therefore \quad a = x + y, \text{ and } \sqrt{b} = 2\sqrt{xy}.$$

Subtracting, $\qquad a - \sqrt{b} = x - 2\sqrt{xy} + y\,;$

$$\therefore \quad \sqrt{a - \sqrt{b}} = \sqrt{x} - \sqrt{y}.$$

128. Square Root of a Binomial Surd. — *The square root of a binomial surd, one of whose terms is rational, may sometimes be expressed by a binomial, one or each of whose terms is a quadratic surd.*

Let $a + \sqrt{b}$ be the given binomial surd.

Assume $\qquad \sqrt{a + \sqrt{b}} = \sqrt{x} + \sqrt{y}.$ . . . (1)

By (3) of Art. 131, $\sqrt{a - \sqrt{b}} = \sqrt{x} - \sqrt{y}.$ . . . (2)

Multiplying (1) by (2), $\sqrt{a^2 - b} = x - y.$ . . . . . (3)

Squaring (1) $\qquad a + \sqrt{b} = x + 2\sqrt{xy} + y.$. (4)

Therefore by (2) of Art. 127, $a = x + y.$ . . . . . (5)

Hence, from (3) and (5), by addition and subtraction, we have

$$x = \frac{a + \sqrt{a^2 - b}}{2}, \quad \ldots \ldots (6)$$

$$y = \frac{a - \sqrt{a^2 - b}}{2}, \quad \ldots \ldots (7)$$

which substituted for x and y in (1) and (2) will give the values of $\sqrt{a + \sqrt{b}}$ and $\sqrt{a - \sqrt{b}}$.

Find the square root of $16 + 2\sqrt{55}$.

Here $a = 16$, and $\sqrt{b} = 2\sqrt{55}$.

Then $\qquad a^2 - b = 256 - 220 = 36,$

which in (6) and (7) gives

$$x = \tfrac{1}{2}(16 + 6) = 11.$$

$$y = \tfrac{1}{2}(16 - 6) = 5.$$

Hence $\qquad \sqrt{16 + 2\sqrt{55}} = \sqrt{11} + \sqrt{5}.$

From the values of x and y in (6) and (7), it is clear that each of them is itself a complex surd unless $\sqrt{a^2 - b}$ is rational; and the expression $\sqrt{x} + \sqrt{y}$ will be more complicated than $\sqrt{a + \sqrt{b}}$ itself. Hence the above method for finding the square root of $a + \sqrt{b}$ fails entirely unless $a^2 - b$ is a square number; and as this condition is not often satisfied, the process has no great practical utility.

The square root of a binomial surd may often be found by inspection. For we see from (4) and (5) that we have to find two numbers whose sum is a and whose product is b; and if two *rational* numbers satisfy these conditions, they can generally be found at once by inspection. Thus

1. Find the square root of $11 + 2\sqrt{30}$.

We have only to find two numbers whose sum is 11, and whose product is 30; and these are evidently 6 and 5.

Hence $11 + 2\sqrt{30} = 6 + 2\sqrt{6 \times 5} + 5$

$$= (\sqrt{6} + \sqrt{5})^2.$$

$\therefore \quad \sqrt{6} + \sqrt{5} =$ the square root of $11 + 2\sqrt{30}$.

2. Find the square root of $53 - 12\sqrt{10}$.

We must write the binomial so that the coefficient of the surd is 2. Thus

$$53 - 12\sqrt{10} = 53 - 2\sqrt{360}.$$

The two numbers whose sum is 53 and whose product is 360 are 45 and 8.

Hence $53 - 2\sqrt{360} = 45 - 2\sqrt{45 \times 8} + 8$

$$= (\sqrt{45} - \sqrt{8})^2.$$

$\therefore \sqrt{53 - 12\sqrt{10}} = \sqrt{45} - \sqrt{8} = 3\sqrt{5} - 2\sqrt{2}.$

EXAMPLES.

Find the square root ot

1. $7 + 2\sqrt{10}.$ *Ans.* $\sqrt{5} + \sqrt{2}.$
2. $13 + 2\sqrt{30}.$ $\sqrt{10} + \sqrt{3}.$
3. $5 + 2\sqrt{6}.$ $\sqrt{3} + \sqrt{2}.$
4. $47 - 4\sqrt{33}.$ $2\sqrt{11} - \sqrt{3}.$
5. $15 + 2\sqrt{56}.$ $\sqrt{8} + \sqrt{7}.$

The *cube* root of a binomial surd may sometimes be found by a method similar to the one just given for obtaining the *square* root. But the method is very imperfect, and is of no practical importance.

129. Equations Involving Surds. — Equations sometimes occur in which the unknown quantity appears under the radical sign. In the solution of such equations, special artifices are often required. We shall here consider only a few of the easier cases, which reduce to simple equations. These can generally be solved by the following

RULE. *Transpose to one member of the equation a single radical term so it will stand by itself; then on raising each member to a power of the same degree as the radical, it will disappear. If there are still radical terms remaining, repeat the process till all are removed.*

EXAMPLES.

1. Solve $2\sqrt{x} - \sqrt{4x - 11} = 1.$

Transposing, $2\sqrt{x} - 1 = \sqrt{4x - 11}.$

Squaring, $4x - 4\sqrt{x} + 1 = 4x - 11.$

Transposing and dividing by -4, $\sqrt{x} = 3.$ $\quad \therefore \quad x = 9.$

2. Solve $\sqrt{x - \sqrt{1 - x}} + \sqrt{x} = 1.$

Transposing, $\sqrt{x - \sqrt{1 - x}} = 1 - \sqrt{x}.$

Squaring, $x - \sqrt{1 - x} = 1 - 2\sqrt{x} + x.$

Canceling x and squaring, $1 - x = 1 - 4\sqrt{x} + 4x.$

Transposing and squaring, $25x^2 = 16x.$

Dividing by $25x$, $x = \frac{16}{25}.$

When radicals appear in a fractional form, the equation should be first cleared of fractions in the usual way before performing the involution.

3. Solve $\dfrac{6\sqrt{x} - 11}{3\sqrt{x}} = \dfrac{2\sqrt{x} + 1}{\sqrt{x} + 6}.$

Clearing of fractions

$$6x + 25\sqrt{x} - 66 = 6x + 3\sqrt{x}.$$

$$\therefore \quad 22\sqrt{x} = 66. \quad \therefore \quad x = 9.$$

Solve the following equations.

4. $\sqrt{x - 5} = 3.$ $\qquad\qquad$ *Ans.* 14.

5. $\sqrt[3]{4x - 7} = 5.$ $\qquad\qquad$ 33.

6. $\sqrt{5x - 1} = 2\sqrt{x + 3}.$ $\qquad\qquad$ 13.

7. $13 - \sqrt[3]{5x - 4} = 7.$ $\qquad\qquad$ 44.

8. $2\sqrt{3 - 7x} - 3\sqrt{8x - 12} = 0.$ $\qquad\qquad$ $\frac{6}{5}.$

9. $\sqrt{1 + \sqrt{3 + \sqrt{6x}}} = 2.$ $\qquad\qquad$ 6.

EXAMPLES.

1. Multiply $a^{-\frac{3}{4}}$ by $a^{\frac{5}{6}}$.   *Ans.* $a^{\frac{1}{12}}$.

2. Multiply $3a^{\frac{3}{4}}b^{\frac{1}{2}}$ by $4a^{\frac{1}{6}}b^{\frac{5}{6}}$. $12a^{\frac{5}{6}}b^{\frac{4}{3}}$.

3. Divide $x^{\frac{2}{3}}$ by $x^{\frac{1}{4}}$. $x^{\frac{5}{12}}$.

4. Divide $a^{-2}x^{-\frac{1}{2}}$ by a^{-3}. $ax^{-\frac{1}{2}}$.

Find the numerical values of

5. $16^{-\frac{1}{4}}$. *Ans.* $\frac{1}{2}$.	8. $4^{-\frac{3}{2}}$. *Ans.* $\frac{1}{8}$.
6. $27^{\frac{4}{3}}$. 81.	9. $16^{-\frac{1}{2}}$. $\frac{1}{4}$.
7. $8^{\frac{2}{3}}$. 4.	10. $(\frac{1}{125})^{-\frac{2}{3}}$. 25.

Express with positive exponents:

11. $\sqrt{a^3y} + \sqrt[3]{ay^3}$.   *Ans.* $a^{\frac{3}{2}}y^{\frac{1}{2}} + a^{\frac{1}{3}}y$.

12. $\sqrt[3]{a^2x^5} + \sqrt{a^3y}$. $a^{\frac{2}{3}}x^{\frac{5}{3}} + a^{\frac{3}{2}}y^{\frac{1}{2}}$.

13. $\dfrac{\sqrt[4]{x^3}}{\sqrt{x^{-1}}} + \dfrac{\sqrt[3]{a-1}}{\sqrt[3]{a}} + \dfrac{1}{4\sqrt[5]{x^{-3}}}$. $x^{\frac{5}{4}} + \dfrac{1}{a^{\frac{1}{3}}} + \dfrac{x^{\frac{3}{5}}}{4}$.

14. $(x \div \sqrt[n]{x})^n + \left(x \times \sqrt[n]{x^{-\frac{1}{n}}}\right)^{\frac{n^2}{1-n}}$. $x^{n-1} + \dfrac{1}{x^{n+1}}$.

15. $\left(\dfrac{a^{-2}b}{a^3b^{-4}}\right)^{-3} \div \left(\dfrac{ab^{-1}}{a^{-8}b^2}\right)^5$. $\dfrac{1}{a^5}$.

16. $\left(\dfrac{a^{-3}}{b^{-\frac{2}{3}}c}\right)^{-\frac{3}{2}} \div \left(\dfrac{\sqrt{a^{-\frac{1}{2}}} \cdot \sqrt[6]{b^3}}{a^2c^{-1}}\right)^{-2}$. $c^{\frac{5}{2}}$.

17. $\left(\sqrt[5]{\dfrac{a^{\frac{1}{2}}x^{-2}}{x^{\frac{1}{2}}a^{-2}}} \times \sqrt[3]{\dfrac{a\sqrt{x}}{x^{-1}\sqrt{a}}}\right)^{-4}$. $\dfrac{1}{a^{\frac{n}{1}}}$.

18. $\dfrac{2^{n+1}}{(2^n)^{n-1}} \div \dfrac{4^{n+1}}{(2^{n-1})^{n+1}}$. $\frac{1}{4}$.

Express with radical signs and positive exponents:

19. $\dfrac{2a^{-2}}{a^{-\frac{3}{2}}} + \dfrac{a^{-\frac{1}{2}}}{3a} + \dfrac{4x^{-1}}{x^{-\frac{1}{3}}}$.   *Ans.* $\dfrac{2}{\sqrt{a}} + \dfrac{1}{3\sqrt{a^3}} + \dfrac{4}{\sqrt[3]{x^2}}$.

20. $a^{\frac{2}{3}} \times a^{\frac{3}{2}} + a^{-\frac{x}{5}} \div a^{-\frac{2x}{5}}$. $\sqrt[6]{a^{13}} + \sqrt[5]{a^x}$.

21. $a^{\frac{3}{2}}b^{-1} - a^{-\frac{3}{2}}b.$ *Ans.* $\dfrac{\sqrt[3]{a^2}}{b} - \dfrac{b}{\sqrt[3]{a^{-1}}}.$

22. $a^{-3}b^{-\frac{1}{2}} + 3a^{\frac{3}{2}}b^{-\frac{2}{3}}.$ $\dfrac{1}{a^3\sqrt{b}} + 3\sqrt[3]{\dfrac{a^2}{b^2}}.$

Multiply

23. $1 + x^{\frac{1}{4}} + x^{\frac{1}{2}}$ by $1 - x^{\frac{1}{4}}.$ $1 - x^{\frac{3}{4}}.$

24. $x^{\frac{5}{4}} - x^{\frac{3}{4}} + x^{\frac{1}{4}} - x^{-\frac{1}{4}}$ by $x^{\frac{3}{4}} + x^{\frac{1}{4}}.$ $x^2 - 1.$

25. $x^n + x^{\frac{n}{2}} + 1$ by $x^{-n} + x^{-\frac{n}{2}} + 1.$

     Ans. $x^n + 2x^{\frac{n}{2}} + 3 + 2x^{-\frac{n}{2}} + x^{-n}.$

26. $1 - 2\sqrt[3]{x} - 2x^{\frac{1}{2}}$ by $1 - \sqrt[6]{x}.$ $1 - x^{\frac{1}{6}} - 2x^{\frac{1}{3}} + 2x^{\frac{2}{3}}.$

27. $2\sqrt[3]{a^5} - a^{\frac{1}{3}} - \dfrac{3}{a}$ by $2a - 3\sqrt[3]{\dfrac{1}{a}} - a^{-\frac{4}{3}}.$

     Ans. $4a^{\frac{8}{3}} - 8a^{\frac{4}{3}} - 5 + 10a^{-\frac{4}{3}} + 3a^{-\frac{8}{3}}.$

Divide

28. $16a^{-3} + 6a^{-2} + 5a^{-1} - 6$ by $2a^{-1} - 1.$

     Ans. $8a^{-2} + 7a^{-1} + 6.$

29. $21a^{3x} + 20 - 27a^x - 26a^{2x}$ by $3a^x - 5.$

     Ans. $7a^{2x} + 3a^x - 4.$

30. $8c^{-n} - 8c^n + 5c^{3n} - 3c^{-3n}$ by $5c^n - 3c^{-n}.$

     Ans. $c^{2n} - 1 + c^{-2n}.$

31. $1 - \sqrt{a} - 2a + 2a^2$ by $1 - a^{\frac{1}{2}}.$ $1 - 2a - 2a^{\frac{3}{2}}.$

32. $x^{\frac{3}{2}} - xy^{\frac{1}{2}} + x^{\frac{1}{2}}y - y^{\frac{3}{2}}$ by $x^{\frac{1}{2}} - y^{\frac{1}{2}}.$ $x + y.$

33. $a^{\frac{2}{3}} + b^{\frac{2}{3}} - c^{\frac{2}{3}} + 2a^{\frac{1}{3}}b^{\frac{1}{3}}$ by $a^{\frac{1}{3}} + b^{\frac{1}{3}} + c^{\frac{1}{3}}.$ $a^{\frac{1}{3}} + b^{\frac{1}{3}} - c^{\frac{1}{3}}.$

Find the values of

34. $\left[\left(\dfrac{1+x}{2}\right)^{\frac{1}{2}} + \left(\dfrac{1-x}{2}\right)^{\frac{1}{2}}\right]^2.$ *Ans.* $1 + (1 - x^2)^{\frac{1}{2}}.$

35. $(a^{\frac{1}{3}}x^{-1} + a^{-\frac{1}{3}}x)^3.$ $ax^{-3} + 3a^{\frac{1}{3}}x^{-1} + 3a^{-\frac{1}{3}}x + a^{-1}x^3.$

36. $(\frac{1}{3}a^{\frac{3}{2}} - a^{-\frac{1}{3}})^2.$ $\frac{1}{9}a^3 - \frac{2}{3} + a^{-\frac{2}{3}}.$

37. $\dfrac{a^{\frac{4}{3}} - 8a^{\frac{1}{3}}b}{a^{\frac{2}{3}} + 2\sqrt[3]{ab} + 4b^{\frac{2}{3}}}.$ $a^{\frac{1}{3}}(a^{\frac{1}{3}} - 2b^{\frac{1}{3}}).$

38. $\dfrac{x - 7x^{\frac{1}{2}}}{x - 5\sqrt{x} - 14} \div \left(1 + \dfrac{2}{\sqrt{x}}\right)^{-1}.$ 1.

Find the square root of

39. $4x^2a^{-2} - 12xa^{-1} + 25 - 24x^{-1}a + 16x^{-2}a^2$.

$\qquad$ *Ans.* $2xa^{-1} - 3 + 4x^{-1}a$.

40. $x^{\frac{2}{3}} - 4x^{\frac{5}{6}} + 4x + 2x^{\frac{7}{6}} - 4x^{\frac{4}{3}} + x^{\frac{5}{3}}$. $x^{\frac{1}{3}} - 2x^{\frac{1}{2}} + x^{\frac{5}{6}}$.

41. $4a - 12a^{\frac{1}{2}}b^{\frac{1}{3}} + 9b^{\frac{2}{3}} + 16a^{\frac{1}{2}}c^{\frac{1}{4}} - 24b^{\frac{1}{3}}c^{\frac{1}{4}} + 16c^{\frac{1}{2}}$.

Express as entire surds $\qquad$ *Ans.* $2a^{\frac{1}{2}} - 3b^{\frac{1}{3}} + 4c^{\frac{1}{4}}$.

42. $\dfrac{3ab}{2c}\sqrt{\dfrac{20c^2}{9a^2b}}$. *Ans.* $\sqrt{5b}$. | 44. $a\sqrt[n]{\dfrac{b^2}{a^{n-2}}}$. *Ans.* $\sqrt[n]{a^2b^2}$.

43. $\dfrac{2a}{3x}\sqrt[3]{\dfrac{27x^4}{a^2}}$. $\sqrt[3]{8ax}$. | 45. $\dfrac{y}{x^n}\sqrt{\dfrac{x^{2n+1}}{y^3}}$. $\sqrt{\dfrac{x}{y}}$.

Express in the simplest form :

46. $3\sqrt{150}$ and $2\sqrt{720}$. $\qquad$ *Ans.* $15\sqrt{6}$ and $24\sqrt{5}$.

47. $\sqrt[3]{-108x^4y^3}$. $\qquad$ $-3xy\sqrt[3]{4x}$.

48. $\sqrt{a^3 + 2a^2b + ab^2}$. $\qquad$ $(a + b)\sqrt{a}$.

49. $\sqrt[3]{8x^4y - 24x^3y^2 + 24x^2y^3 - 8xy^4}$. $\quad 2(x - y)\sqrt[3]{xy}$.

50. $\sqrt[4]{768}, \sqrt[4]{1250}, \sqrt[4]{3888}$. $\qquad 4\sqrt[4]{3}, 5\sqrt[4]{2}, 6\sqrt[4]{3}$.

Express as surds of the same lowest order :

51. $\sqrt{ax^2}, \sqrt[30]{a^9x^6}$. $\qquad$ *Ans.* $\sqrt[26]{a^{13}x^{26}}, \sqrt[26]{a^6x^4}$.

52. $\sqrt{5}, \sqrt[3]{11}, \sqrt[6]{13}$. $\qquad \sqrt[6]{125}, \sqrt[6]{121}, \sqrt[6]{13}$.

53. $\sqrt[4]{8}, \sqrt{3}, \sqrt[8]{6}$. $\qquad \sqrt[8]{64}, \sqrt[8]{81}, \sqrt[8]{6}$.

54. $3^{\frac{3}{4}}, 2^{\frac{3}{4}}, 5^{\frac{1}{2}}$. $\qquad \sqrt[12]{6561}, \sqrt[12]{512}, \sqrt[12]{15625}$.

Find the value of

55. $\sqrt{243} + \sqrt{27} + \sqrt{48}$. $\qquad$ *Ans.* $16\sqrt{3}$.

56. $2\sqrt[3]{189} + 3\sqrt[3]{875} - 7\sqrt[3]{56}$. $\qquad 7\sqrt[3]{7}$.

57. $5\sqrt[3]{81} - 7\sqrt[3]{192} + 4\sqrt[3]{648}$. $\qquad 11\sqrt[3]{3}$.

58. $3\sqrt[4]{162} - 7\sqrt[4]{32} + \sqrt[4]{1250}$. $\qquad 0$.

59. $3\sqrt{2} + 4\sqrt{8} - \sqrt{32}$. $\qquad 7\sqrt{2}$.

60. $2\sqrt[3]{4} + 5\sqrt[3]{32} - \sqrt[3]{108}$. $\qquad 9\sqrt[3]{4}$.

61. $4\sqrt{128} + 4\sqrt{75} - 5\sqrt{162}$. $\qquad 20\sqrt{3} - 13\sqrt{2}$.

62. $5\sqrt[3]{128} \times 2\sqrt[3]{432}$.   *Ans.* $240\sqrt[3]{4}$.

63. $3\sqrt[3]{b} \times 4\sqrt[4]{a}$. $12\sqrt[12]{a^3b^4}$.

64. $\sqrt{2} \times \sqrt[3]{3} \times \sqrt[4]{5}$. $\sqrt[12]{648000}$.

65. $4\sqrt{5} \times 2\sqrt[3]{11}$. $8\sqrt[6]{15125}$.

66. $\dfrac{3}{x}\sqrt{\dfrac{a^2}{x}} \times \tfrac{4}{3}\sqrt{\dfrac{x^3}{2a^4}}$. $\dfrac{2\sqrt{2}}{a}$.

67. $(2\sqrt{3} + 3\sqrt{2})^2$. $30 + 12\sqrt{6}$.

68. $(\sqrt{x} + \sqrt{x-1})\sqrt{x-1}$. $x - 1 + \sqrt{x^2 - x}$.

69. $(\sqrt{x+a} - \sqrt{x-a})\sqrt{x+a}$. $x + a - \sqrt{x^2 - a^2}$.

70. $(\sqrt{2} + \sqrt{3} - \sqrt{5})(\sqrt{2} + \sqrt{3} + \sqrt{5})$. $2\sqrt{6}$.

71. $(\sqrt[3]{12} + \sqrt{19})(\sqrt[3]{12} - \sqrt{19})$. 5.

72. $(x^2 + x\sqrt{2} + 1)(x^2 - x\sqrt{2} + 1)$. $x^4 + 1$.

Rationalize the denominator of

73. $\dfrac{2\sqrt{3} + 3\sqrt{2}}{5 + 2\sqrt{6}}$.   *Ans.* $3\sqrt{2} - 2\sqrt{3}$.

74. $\dfrac{2\sqrt{5}}{\sqrt{5} + \sqrt{3}}$. $5 - \sqrt{15}$.

75. $\dfrac{3 + \sqrt{5}}{3 - \sqrt{5}}$. $\tfrac{1}{2}(7 + 3\sqrt{5})$.

76. $\dfrac{y^2}{x + \sqrt{x^2 - y^2}}$. $x - \sqrt{x^2 - y^2}$.

77. $\dfrac{x^2}{\sqrt{x^2 + a^2} + a}$. $\sqrt{x^2 + a^2} - a$.

78. $\dfrac{\sqrt{1 + x^2} - \sqrt{1 - x^2}}{\sqrt{1 + x^2} + \sqrt{1 - x^2}}$. $\dfrac{1 - \sqrt{1 - x^4}}{x^2}$.

Find the numerical value of the following to five places of decimals :

79. $\dfrac{14}{\sqrt{2}}, \dfrac{10}{\sqrt{7}}$.   *Ans.* 9.8995, 3.77964.

80. $\dfrac{\sqrt{2}}{\sqrt{3}}, \dfrac{1}{2\sqrt{3}}$.   .81649, .28867.

81. $\dfrac{\sqrt{5}+\sqrt{3}}{4+\sqrt{15}}$. *Ans.* $\sqrt{5}-\sqrt{3}=.50402$.

Find the value of

82. $\dfrac{3\sqrt{11}}{2\sqrt{98}} \div \dfrac{5}{7\sqrt{22}}$. $\tfrac{33}{10}$.

83. $\dfrac{3\sqrt{48}}{5\sqrt{112}} \div \dfrac{6\sqrt{84}}{\sqrt{392}}$. $\tfrac{1}{10}\sqrt{2}$.

84. $(3+\sqrt{5})(\sqrt{5}-2) \div (5-\sqrt{5})$. $\dfrac{\sqrt{5}}{5}$.

85. $\dfrac{\sqrt{a}}{\sqrt{a}-\sqrt{x}} \div \dfrac{\sqrt{a}+\sqrt{x}}{\sqrt{x}}$. $\dfrac{\sqrt{ax}}{a-x}$.

Find the square root of

86. $41-24\sqrt{2}$. *Ans.* $4\sqrt{2}-3$.
87. $83+12\sqrt{35}$. $2\sqrt{5}+3\sqrt{7}$.
88. $101-28\sqrt{13}$. $2\sqrt{13}-7$.
89. $117-36\sqrt{10}$. $6\sqrt{2}-3\sqrt{5}$.
90. $280+56\sqrt{21}$. $14+2\sqrt{21}$.
91. $8+4\sqrt{3}$. $\sqrt{6}+\sqrt{2}$.
92. $4-\sqrt{15}$. $\sqrt{\tfrac{5}{2}}-\sqrt{\tfrac{3}{2}}$.
93. $75-12\sqrt{21}$. $3\sqrt{7}-2\sqrt{3}$.

Solve the following equations:

94. $8-2\sqrt{x}=4$. *Ans.* 4.
95. $6+\sqrt{x}=2\sqrt{12+x}$. 4.
96. $\sqrt{x-3}-\sqrt{x+12}=-3$. 4.
97. $\sqrt{3x+10}-\sqrt{3x+25}+3=0$. -3.
98. $\sqrt{3x}+\sqrt{3x+13}=\dfrac{91}{\sqrt{3x+13}}$. 12.
99. $2+\sqrt[3]{x-5}=13$. 1336.
100. $\sqrt{x}-\sqrt{x-8}=\dfrac{2}{\sqrt{x-8}}$. 9.

CHAPTER XIII.

QUADRATIC EQUATIONS OF ONE UNKNOWN QUANTITY.

130. Quadratic Equations. — An equation which contains the *square* of the unknown quantity, but no higher power, is called a *quadratic equation*, or an *equation of the second degree*.

A *Pure quadratic equation* is one which contains only the *square* of the unknown quantity; it is sometimes called an *incomplete quadratic equation*. An *Adfected*, or *Affected*,[*] *quadratic equation* is one which contains both the *square* and the *first power* of the unknown quantity; it is also called a *complete quadratic equation*.

Thus, $2x^2 = 50$, and $ax^2 + b = 0$,

are pure quadratic equations; and

$2x^2 - 5x = 4$, and $ax^2 + bx + c = 0$,

are affected quadratic equations.

131. Pure Quadratic Equations. — A pure quadratic may be solved for the *square* of the unknown quantity by the method of solving a simple equation.

Let it be required to solve

$$\frac{x^2 - 13}{3} + \frac{x^2 - 5}{10} = 6.$$

Clear of fractions,

$$10x^2 - 130 + 3x^2 - 15 = 180.$$
$$\therefore\ 13x^2 = 325.$$
$$\therefore\ x^2 = 25.$$

Extracting the square root $\qquad x = \pm 5$.

---

* The term *adfected*, or *affected*, was introduced by Vieta, about the year 1600, to distinguish equations which involve, or are *affected* with, different powers of the unknown quantity from those which contain one power only.

In this example we find that $x^2 = 25$. Therefore x must be such a number that if multiplied by itself the product is 25; i.e., x must be the square root of 25; we prefix the double sign to 5 because the square root of a quantity may be either positive or negative. [Art. 106, (1)].

NOTE. — In extracting the square root of the two members of the equation $x^2 = 25$, it might seem at first that we ought to prefix the double sign to the square root of each side, and write $\pm x = \pm 5$. An examination however of the various cases shows this to be unnecessary, because we obtain no new results in so doing. Thus, if we write $\pm x = \pm 5$, we have the four cases:

$$+x = +5, \ +x = -5, \ -x = +5, \ -x = -5;$$

but the last two are equivalent to the first two, and become identical with them on changing the signs. Hence there are no new results obtained, and therefore when we extract the square root of the two members of an equation, it is sufficient to put the double sign before *one* member only. Thus the equation has two roots, and only two.

A pure quadratic equation can always be reduced to the form

$$ax^2 + b = 0;$$

for all the terms containing x^2 may be reduced to one term, as ax^2; and the known terms to another, as b.

By transposing b, dividing by a, and putting $q = -\dfrac{b}{a}$, the equation may be written

$$x^2 = q.$$

Such an equation is called a *binomial equation*, because it has but two terms.

Solving this equation by extracting the square root of each member, we have

$$x = \pm\sqrt{q}.$$

That is, *Every pure quadratic equation has two roots, numerically equal, but with contrary signs.*

Hence, for the solution of a *pure* quadratic equation we have the following

RULE.

Find the value of the square of the unknown quantity by the rule for solving a simple equation, and then extract the square root of both members.

EXAMPLES.

Solve

1. $11x^2 - 44 = 5x^2 + 10$. 　　　　*Ans.* $x = \pm 3$.

2. $(x + 2)^2 = 4x + 5$. 　　　　　　$x = \pm 1$.

3. $\dfrac{8}{1 - 2x} + \dfrac{8}{1 + 2x} = 25$. 　　$x = \pm.3$

4. $14 - \sqrt{x^2 - 36} = 6$. 　　　　　$x = \pm 10$.

132. Affected Quadratic Equations. — *An affected quadratic equation* can always be reduced to the form

$$ax^2 + bx + c = 0;$$

for all the terms containing x^2 may be reduced to one term, as ax^2; those containing x to one, as bx; and the known terms to another, as c.

If we divide by a, and put $p = \dfrac{b}{a}$, and $q = -\dfrac{c}{a}$, the equation may be written

$$x^2 + px = q,$$

where p and q are positive or negative. This is called the *General Quadratic Equation.*

Let it be required to solve this equation. If the first member of this equation were a perfect square, we might solve it by extracting the square root, as in Art. 131. To ascertain what must be done to make the first member a perfect square, let us compare it with the square of the binomial,

$$x + \frac{p}{2}, \text{ which is } x^2 + px + \frac{p^2}{4}.$$

Thus, we see that $x^2 + px$ is rendered a perfect square by the addition of $\dfrac{p^2}{4}$; i.e., *by the addition of the square of half the coefficient of* x. Hence, adding $\dfrac{p^2}{4}$ to both members, to preserve the equality, we have

$$x^2 + px + \frac{p^2}{4} = q + \frac{p^2}{4}.$$

This is called *completing the square.*

Extracting the square root of each member, we have

$$x + \frac{p}{2} = \pm\sqrt{q + \frac{p^2}{4}}.$$

$$\therefore \quad x = -\frac{p}{2} \pm \sqrt{q + \frac{p^2}{4}}. \quad \cdots \quad (1)$$

$$x = -\frac{p}{2} + \sqrt{q + \frac{p^2}{4}}, \text{ or } -\frac{p}{2} - \sqrt{q + \frac{p^2}{4}}.$$

Thus there are *two* roots of a quadratic equation.

Note 1. — When an expression is a perfect square, the *square terms* are always *positive.* Hence, the coefficient of x^2 must be made $+1$, if necessary, before completing the square.

1. Solve $x^2 + 6x = 27$.

Here half the coefficient of x is 3 ; add 3^2,

$$x^2 + 6x + 3^2 = 27 + 9 = 36.$$

Extracting the square root,

$$x + 3 = \pm 6.$$

$$\therefore \quad x = -3 \pm 6 = 3, \text{ or } -9.$$

We may verify these values as follows :

Putting 3 for x in the given equation, $\quad 9 + 18 = 27$.

Putting -9 for x in the given equation, $81 - 54 = 27$.

These results being identical, the values of x are verified.

It will be well for the student thus to verify his results.

2. Solve $7x = x^2 - 8$.

Transposing, so that the terms which involve x are alone in the first member, and the coefficient of x^2 is $+1$, we have

$$x^2 - 7x = 8.$$

Here half the coefficient of x is $-\frac{7}{2}$;
completing the square,

$$x^2 - 7x + (\tfrac{7}{2})^2 = 8 + \tfrac{49}{4} = \tfrac{81}{4}.$$
$$\therefore \quad x - \tfrac{7}{2} = \pm\tfrac{9}{2}.$$
$$\therefore \quad x = \tfrac{7}{2} \pm \tfrac{9}{2} = 8, \text{ or } -1.$$

NOTE 2. — We indicate $(\tfrac{7}{2})^2$ in the first member.

3. Solve $32 - 3x^2 = 10x$.

Transposing, changing signs, and dividing by 3, so as to make the coefficient of x^2 unity and positive,

$$x^2 + \tfrac{10}{3}x = \tfrac{32}{3};$$

completing the square,

$$x^2 + \tfrac{10}{3}x + (\tfrac{5}{3})^2 = \tfrac{32}{3} + \tfrac{25}{9} = \tfrac{121}{9}.$$
$$\therefore \quad x + \tfrac{5}{3} = \pm\tfrac{11}{3}.$$
$$\therefore \quad x = -\tfrac{5}{3} \pm \tfrac{11}{3} = 2, \text{ or } -5\tfrac{1}{3}.$$

NOTE 3. — We add $(\tfrac{5}{3})^2$ and not $(\tfrac{10}{6})^2$, to complete the square.

4. Solve $5x^2 + 11x = 12$.

Dividing by 5, $x^2 + \tfrac{11}{5}x = \tfrac{12}{5}$;
completing the square,

$$x^2 + \tfrac{11}{5}x + (\tfrac{11}{10})^2 = \tfrac{12}{5} + \tfrac{121}{100} = \tfrac{361}{100}.$$
$$\therefore \quad x + \tfrac{11}{10} = \pm\tfrac{19}{10}.$$
$$\therefore \quad x = -\tfrac{11}{10} \pm \tfrac{19}{10} = \tfrac{4}{5}, \text{ or } -3.$$

Hence, for solving affected quadratic equations, we have the

RULE I.

Reduce the equation so that the terms involving the unknown quantity are alone in one member, and the coefficient of x^2 is $+1$; complete the square by adding to each member of the equation the square of half the coefficient of x; extract the square root of both members, and solve the resulting simple equation.

NOTE 4. — There are other ways of completing the square of an affected quadratic, which are convenient in *special cases*, and some of which will be given as we proceed; but the method just explained is the most important, and will solve every case.

Instead of going through the process of completing the square in every particular example, it is more convenient to apply the following rule deduced from formula (1) of this Article :

RULE II.

Reduce the equation to the general form, $x^2 + px = q$. Then the value of x is half the coefficient of the first power of x with a contrary sign, plus or minus the square root of the second member increased by the square of half the same coefficient.

NOTE 5. — The student should use this method in practice, and become familiar with it, but at the same time be careful that he does not lose sight of the complete method.

5. Solve $36x - 3x^2 = 105$.

Transposing, changing signs, and dividing by 3,

$$x^2 - 12x = -35.$$

Therefore by Rule II, $x = 6 \pm \sqrt{-35 + 36} = 1$.

$$\therefore \quad x = 6 \pm 1 = 7, \text{ or } 5.$$

6. Solve $\dfrac{3x - 2}{2x - 3} = \dfrac{5x}{x + 4} - 2$.

Simplifying, $\dfrac{3x - 2}{2x - 3} = \dfrac{3x - 8}{x + 4}$.

Clearing of fractions,

$$3x^2 + 10x - 8 = 6x^2 - 25x + 24.$$

Reducing, $x^2 - \tfrac{35}{3}x = -\tfrac{32}{3}$.

Therefore, Rule II, $x = \tfrac{35}{6} \pm \sqrt{-\tfrac{32}{3} + \tfrac{1225}{36}} = \tfrac{841}{36}$.

$$\therefore \quad x = \tfrac{35}{6} \pm \tfrac{29}{6} = 10\tfrac{2}{3}, \text{ or } 1.$$

7. Solve $x^2 - 4x = 1$.

Rule II, $x = 2 \pm \sqrt{1 + 4} = 5$.

$$\therefore \quad x = 2 \pm 2.236 = 4.236, \text{ or } -0.236.$$

These values of x are correct only to three places of decimals, and neither of them will be found to satisfy the equation *exactly*.

If the *numerical* values of the unknown quantity are not required, it is usual to leave the roots in the form

$$2 + \sqrt{5}, \text{ and } 2 - \sqrt{5}.$$

8. Solve $x^2 - 10x = -32$.

Rule II, $\qquad x = 5 \pm \sqrt{-32 + 25} = -7.$

$\qquad \therefore \quad x = 5 \pm \sqrt{-7}.$

But -7 has no square root, either exact or approximate (Art. 106) ; so that no real value of x can be found to satisfy the given equation. In such a case the quadratic equation has no real roots ; the roots are said to be *imaginary* or *impossible*.

In the examples hitherto considered, the quadratic equations have had *two different roots*. Sometimes however, there is only *one* root. Take, for example, the equation, $x^2 - 10x + 25 = 0$; by extracting the square root we have $x - 5 = 0$; therefore $x = 5$. It is found convenient however in this and similar cases to say that the quadratic has *two equal roots*.

EXAMPLES.

Solve

9. $x^2 = x + 72$.	*Ans.* 9, — 8.
10. $9x - x^2 + 220 = 0$.	20, —11.
11. $x^2 - \frac{2}{3}x = 32$.	$6, -\frac{16}{3}$.
12. $\frac{19}{5}x = \frac{4}{5} - x^2$.	$\frac{1}{5}, - 4.$
13. $5x^2 = 8x + 21$.	$3, - \frac{7}{5}.$
14. $\dfrac{5x + 7}{x - 1} = 3x + 2$.	3, — 1.
15. $\dfrac{3x - 8}{x - 2} = \dfrac{5x - 2}{x + 5}$.	4, $1\frac{1}{2}$.

133. Condition for Equal Roots. — *To find the relation that must exist between the known quantities of a quadratic equation in order that the two roots may be equal.*

Take the general equation

$$ax^2 + bx + c = 0.$$

Transpose c and divide by a,

$$x^2 + \frac{b}{a}x = -\frac{c}{a}.$$

Rule II, $x = -\dfrac{b}{2a} \pm \sqrt{-\dfrac{c}{a} + \dfrac{b^2}{4a^2}} = \dfrac{b^2 - 4ac}{4a^2}.$

$\therefore$ $x = \dfrac{-b \pm \sqrt{b^2 - 4ac}}{2a}.$ . . . . (1)

Denoting the root corresponding to the positive surd by x_1, and that corresponding to the negative surd by x_2, we have

$$x_1 = \frac{-b + \sqrt{b^2 - 4ac}}{2a}.$$

$$x_2 = \frac{-b - \sqrt{b^2 - 4ac}}{2a}.$$

Now we see that if $b^2 - 4ac = 0$, these two roots are *equal*, and each of them is $-\dfrac{b}{2a}$.

Hence the relation, $b^2 - 4ac = 0$, *is the condition that the two roots of the equation* $ax^2 + bx + c = 0$ *may be equal.*

The two roots are *real* and *unequal* if $b^2 - 4ac$ is *positive*, i.e., if b^2 is Algebraically greater than $4ac$.

The two roots are *imaginary* if $b^2 - 4ac$ is *negative*, i.e., if b^2 is Algebraically less than $4ac$.

Hence the two roots of this equation are *real* and *unequal*, *equal*, or *imaginary*, according as b^2 is greater than, equal to, or less than $4ac$.

NOTE 1. — If either of the roots of a quadratic equation is imaginary, they are *both* imaginary.

By applying these tests, the nature of the roots of any quadratic may be determined without solving the equation.

1. Show that the equation $2x^2 - 6x + 7 = 0$ cannot be satisfied by any real values of x.

Here $a = 2$, $b = -6$, $c = 7$.

$\therefore$ $b^2 - 4ac = 36 - 4 \cdot 2 \cdot 7 = -20.$

Hence the roots are imaginary.

Determine the nature of the roots of

2. $x^2 + 3x + 1 = 0.$ *Ans.* Real and surds.

3. $3x^2 - 4x - 4 = 0.$ Rational.

4. If the equation $x^2 + 2(k + 2)x + 9k = 0$ has equal roots, find k.   *Ans.* $k = 4$, or 1.

When an equation is in the general form $ax^2 + bx + c = 0$, instead of solving it by either of the rules in Art. 136, we may make use of formula (1) above as follows:

5. Solve $5x^2 + 11x = 12.$

Here $a = 5$, $b = 11$, $c = -12$; substituting these values in (1)

$$x = \frac{-11 \pm \sqrt{(11)^2 - 4 \cdot 5(-12)}}{10}$$

$$= \frac{-11 \pm \sqrt{361}}{10} = \frac{-11 \pm 19}{10} = \tfrac{4}{5}, \text{ or } -3,$$

which agrees with the solution of Ex. 4, (Art. 132).

Solve by this method the following:

6. $3x^2 = 15 - 4x.$ *Ans.* $\tfrac{5}{3}, -3.$

7. $2x^2 + 7x = 15.$ $\tfrac{3}{2}, -5.$

8. $5x^2 + 4 + 21x = 0.$ $-4, -\tfrac{1}{5}.$

9. $8x^2 = x + 7.$ $1, -\tfrac{7}{8}.$

10. $35 + 9x - 2x^2 = 0.$ $7, -\tfrac{5}{2}.$

NOTE 2. — Though we can always find the roots of a given quadratic equation by substituting in formula (1), yet the student is advised to solve each separate equation either by the method given in Art. 132, and embodied in *Rule II*, or by one of the two following.

134. Hindoo Method of Completing the Square.

— When a quadratic equation appears in the general form $ax^2 + bx + c = 0$, the first member may be made a complete square, without dividing by the coefficient of x^2, thus avoiding fractions, by another method (called the Hindoo method), as follows:

Transpose c, and multiply by $4a$,

$$4a^2x^2 + 4abx = -4ac.$$

Now since the middle term of any trinomial square is twice the product of the square roots of the other two (Art. 41), the square root of the third term must be equal to the second term divided by twice the square root of the first term. Hence, dividing $4abx$ by twice the square root of $4a^2x^2$, i.e., by $4ax$, and adding the square of the quotient, b^2, to both members, the first becomes a perfect square. Thus,

$$4a^2x^2 + 4abx + b^2 = b^2 - 4ac.$$

Extracting the square root,

$$2ax + b = \pm \sqrt{b^2 - 4ac}.$$

$$\therefore \quad x = \frac{-b \pm \sqrt{b^2 - 4ac}}{2a},$$

which are the same values we obtained in (1) of Art. 133.

RULE.

Reduce the equation to the form $ax^2 + bx + c = 0$. Multiply it by four times the coefficient of x^2; add to each member the square of the coefficient of x in the given equation; extract the square root of both members, and solve the resulting simple equation.

NOTE. — This method may be used to advantage when we wish to avoid *fractions* in completing the square, and it is often preferred in solving *literal* equations. (See Note 4 of Art. 132.)

1. Solve $2x^2 - 5x = 3$.

Multiply by four times 2, or 8,

$$16x^2 - 40x = 24.$$

Add to each side 5^2, or 25,

$$16x^2 - 40x + 25 = 49.$$

Extract the square root,

$$4x - 5 = \pm 7.$$

$$\therefore \quad x = \frac{5 \pm 7}{4} = 3, \text{ or } -\tfrac{1}{2}.$$

Solve by the Hindoo method the following:

2. $3x^2 + 5x = 2$.   *Ans.* $\frac{1}{3}$, -2.

3. $6x^2 - 12 = x$. $1\frac{1}{2}$, $-1\frac{1}{3}$.

4. $3x^2 + 2x = 85$. 5, $-5\frac{2}{3}$.

5. $acx^2 - bcx + adx = bd$. $\dfrac{b}{a}$, $-\dfrac{d}{c}$.

135. Solving a Quadratic by Factoring. — There is still one method of solving a quadratic which is often shorter than either of the methods already given.

1. Consider the equation $x^2 - 2x - 15 = 0$. Resolving this into factors (Art. 65), we have

$$(x - 5)(x + 3) = 0.$$

Now it is clear that a product is zero when any one of its factors is zero; and it is also clear that no product can be zero unless one of the factors is zero. Thus ab is zero if a is zero, or if b is zero; and, if we know that ab is zero, we know that *either* a or b must be zero; and so on for any number of factors.

Similarly the product $(x - 5)(x + 3)$ is zero, when *either* of the factors, $x - 5$, $x + 3$, is zero, and in no other case.

Hence the equation

$$(x - 5)(x + 3) = 0,$$

is satisfied if $x - 5 = 0$, or if $x + 3 = 0$; i.e., if $x = 5$, or if $x = -3$, and in no other case.

Therefore the *roots* of the equation are 5, and -3.

2. Solve $x^2 - 5x + 6 = 0$. Resolving this into factors, we have

$$(x - 2)(x - 3) = 0.$$

The first member is zero either when $x - 2 = 0$, or when $x - 3 = 0$; and in no other case. Hence the equation is satisfied by $x = 2$, or 3; and by no other values; thus, 2 and 3 are the *roots* of the equation.

From these examples it appears that when a quadratic equation has been simplified, and has all its terms in the first member, its solution can always be readily obtained if the expression can be resolved into factors. Hence for the solution of such an equation, we have the following

Rule.

Reduce the equation to its simplest form, with all its terms in the first member; then resolve the whole expression into factors, and the values obtained by equating each of these factors separately to zero will be the required roots.

3. Solve $x^2 - 4x = 0$.

Factoring, we have $x(x - 4) = 0$.

The equation is satisfied if $x = 0$, or if $x - 4 = 0$, and in no other case. Hence we must have either

$$x = 0, \text{ or } x - 4 = 0.$$

$$\therefore \quad x = 0, \text{ or } 4.$$

Note 1. — In this example we might have divided the given equation by x and obtained the simple equation $x - 4 = 0$, whence $x = 4$, which is *one* of the solutions. But the student must be particularly careful to notice that whenever we divide every term of an equation by x, it must not be neglected, since the equation is satisfied by $x = 0$, which is therefore one of the roots.

Note 2. — When the factors can be written down *by inspection*, the student should always solve the example in this way, as he will thus save himself a great deal of unnecessary work.

Solve the following by resolution into factors :

4. $(3x - 1)(3x + 1) = 0$.   *Ans.* $\pm \frac{1}{3}$.

5. $x^2 - 11x = 0$.   0, 11.

6. $x^2 - 3x + 2 = 0$.   1, 2.

7. $x^2 - 2x = 8$.   4, -2.

8. $x^2 - 2ax + 4ab = 2bx$.   2a, 2b.

9. $x^2 - 2ax + 8x = 16a$.   2a, -8.

Note 3. — When the student cannot factor the equation readily by inspection, he should solve it by Rule II, Art. 132, or by Art. 134.

136. To Form a Quadratic when the Roots are Given.

— We have seen (Art. 135) that if

$$x^2 + px + q = (x - a)(x - b),$$

then a and b are the roots of the equation

$$x^2 + px + q = 0 . \quad . \quad . \quad . \quad . \quad (1)$$

Conversely, if a and b are roots of (1), then $x - a$ and $x - b$ are factors of the expression $x^2 + px + q$, which may be proved as follows:

Since a is a root of (1), we have

$$a^2 + pa + q = 0 . \quad . \quad . \quad . \quad . \quad (2)$$

Hence $x^2 + px + q = x^2 + px + q - (a^2 + pa + q)$

$$= (x - a)(x + a + p).$$

$\therefore$ $x - a$ is a factor of $x^2 + px + q$.

Hence, if a is a root of (1), $x - a$ will be a factor of $x^2 + px + q$.

Similarly it may be shown that if b is a root of (1), then $x - b$ will be a factor of $x^2 + px + q$.

Now $x^2 + px + q$ cannot have more than *two* factors of the form $x - a$, for the product of any number of factors of the form $x - a$ must be of the same degree in x as the number of the factors; also $x^2 + px + q$ clearly has no factors not containing x.

Hence $\qquad x^2 + px + q = (x - a)(x - b) \quad . \quad . \quad . \quad (3)$

Performing the multiplication in (3), we have

$$x^2 + px + q = x^2 - (a + b)x + ab.$$

Hence we have $\quad a + b = -p \atop ab = \quad q.$ $\left.\vphantom{\begin{matrix}a\\a\end{matrix}}\right\}$ $\quad . \quad . \quad . \quad . \quad . \quad . \quad (4)$

That is, *in a quadratic equation where the coefficient of x^2 is unity and all the terms are in the first member, the sum of the roots is equal to the coefficient of x with its sign changed, and the product of the roots is equal to the third term.*

Thus, dividing the general equation by a, it becomes

$$x^2 + \frac{b}{a}x + \frac{c}{a} = 0 . \quad . \quad . \quad . \quad . \quad (5)$$

Adding together the two roots of (1), Art. 133, we have

$$x_1 + x_2 = -\frac{b}{a};$$

and by multiplication we have

$$x_1 x_2 = \frac{b^2 - (b^2 - 4ac)}{4a^2} = \frac{c}{a},$$

which confirms the proposition.

Hence any quadratic may be expressed in the form

$$x^2 - \text{(sum of roots)} \; x + \text{product of roots} = 0 \; . \; (6)$$

By this principle we may easily form a quadratic with given roots. Although we cannot in all cases find the roots of a *given equation*, it is easy to solve the converse problem, namely, the problem of finding an equation which has *given roots*.

These relations are useful in verifying the solution of a quadratic equation. If the roots obtained do not satisfy these relations, we know there is some error in the work.

Relations analogous to those above hold good for equations of the third and of higher degrees. But we defer the proof to a subsequent chapter.

When we know one root of an equation, we may by division lower the degree of the equation. Thus if a is one root of an equation, we may divide it by the factor $x - a$.

NOTE 1. — In any equation the term which does not contain the unknown quantity is frequently called the *absolute term*.

A quadratic equation cannot have more than two roots. For no other value of x besides a and b can make $(x - a)(x - b)$ in (3) equal to 0, since the product of two factors cannot equal 0 if neither factor is equal to 0.

It may therefore be inferred that a *cubic* equation has *three* and *only three* roots; and that in *any* equation the number of roots is equal to the *degree* of the equation.

NOTE 2. — The student must carefully distinguish between a *quadratic equation* and a *quadratic expression*. Thus, in the quadratic equation $x^2 + px + q = 0$, we must suppose x to have one of two definite values, or roots, but when we speak of the quadratic *expression* $x^2 + px + q$, without saying that it is to be equal to zero, we may suppose x to have any value we please.

EXAMPLES.

Form the quadratic equation whose roots are

1. 2 and 3.   *Ans.* $x^2 - 5x + 6 = 0$.

2. 3 and -2. $x^2 - x - 6 = 0$.

3. $2 + \sqrt{3}$ and $2 - \sqrt{3}$. $x^2 - 4x + 1 = 0$.

4. 6 and 8. $x^2 - 14x + 48 = 0$.

5. 4 and 5. $x^2 - 9x + 20 = 0$.

137. Equations Having Imaginary Roots. — It was shown (Art. 133) that when b^2 is less than $4ac$, i.e., when $\dfrac{b^2}{4a^2}$ is less than $\dfrac{c}{a}$, the two roots are *imaginary*. Hence, from (5) and (6) of Art. 136, *the roots are imaginary when the square of half their sum is less than their product.* Now it is *impossible* to have two numbers such that the square of half their sum is less than their product, which may be shown as follows :

Let a represent any number ; and suppose it to be divided into two parts $\dfrac{a}{2} + x$ and $\dfrac{a}{2} - x$. Then the product is

$$\frac{a^2}{4} - x^2,$$

which is evidently the greatest when x^2 is the least. But when x^2 or $x = 0$ the parts are each $\dfrac{a}{2}$; thus the product of two *unequal* numbers is *less* than the square of half their sum. Hence,

The square of half the sum of two numbers can never be less than their product.

If then any problem furnishes an equation of the general quadratic form $x^2 + px + q = 0$, in which q is positive and *greater* than the square of $\dfrac{p}{2}$, we infer that the conditions of the problem are *incompatible* with each other, and hence the problem is *impossible.* Thus,

Let it be required to divide 6 into two parts whose product shall be 10.

Let x = one of the parts,

then $\qquad\qquad 6 - x$ = the other.

$\qquad\therefore\ x(6 - x) = 10,$

whence $\qquad\qquad x = 3 \pm \sqrt{-1}.$

Thus, the roots are imaginary. Now we know from the preceding proposition, that the number 6 cannot be divided into any two parts whose product will be greater than 9. Hence when we are required to divide 6 into two parts whose product is 10, we are required to solve an impossible problem. Thus, *the imaginary root shows that the problem is impossible.*

138. Equations of Higher Degree than the Second. — There are many equations which are not really quadratic, but which may be reduced to the quadratic form, and solved by the methods explained in this Chapter. An equation is in the *quadratic form* when the unknown quantity is found in *two terms, and its exponent in one term is twice as great as in the other.* Thus,

$$x^4 - 9x^2 = -20,\ (x^2 + x)^2 + 4(x^2 + x) = 12,$$
$$ax^{2n} + bx^n + c = 0,\ \text{etc.},$$

are in the quadratic form, and may be solved by either of the preceding rules; care however should be taken to use the one best adapted to the example considered.

1. Solve $x^4 - 9x^2 = -20$.

Here we may complete the square and solve by Rule I, Art. 132, or we may write out the value of x^2 at once by Rule II, Art. 132 as follows:

$$x^2 = \tfrac{9}{2} \pm \sqrt{-20 + \tfrac{81}{4}} = \tfrac{1}{4}$$
$$= \tfrac{9}{2} \pm \tfrac{1}{2} = 5,\ \text{or } 4.$$
$$x = \pm\sqrt{5},\ \text{or } \pm 2.$$

Thus there are four roots, $\pm\sqrt{5},\ \pm 2.$

Otherwise thus.

Transposing and factoring the first member,

$$(x^2 - 5)(x^2 - 4) = 0.$$

$$\therefore \quad x^2 - 5 = 0, \text{ giving } x = \pm\sqrt{5},$$

or $\qquad x^2 - 4 = 0$, giving $x = \pm 2$.

2. Solve $ax^{2n} + bx^n + c = 0$.

Transpose and divide by a,

$$x^{2n} + \frac{b}{a}x^n = -\frac{c}{a}.$$

Art. 132, Rule II, $x^n = -\dfrac{b}{2a} \pm\sqrt{-\dfrac{c}{a} + \dfrac{b^2}{4a^2}} = \dfrac{b^2 - 4ac}{4a^2}$

$$= \frac{-b \pm \sqrt{b^2 - 4ac}}{2a},$$

from which x may be found by taking the nth root of both members.

NOTE 1. — If the student prefers, he may let $x^n = y$; then $x^{2n} = y^2$. Substituting, the equation becomes

$$ay^2 + by + c = 0.$$

After solving for y, he may replace the value of y.

3. Solve $x^6 - 6x^3 = 16$.

Art. 132, Rule II, $x^3 = 3 \pm\sqrt{25} = 3 \pm 5 = 8$, or -2.

$$\therefore \quad x = 2, \text{ or } -\sqrt[3]{2}.$$

4. Solve $x^{-\frac{1}{2}} + x^{-\frac{1}{4}} = 6$.

Solving for $x^{-\frac{1}{4}}$, $x^{-\frac{1}{4}} = -\frac{1}{2} \pm \sqrt{6 + \frac{1}{4}} = 2$, or -3.

$$\therefore \quad x^{-1} = 16, \text{ or } 81.$$

$$\therefore \quad x = \tfrac{1}{16}, \text{ or } \tfrac{1}{81}.$$

5. Solve $\sqrt{x^2 + 12} + \sqrt[4]{x^2 + 12} = 6$.

Solving for $\sqrt[4]{x^2 + 12}$,

$$\sqrt[4]{x^2 + 12} = -\tfrac{1}{2} \pm \sqrt{6 + \tfrac{1}{4}} = -\tfrac{1}{2} \pm \tfrac{5}{2} = 2, \text{ or } -3.$$

$$\therefore \quad x^2 + 12 = 16, \text{ or } 81.$$

$$\therefore \quad x^2 = 4, \text{ or } 69.$$

$$\therefore \quad x = \pm 2, \text{ or } \pm\sqrt{69}.$$

6. Solve $x + \sqrt{5x + 10} = 8$.

By transposing x, and squaring,

$$5x + 10 = 64 - 16x + x^2.$$

$$\therefore x^2 - 21x = -54.$$

Solving, we get $\qquad x = \frac{21}{2} \pm \frac{15}{2} = 18$, or 3.

If we proceed to verify these values of x by substituting them in the *given* equation, we shall find that 3 satisfies the equation, but that 18 does not, while it *does* satisfy the equation

$$x - \sqrt{5x + 10} = 8.$$

Now the reason is this: the equation $x^2 - 21x = -54$, which we obtained from the given equation by transposing and squaring, might have been obtained as well from $x - \sqrt{5x + 10} = 8$, since the square root of a quantity may have either the sign $+$ or $-$ prefixed to it; i.e., the resulting equation $x^2 - 21x = -54$, of which 18 and 3 are the roots, would be obtained, whether the sign of the radical be $+$ or $-$. Hence we see that when an equation has been reduced to a rational form by squaring, we cannot be certain without trial whether the values which are finally obtained for the unknown quantity are roots of the given equation.

7. Solve $x^2 - 7x + \sqrt{x^2 - 7x + 18} = 24$.

Add 18 to both members in order that the equation may be in the quadratic form.

$x^2 - 7x + 18 + \sqrt{x^2 - 7x + 18} = 42.$

Solving, $\qquad \sqrt{x^2 - 7x + 18} = -\frac{1}{2} \pm \sqrt{42 + \frac{1}{4}}$

$$= -\frac{1}{2} \pm \frac{13}{2} = 6, \text{ or } -7.$$

$$\therefore x^2 - 7x + 18 = 36, \text{ or } 49.$$

Solving the first quadratic, we obtain $\qquad x = 9$, or -2.

Solving the second quadratic, we obtain $x = \frac{1}{2}(7 \pm \sqrt{173})$.

Only the first two values are roots of the given equation; the other two are roots of the equation

$$x^2 - 7x - \sqrt{x^2 - 7x + 18} = 24.$$

8. Solve $x^4 - 4x^3 - 2x^2 + 12x - 16 = 0$.

We proceed to form a perfect trinomial square with the first two terms and a part of the third. The square root of this square is evidently $(x^2 - 2x)$, the square of which is $x^4 - 4x^3 + 4x^2$; having *added* $4x^2$ to the equation we must now *subtract* $4x^2$. Hence the equation becomes

$$x^4 - 4x^3 + 4x^2 - 6x^2 + 12x - 16 = 0,$$

or $\qquad (x^2 - 2x)^2 - 6(x^2 - 2x) = 16,$

which is in the quadratic form, and may be solved as those above.

Hence $x^2 - 2x = 3 \pm \sqrt{16 + 9} = 8$, or -2.

$$\therefore \quad x = 1 \pm \sqrt{8 + 1}, \text{ or } 1 \pm \sqrt{-2 + 1}.$$

$$\therefore \quad x = 4, \text{ or } -2, \text{ or } 1 \pm \sqrt{-1}.$$

Solve the following:

9. $x^4 - 13x^2 + 36 = 0$. $\hfill Ans. \pm 2, \pm 3.$

10. $x^2 + \sqrt{x^2 + 9} = 21$. $\hfill \pm 4.$

11. $9\sqrt{x^2 - 9x + 28} + 9x = x^2 + 36$. $\hfill 12, -3.$

12. $x^{\frac{5}{6}} + x^{\frac{5}{3}} = 1056$. $\hfill 64, (-33)^{\frac{6}{5}}.$

13. $(x^2 - 9)^2 - 11(x^2 - 2) = 3$. $\hfill \pm 5, \pm 2.$

14. $x^4 - 8x^3 + 10x^2 + 24x + 5 = 0$. $\hfill 5, -1, 2 \pm \sqrt{5}.$

139. Solutions by Factoring. — By the principles of Art. 135, many equations of a higher degree than the second may be solved, which cannot be reduced to the quadratic form. If an equation can be reduced to the form

$$(x - a)X = 0,$$

in which X represents an expression involving x, we have either

$$x - a = 0, \text{ or } X = 0;$$

therefore $x = a$, is one value of x; and if we solve the equation $X = 0$, we shall have the other values of x. Hence whenever we have one factor of an equation, we have at least *one* root, and by division we may lower the degree of the equation by one (Art. 136). Thus

1. Solve $(x - 5)(x^2 - 3x + 2) = 0$.

Here the first member is zero either when $x - 5 = 0$, or when $x^2 - 3x + 2 = 0$; and in no other case. Hence we have

$$x - 5 = 0, \text{ or } x^2 - 3x + 2 = 0.$$

From the first we have $x = 5$; and the other roots of the equation are those given by

$$x^2 - 3x + 2 = 0,$$

that is, $(x - 2)(x - 1) = 0.$

Thus the cubic equation $(x - 5)(x^2 - 3x + 2) = 0$ has the three roots 5, 2, and 1.

The difficulty to be overcome in this method consists in resolving the equation into factors; and facility in separating expressions into factors can be acquired only by experience.

2. Solve $x^3 - 1 = 0$.

Since $x^3 - 1 = (x - 1)(x^2 + x + 1),$

we have $(x - 1)(x^2 + x + 1) = 0.$

   ∴ $x - 1 = 0, \text{ or } x^2 + x + 1 = 0,$

the roots of which are 1, or $-\frac{1}{2} \pm \sqrt{-\frac{3}{4}}$.

Hence there are three roots of the equation $x^3 = 1$, one being real and the other two imaginary. Thus there are three numbers whose cubes are equal to 1; that is, there are *three cube roots* of 1.

3. Solve $x - 1 = 2 + \dfrac{2}{\sqrt{x}}$.   *Ans.* 4.

4. " $2x^3 - x^2 - 6x = 0$. $0, 2, -\frac{3}{2}$.

5. " $x^3 + x^2 - 4x = 4$. $-1, 2, -2$.

6. " $x^3 - 3x = 2$. $-1, 2$.

7. " $x^2 - \dfrac{2}{3x} = 1\frac{4}{9}$. $-\frac{2}{3}, \frac{1}{3}(1 \pm \sqrt{10})$.

140. Problems Leading to Quadratic Equations of One Unknown Quantity. — We shall now give some examples of problems which lead to pure or affected quadratic equations of one unknown quantity.

In the solution of such problems, the equations are found on the same principles as in problems producing simple equations (Art. 61).

EXAMPLES.

1. Find two numbers such that their sum is 15, and their product is 54.

Let $x =$ one of the numbers,

then $15 - x =$ the other number.

Hence from the conditions, we have

$$x(15 - x) = 54,$$

or $x^2 - 15x = -54.$

Solving, we get $x = 9$, or 6.

If we take $x = 9$ we have $15 - x = 6$; and if we take $x = 6$ we have $15 - x = 9$. Thus, whichever value of x we take, we get for the two numbers 6 and 9. Hence, although the equation gives two values of x, yet there is really only one solution of the problem.

2. A man buys a horse which he sells again for $96; he finds that he thus loses one-fourth as much per cent as the horse cost him : find the price of the horse.

Let $x =$ the price of the horse in dollars,

then $\dfrac{x^2}{400} =$ the man's loss in dollars.

Hence from the conditions, we have

$$\frac{x^2}{400} = x - 96.$$

Solving, we get $x = 240$ or 160.

That is, the price was either $240 or $160, for each of these values satisfies the conditions of the problem.

3. A train travels 300 miles at a uniform rate ; if the rate had been 5 miles an hour more, the journey would have taken two hours less : find the rate of the train.

Let $x =$ the rate the train runs in miles per hour ;

then $300 \div x =$ the time of running on the first supposition ;

and $300 \div (x + 5) =$ the time of running on the second supposition.

$$\therefore \quad \frac{300}{x + 5} = \frac{300}{x} - 2.$$

Solving, we get $x = 25$, or -30.

Only the positive value of x is admissible, and thus the train runs 25 miles per hour.

NOTE. — In the solutions of problems it often happens that the roots of the equation, which is the Algebraic statement of the relation between the magnitudes of the known and unknown quantities, do not all satisfy the conditions of the problem. The reason of this is that the Algebraic statement is more general than ordinary language; and the equation, which is a proper representation of the conditions, will also express other conditions. Thus, the roots of the *equation* are the numbers, whether *positive, negative, integral,* or *fractional,* which *satisfy that equation;* but in the *problem* there may be *restrictions* on the numbers, expressed or implied, which cannot be retained in the equation. If for instance, one of the roots of an equation is a *fraction,* it cannot be a solution of a problem which refers to a number of men, for such a number must be *integral.* Thus

4. Eleven times the number of men in a group is greater by twelve than twice the square of the number: find the number of men in the group.

Let $x =$ the number of men ; then we have

$$11x = 2x^2 + 12,$$

or $\qquad\qquad 2x^2 - 11x = -12.$

Solving, we get $\qquad\qquad x = 4$, or $1\frac{1}{2}$.

Thus, there are 4 men ; the value $1\frac{1}{2}$ is plainly inadmissible.

5. Eleven times the number of feet in the length of a rod is greater by twelve than twice the square of the number of feet: how long is the rod?

This question leads to the same equation as Ex. 4, only here we cannot reject the fractional result, since the rod may be either 4 feet long or $1\frac{1}{2}$ feet long.

6. The square of the number of dollars a man possesses is greater by 600 than ten times the number: how much has the man?

Let $x =$ the number of dollars the man has.

Then $x^2 = 10x + 600.$

Solving, we get $x = 30,$ or $-20.$

Both these values are admissible, since a negative possession is a debt (Art. 20).

7. The sum of the ages of a father and his son is 80 years; also one-fourth of the product of their ages, in years, exceeds the father's age by 240: how old are they?

Let $x =$ the father's age in years;

then $80 - x =$ the son's age in years.

Hence $\frac{1}{4}x(80 - x) = x + 240,$

or $x^2 - 76x = -960.$

$\therefore x = 60,$ or 16.

Thus the father is 60 and the son 20 years old.

The second solution 16 is not admissible, since it would make the father younger than his son.

Note. — The student should examine each root of every equation to see if it satisfies the conditions of the problem, and reject those which do not.

8. A cistern can be filled by two pipes in $33\frac{1}{3}$ minutes; if the larger pipe takes 15 minutes less than the smaller to fill the cistern, find in what time it will be filled by each pipe singly. *Ans.* 75 and 60 minutes.

9. A person selling a horse for \$72, finds that his loss per cent is one-eighth of the number of dollars that he paid for the horse: what was the cost price? *Ans.* \$80, or \$720.

10. Divide the number 10 into two parts such that their product added to the sum of their squares may make 76.

Ans. 4, 6.

11. Find the number which added to its square root will make 210. *Ans.* 196.

12. A and B together can do a piece of work in $14\frac{2}{5}$ days; and A alone can do it in 12 days less than B alone: find the time in which A alone can do the work. *Ans.* 24 days.

13. A company dining together at an inn, find their bill amounts to \$35; two of them were not allowed to pay, and the rest found that their shares amounted to \$2 a man more than if all had paid: find the number of men in the company.

Ans. 7.

14. The side of a square is 110 inches long: find the length and breadth of a rectangle which shall have its perimeter 4 inches longer than that of the square, and its area 4 square inches less than that of the square. *Ans.* 126, 96.

NOTE. — We will conclude this chapter with the following examples. In solving them care must be taken to select the method best adapted to the example considered. Many of them may be solved by special methods (Arts. 133, 134, 135); but the methods of Art. 132 are the most important, and will solve every example.

EXAMPLES OF QUADRATICS.

1. $x\sqrt{6 + x^2} = 1 + x^2.$ *Ans.* $\pm\frac{1}{2}.$

2. $\dfrac{7x^2 + 8}{21} - \dfrac{x^2 + 4}{8x^2 - 11} = \dfrac{x^2}{3}.$ $\pm 2.$

3. $x + \sqrt{1 + x^2} = \dfrac{2}{\sqrt{1 + x^2}}.$ $\pm\frac{1}{3}\sqrt{3}.$

4. $\dfrac{1}{x + \sqrt{2 - x^2}} + \dfrac{1}{x - \sqrt{2 - x^2}} = \dfrac{x}{2}.$ $\pm\sqrt{3}.$

5. $x^2 - 5x = -6.$ 2, 3.

6. $x^2 - 10x = -9.$ 1, 9.

7. $x^2 + 12x = 13.$ 1, −13.

8. $x^2 + 9x = 22.$　　　　　　　　　　*Ans.* 2, $-11.$

9. $x^2 + 2x = 143.$　　　　　　　　　　11, $-13.$

10. $15x^2 - 2ax = a^2.$　　　　　　　　$\dfrac{a}{3},\ -\dfrac{a}{5}.$

11. $21x^2 = 2ax + 3a^2.$　　　　　　　$\frac{3}{7}a,\ -\dfrac{a}{3}.$

12. $\dfrac{x + 3}{2x - 7} = \dfrac{2x - 1}{x - 3}.$　　　　　　　4, $\frac{4}{3}.$

13. $\dfrac{1}{1 + x} - \dfrac{1}{3 - x} = \frac{6}{35}.$　　　　　　13, $\frac{2}{3}.$

14. $\dfrac{4}{x - 1} - \dfrac{5}{x + 2} = \dfrac{3}{x}.$　　　　　　3, $-\frac{1}{2}.$

15. $(x + 1)(2x + 3) = 4x^2 - 22.$　　　5, $-2\frac{1}{2}.$

16. $\dfrac{2 + x^2}{3} - \dfrac{x - x^2}{2} = 1 - x + x^2.$　　**1, 2.**

17. $8x + 11 + \dfrac{7}{x} = \dfrac{68x}{7}.$　　　　　**7, $-\frac{7}{12}.$**

18. $\dfrac{3(x - 1)}{x + 1} - \dfrac{2(x + 1)}{x - 1} = 5.$　　**$\frac{1}{2}$, $-3.$**

19. $\dfrac{x + 1}{x - 1} - \dfrac{x - 2}{x + 2} = \frac{9}{5}.$　　　**3, $-\frac{1}{2}.$**

20. $\dfrac{1}{x - 2} - \dfrac{2}{x + 2} = \frac{3}{5}.$　　　　**3, $-4\frac{2}{3}.$**

21. $\dfrac{3}{2(x^2 - 1)} - \dfrac{1}{4(x + 1)} = \frac{1}{8}.$　　**3, $-5.$**

22. $\dfrac{5}{x + 2} + \dfrac{3}{x} = \dfrac{14}{x + 4}.$　　　**3, $-1\frac{1}{3}.$**

23. $\dfrac{4}{x + 1} + \dfrac{5}{x + 2} = \dfrac{12}{x + 3}.$　　**3, $-1\frac{2}{3}.$**

24. $\dfrac{x + 1}{x + 2} + \dfrac{x - 1}{x - 2} = \dfrac{2x - 1}{x - 1}.$　　4, 0.

25. $\dfrac{x - 2}{x + 2} + \dfrac{x + 2}{x - 2} = 2\!\left(\dfrac{x + 3}{x - 3}\right).$　　**$1\frac{1}{3}$, 0.**

26. $\dfrac{x-1}{x+1} + \dfrac{x-2}{x+2} = \dfrac{2x+13}{x+16}.$ *Ans.* $5, -1\frac{5}{13}.$

27. $\dfrac{x+1}{x-1} + \dfrac{x+2}{x-2} = \dfrac{2x+13}{x+1}.$ $5, 1\frac{1}{5}.$

28. $\dfrac{2x-1}{x+1} + \dfrac{3x-1}{x+2} = \dfrac{5x-11}{x-1}.$ $5, -1\frac{1}{4}.$

29. $x - \dfrac{14x-9}{8x-3} = \dfrac{x^2-3}{x+1}.$ $2\frac{2}{3}, 0.$

30. $\dfrac{x+\dfrac{1}{x}}{x-\dfrac{1}{x}} + \dfrac{1+\dfrac{1}{x}}{1-\dfrac{1}{x}} = 3\frac{1}{4}.$ $3, -1\frac{2}{5}.$

31. $\dfrac{\sqrt{4x}+2}{4+\sqrt{x}} = \dfrac{4-\sqrt{x}}{\sqrt{x}}.$ $4.$

32. $a^2x^2 - 2a^3x + a^4 - 1 = 0.$ $a \pm \dfrac{1}{a}.$

33. $4a^2x = (a^2 - b^2 + x)^2.$ $(a \pm b)^2.$

34. $\dfrac{x}{a} + \dfrac{a}{x} = \dfrac{x}{b} + \dfrac{b}{x}.$ $\pm\sqrt{ab}.$

35. $(3a^2 + b^2)(x^2 - x + 1) = (3b^2 + a^2)(x^2 + x + 1).$

      Ans. $\dfrac{a+b}{a-b}, \dfrac{a-b}{a+b}.$

36. $\dfrac{1}{a+b+x} = \dfrac{1}{a} + \dfrac{1}{b} + \dfrac{1}{x}.$ $-a, -b.$

37. $\dfrac{a+c(a+x)}{a+c(a-x)} + \dfrac{a+x}{x} = \dfrac{a}{a-2cx}.$ $-a, \dfrac{a(1+c)}{c(2c+3)}.$

38. $\dfrac{x}{a} + \dfrac{a}{x} - \dfrac{2}{a} = 0.$ $1 \pm \sqrt{1-a^2}.$

39. $\dfrac{x^2}{\sqrt{a}+\sqrt{b}} - (a^{\frac{1}{2}} - b^{\frac{1}{2}})x = \dfrac{1}{(ab^2)^{-\frac{1}{2}} + (a^2b)^{-\frac{1}{2}}}.$ $a, -b.$

Form the equation whose roots are

40. 1 ± 5.   *Ans.* $x^2 - 2x - 4 = 0$.

41. $-\frac{4}{5}, \frac{3}{7}$. $35x^2 + 13x - 12 = 0$.

42. $\dfrac{p - q}{p + q}, -\dfrac{p + q}{p - q}$. $(p^2 - q^2)x^2 + 4pqx - p^2 + q^2 = 0$.

43. $7 \pm 2\sqrt{5}$. $x^2 - 14x + 29 = 0$.

44. $\pm 2\sqrt{3} - 5$. $x^2 + 10x + 13 = 0$.

45. $-p \pm 2\sqrt{2q}$. $x^2 + 2px + p^2 - 8q = 0$.

Show that the roots of the following equations are real :

46. $x^2 - 2ax + a^2 - b^2 - c^2 = 0$.

47. $(a - b + c)x^2 + 4(a - b)x + (a - b - c) = 0$.

For what values of m will the following equations have equal roots?

48. $x^2 - 15 - m(2x - 8) = 0$.   *Ans.* $3, 5$.

49. $x^2 - 2x(1 + 3m) + 7(3 + 2m) = 0$. $2, -\frac{10}{9}$.

EXAMPLES OF EQUATIONS REDUCIBLE TO QUADRATICS.

50. $x^4 - 5x^2 + 4 = 0$.   *Ans.* $\pm 1, \pm 2$.

51. $(x^2 - x)^2 - 8(x^2 - x) + 12 = 0$. $3, -1, \pm 2$.

52. $(x^2 + x)(x^2 + x + 1) = 42$. $2, -3, \dfrac{-1 \pm \sqrt{-27}}{2}$.

53. $\left(x + \dfrac{1}{x}\right)^2 + 4\left(x + \dfrac{1}{x}\right) = 12$. $1, -3 \pm 2\sqrt{2}$.

54. $x^2 + 3 - \sqrt{2x^2 - 3x + 2} = \frac{3}{2}(x + 1)$. $1, \frac{1}{2}$.

55. $x^4 + 2x^3 - 11x^2 + 4x + 4 = 0$. $1, 2$.

56. $x^4 + 4a^3x = a^4$. $-\dfrac{a}{\sqrt{2}} \pm \dfrac{a\sqrt{2\sqrt{2} - 1}}{\sqrt{2}}$.

57. $\dfrac{1 + x^4}{(1 + x)^4} = \frac{1}{2}$.

$\qquad$ *Ans.* $1 + \sqrt{3} \pm \sqrt{3 + 2\sqrt{3}}, 1 - \sqrt{3} \pm \sqrt{3 - 2\sqrt{3}}$.

58. $27x^2 - \dfrac{841}{3x^2} + 1\tfrac{7}{3} = \dfrac{232}{3x} - \dfrac{1}{3x^2} + 5.$

$\qquad\qquad$ *Ans.* $2, -1\tfrac{5}{9}, \tfrac{1}{9}\,(-2 \pm \sqrt{-266}).$

Solve as explained in Art. 143 :

59. $2x^3 - x^2 = 1.$ $\qquad\qquad$ *Ans.* $1, \tfrac{1}{4}(-1 \pm \sqrt{-7}).$

60. $x^3 - 6x = 9.$ $\qquad\qquad\qquad 3, \tfrac{1}{2}(-3 \pm \sqrt{-3}).$

61. $8x^3 + 16x = 9.$ $\qquad\qquad\quad \tfrac{1}{2}, \tfrac{1}{4}(-1 \pm \sqrt{-35}).$

62. $x^3 - 3x^2 + x + 2 = 0.$ $\qquad\qquad 2, \tfrac{1}{2}(1 \pm \sqrt{5}).$

63. $3x^6 + 8x^4 - 8x^2 = 3.$ $\qquad \pm 1, [\tfrac{1}{6}(-11 \pm \sqrt{85})]^{\frac{1}{2}}.$

64. $x^3 + x^2 - 4x - 4 = 0.$ $\qquad\qquad\qquad -1, \pm 2.$

65. $x^3 + ax^2 + \left(a - 1 + \dfrac{1}{a-1}\right)x + 1 = 0.$

$\qquad\qquad\qquad 1 - a, \tfrac{1}{2}\left(-1 \pm \sqrt{\dfrac{5-a}{1-a}}\right).$

66. Find a number whose square diminished by 119 is equal to ten times the excess of the number over 8. *Ans.* 13.

67. A man is five times as old as his son, and the sum of the squares of their ages is equal to 2106 : find their ages.

$\qquad\qquad\qquad\qquad\qquad\qquad$ *Ans.* 45, 9.

68. If a train traveled 5 miles an hour faster it would take one hour less to travel 210 miles : what time does it take? $\qquad\qquad\qquad\qquad$ *Ans.* 7 hours.

69. The sum of the reciprocals of two consecutive numbers is $\tfrac{15}{56}$: find them. $\qquad\qquad\qquad\qquad$ *Ans.* 7, 8.

70. The perimeter of a rectangular field is 500 yards, and its area is 14400 square yards : find the length of the sides.

$\qquad\qquad\qquad\qquad\qquad$ *Ans.* 90 and 160 yards.

71. A number of two digits is equal to twice the product of the digits, and the digit in the ten's place is less by 3 than the digit in the unit's place : what is the number?

$\qquad\qquad\qquad\qquad\qquad\qquad\qquad$ *Ans.* 36.

72. The sum of a certain number and its square root is 42 : what is the number? $\qquad\qquad\qquad$ *Ans.* 36.

73. A rectangular court is ten yards longer than it is broad ; its area is 1131 square yards : find its length and breadth. *Ans.* 39 and 29.

74. One hundred and ten bushels of coal were divided among a certain number of persons ; if each person had received one bushel more he would have received as many bushels as there were persons : find the number of persons.

Ans. 11.

75. A cistern can be supplied with water by two pipes ; by one of them it would be filled 6 hours sooner than by the other, and by both together in 4 hours : find the time in which each pipe alone would fill it. *Ans.* 6, 12.

76. Two messengers A and B were sent at the same time to a place 90 miles distant ; the former by riding one mile per hour more than the latter arrived at the end of his journey one hour before him : find at what rate per hour each traveled. *Ans.* 10, 9 miles.

77. A person rents a certain number of acres of land for $280 ; he keeps 8 acres in his own possession, and sublets the remainder at $1 per acre more than he gave, and thus he covers his rent and has $8 over : find the number of acres.

Ans. 56.

78. From two places 320 miles apart, two persons A and B set out in order to meet each other. A traveled 8 miles a day more than B ; and the number of days in which they met was equal to half the number of miles B went in a day : find how far each traveled before they met. *Ans.* 192, 128.

79. A certain number is formed by the product of three con-secutive numbers, and if it be divided by each of them in turn, the sum of the quotients is 47 : find the number. *Ans.* 60.

80. A boat's crew row $3\frac{1}{2}$ miles down a river and back again in 1 hour and 40 minutes : supposing the river to have a current of 2 miles per hour, find the rate at which the crew would row in still water. *Ans.* 5 miles per hour.

81. A person rents a certain number of acres of land for $336 ; he cultivates 4 acres himself, and letting the rest for

$2 an acre more than he pays for it, receives for this portion the whole rent, $336 : find the number of acres. *Ans.* 28 acres.

82. A person bought a certain number of sheep for $140 ; after losing two of them he sold the rest at $2 a head more than he paid for them, and by so doing gained $4 by the transaction : find the number of sheep he bought. *Ans.* 14.

83. A man sends a lad to the market to buy 12 cents' worth of oranges ; the lad having eaten a couple, the man pays at the rate of one cent for 15 more than the market price : how many did the man get for his 12 cents? *Ans.* 18.

84. A person drew a quantity of wine from a full vessel which held 81 gallons, and then filled up the vessel with water ; he then drew from the mixture as much as he before drew of pure wine ; and it was found that 64 gallons of pure wine remained : find how much he drew each time.

Ans. 9 gallons.

85. A certain company of soldiers can be formed into a solid square ; a battalion consisting of seven such equal companies can be formed into a hollow square, the men being four deep. The hollow square formed by the battalion is sixteen times as large as the solid square formed by one company : find the number of men in the company. *Ans.* 64.

86. Find that number whose square added to its cube is nine times the next higher number. *Ans.* 3.

87. A courier proceeds from one place P to another place Q in 14 hours ; a second courier starts at the same time as the first from a place 10 miles behind P, and arrives at Q at the same time as the first courier. The second courier finds that he takes half an hour less than the first to go 20 miles : find the distance from P to Q. *Ans.* 70 miles.

88. A vessel can be filled with water by two pipes ; by one of these pipes alone the vessel would be filled 2 hours sooner than by the other ; also the vessel can be filled by both pipes together in $1\frac{7}{8}$ hours : find the time which each pipe alone would take to fill the vessel. *Ans.* 5 and 3 hours.

CHAPTER XIV.

SIMULTANEOUS QUADRATIC EQUATIONS.

141. Simultaneous Quadratic Equations.—We shall now consider some of the most useful methods of solving simultaneous equations, one or more of which may be of a degree higher than the first. It should be remarked that it is, in general, impossible to solve a pair of simultaneous quadratic equations; for, if we eliminate one of the unknown quantities, the resulting equation will be of the fourth degree with respect to the other unknown quantity, and we cannot, in general, solve an equation of a higher degree than the second.

There are several cases however in which a solution of two equations may be effected, when one or both of them are of the second or some higher degree. Various artifices are employed for the solution of such equations, the proper application of which must be learned by experience.

142. Case I. When One of the Equations is of the First Degree. — This case may be solved by the following

RULE. *From the simple equation find the value of one of the unknown quantities in terms of the other, and substitute this value in the other equation.*

1. Solve
$$3x + 4y = 18 \ . \ \ . \ . \ . \ . \ . \ . \ (1)$$
$$5x^2 - 3xy = 2 \ . \ . \ . \ . \ . \ . \ (2)$$

From (1) we have
$$y = \frac{18 - 3x}{4}; \ . \ . \ . \ . \ (3)$$

and substituting in (2),
$$5x^2 - \frac{3x(18 - 3x)}{4} = 2.$$

$$\therefore \ 20x^2 - 54x + 9x^2 = 8,$$

or
$$29x^2 - 54x = 8.$$

Solving, we get $\qquad x = 2,$ or $-\frac{4}{29}$;

and by substituting in (3), $y = 3,$ or $\frac{267}{58}$.

Solve the following:

2. $3x - 4y = 5,$ $\qquad$ $Ans.$ $x = 3,$ or $-1\frac{31}{9},$

$\qquad$ $3x^2 - xy - 3y^2 = 21.$ $\qquad$ $y = 1,$ or $-\frac{38}{3}.$

3. $5x - y = 17,$ $\qquad$ $x = 4,$ or $-\frac{3}{5},$

$\qquad$ $xy = 12.$ $\qquad$ $y = 3,$ or $-20.$

4. $2x - 5y = 3,$ $\qquad$ $x = 4,$ or $-\frac{25}{7},$

$\qquad$ $x^2 + xy = 20.$ $\qquad$ $y = 1,$ or $-\frac{71}{35}.$

5. $4x - 5y = 1,$ $\qquad$ $x = 4,$ or $-\frac{48}{13},$

$\qquad$ $2x^2 - xy + 3y^2 + 3x - 4y = 47.$ $\quad$ $y = 3,$ or $-\frac{41}{13}.$

143. Case II. Equations of the Form $x \pm y = a$, and $xy = b$; or $x^2 + y^2 = a$, and $xy = b$.

1. Solve $\qquad\qquad x + y = 15 \ . \ . \ . \ . \ (1)$

$\qquad\qquad\qquad\qquad xy = 36 \ . \ . \ . \ . \ (2)$

Square (1), $\qquad x^2 + 2xy + y^2 = 225; \ . \ . \ . \ (3)$

multiply (2) by 4, $\qquad\qquad 4xy = 144; \ . \ . \ . \ (4)$

subtract (4) from (3), $x^2 - 2xy + y^2 = 81;$

extract the square root, $\qquad x - y = \pm 9. \ . \ . \ . \ (5)$

Combining (5) with (1), we have the two cases

$$\begin{matrix} x + y = 15, \\ x - y = \ \ 9. \end{matrix} \Big\} \quad \begin{matrix} x + y = \ \ 15, \\ x - y = -9. \end{matrix} \Big\}$$

from which we have $\begin{matrix} x = 12, \\ y = \ \ 3. \end{matrix} \Big\} \qquad \begin{matrix} x = \ \ 3, \\ y = 12. \end{matrix} \Big\}$

2. Solve $\qquad\qquad x^2 + y^2 = 25 \ . \ . \ . \ . \ . \ . \ (1)$

$\qquad\qquad\qquad\qquad xy = 12 \ . \ . \ . \ . \ . \ (2)$

Multiply (2) by 2; then by addition and subtraction we have

$$x^2 + 2xy + y^2 = \ \ 49,$$

$$x^2 - 2xy + y^2 = \ \ 1.$$

$$\therefore \ \ x + y = \pm 7,$$

$$x - y = \pm 1.$$

We have now four cases to consider, namely,

$$x + y = 7 \atop x - y = 1 \Big\} \quad x + y = 7 \atop x - y = -1 \Big\} \quad x + y = -7 \atop x - y = 1 \Big\} \quad x + y = -7 \atop x - y = -1 \Big\}$$

From which the values of x are $\quad 4, 3, -3, -4$;
and the corresponding values of y are $3, 4, -4, -3$.

Thus there are four pairs of values, two of which are given by $x = \pm 4$, $y = \pm 3$, and the other two by $x = \pm 3$, $y = \pm 4$, it being understood that in both cases the upper signs are to be taken together, and the lower signs are to be taken together.

These are the simplest forms that occur, but they are specially important, since the solution of a large number of other forms is dependent upon them. Our object, as a rule, is to solve the given equations *symmetrically*, by finding the values of $x + y$ and $x - y$, which we can always do as soon as we have obtained the product of the unknown quantities and either their sum or their difference.

Any pair of equations of the form

$$x^2 \pm pxy + y^2 = a^2 \quad \cdots \cdots (1)$$

$$x \pm y = b \quad \cdots \cdots (2)$$

where p is any numerical quantity, can be reduced to one of the forms above considered ; for by squaring (2) and combining with (1), we obtain an equation to find xy ; the solution can then be completed by the aid of equation (2). Thus

3. Solve $\qquad x^2 + xy + y^2 = 333 \quad \cdots \cdots (1)$

$$x - y = 3 \quad \cdots \cdots (2)$$

Square (2), $\qquad x^2 - 2xy + y^2 = 9 \quad \cdots \cdots (3)$

Subtract (3) from (1), $\qquad 3xy = 324.$

$$\therefore \quad xy = 108 \quad \cdots \cdots (4)$$

Add (1) and (4) and extract square root,

$$x + y = \pm 21 \quad \cdots \cdots (5)$$

From (2) and (5) $\qquad x = 12, \text{ or } -9. \atop y = 9, \text{ or } -12. \Big\}$

4. Solve $\quad\quad \dfrac{1}{x} - \dfrac{1}{y} = \tfrac{1}{3}$ . . . . (1)

$$\dfrac{1}{x^2} + \dfrac{1}{y^2} = \tfrac{5}{9} \quad\quad \text{. . . . . (2)}$$

Square (1), $\dfrac{1}{x^2} - \dfrac{2}{xy} + \dfrac{1}{y^2} = \tfrac{1}{9}$ . . . . (3)

Subtract (3) from (2), $\quad \dfrac{2}{xy} = \tfrac{4}{9}$ . . . . (4)

Add (2) and (4) and take the square root,

$$\dfrac{1}{x} + \dfrac{1}{y} = \pm 1 \quad\quad \text{. . . . . . . (5)}$$

From (1) and (5), $\quad \dfrac{1}{x} = \tfrac{2}{3}$, or $-\tfrac{1}{3}$.

$$\dfrac{1}{y} = \tfrac{1}{3}, \text{ or } -\tfrac{2}{3}.$$

$\therefore \quad x = \tfrac{3}{2}$, or -3, and $y = 3$, or $-\tfrac{3}{2}$.

Solve the following :

5. $x - y = 12,$ $\quad\quad\quad$ Ans. $x = 17, \quad$ or $- 5,$

$\quad\quad xy = 85.$ $\quad\quad\quad\quad\quad\quad y = 5, \quad$ or $-17.$

6. $x + y = 74,$ $\quad\quad\quad\quad\quad\quad x = 53, \quad$ or $\quad 21,$

$\quad\quad xy = 1113.$ $\quad\quad\quad\quad\quad\quad y = 21, \quad$ or $\quad 53.$

7. $x + y = 84,$ $\quad\quad\quad\quad\quad\quad x = 71, \quad$ or $\quad 13,$

$\quad\quad xy = 923.$ $\quad\quad\quad\quad\quad\quad y = 13, \quad$ or $\quad 71.$

8. $x^2 + y^2 = 97,$ $\quad\quad\quad\quad\quad x = 9, \quad$ or $\quad 4,$

$\quad x + y = 13.$ $\quad\quad\quad\quad\quad\quad y = 4, \quad$ or $\quad 9.$

9. $\quad\quad x + y = 9,$ $\quad\quad\quad\quad\quad x = 5, \quad$ or $\quad 4,$

$\quad x^2 + xy + y^2 = 61.$ $\quad\quad\quad\quad y = 4, \quad$ or $\quad 5.$

144. Case III. When the Two Equations Contain a Common Algebraic Factor.

RULE. *Divide one equation by the other, and cancel the common factor.*

1. Solve
$$xy + y^2 = 133 \quad \cdots \cdots \cdots \quad (1)$$
$$x^2 - y^2 = 95 \quad \cdots \cdots \cdots \quad (2)$$

Divide (2) by (1) and cancel the common factor $x + y$,

$$\frac{x - y}{y} = \tfrac{95}{133} = \tfrac{5}{7}.$$

$$\therefore \quad x = \tfrac{12}{7}y; \quad \cdots \cdots \cdots \quad (3)$$

and substituting in (1)

$$\tfrac{12}{7}y^2 + y^2 = 133.$$

Solving, we get $\qquad y = \pm 7$;
and by substituting in (3), $\quad x = \pm 12.$

NOTE. — This includes the case where one equation is exactly divisible by the other.

2. Solve
$$x^3 + y^3 = 18xy. \quad \cdots \cdots \quad (1)$$
$$x + y = 12 \quad \cdots \cdots \quad (2)$$

Divide (1) by (2), $x^2 - xy + y^2 = \tfrac{3}{2}xy.$

$$\therefore \quad x^2 - \tfrac{5}{2}xy + y^2 = 0. \quad \cdots \cdots \quad (3)$$

Subtract (3) from the square of (2),

$$\tfrac{9}{2}xy = 144.$$

$$\therefore \quad \tfrac{1}{2}xy = 16. \quad \cdots \cdots \cdots \quad (4)$$

Add (3) and (4) and take the square root,

$$x - y = \pm 4 \quad \cdots \cdots \cdots \quad (5)$$

From (2) and (5) $\qquad x = 8, \text{ or } 4,$
$$y = 4, \text{ or } 8.$$

3. Solve
$$x^4 + x^2y^2 + y^4 = 2613. \quad \cdots \cdots \quad (1)$$
$$x^2 + xy + y^2 = \quad 67. \quad \cdots \cdots \quad (2)$$

Divide (1) by (2), $x^2 - xy + y^2 = \quad 39. \quad \cdots \cdots \quad (3)$

Add (2) and (3), $\qquad x^2 + y^2 = \quad 53.$

Subtract (3) from (2), $\qquad xy = \quad 14.$

$$\therefore \quad x = \pm 7, \text{ or } \pm 2,$$
$$y = \pm 2, \text{ or } \pm 7.$$

Solve the following:

4. $x^2 + xy = 84$,   *Ans.* $x = \pm 7$,
   $x^2 - y^2 = 24$. $y = \pm 5$.

5. $x^4 + x^2y^2 + y^4 = 133$, $x = \pm 3$, or ± 2,
   $x^2 + xy + y^2 = 19$. $y = \pm 2$, or ± 3.

6. $x^3 + y^3 = 407$, $x = 7$, or 4,
   $x + y = 11$. $y = 4$, or 7.

7. $x - y = 4$, $x = 11$, or -7,
   $x^3 - y^3 = 988$. $y = 7$, or -11.

145. Case IV. When the Two Equations are of the Second Degree and Homogeneous.

First method.

1. Solve
$$2x^2 + 3xy + y^2 = 70 \quad \cdot \quad \cdot \quad \cdot \quad \cdot \quad (1)$$
$$6x^2 + xy - y^2 = 50 \quad \cdot \quad \cdot \quad \cdot \quad \cdot \quad (2)$$

Divide (1) by (2), $\dfrac{2x^2 + 3xy + y^2}{6x^2 + xy - y^2} = \dfrac{7}{5}$.

Hence $10x^2 + 15xy + 5y^2 = 42x^2 + 7xy - 7y^2$;

or $32x^2 - 8xy - 12y^2 = 0$.

This is a quadratic equation which we may solve, and find the value of one unknown quantity in terms of the other. Thus, solving for x,
$$x^2 - \tfrac{1}{4}xy = \tfrac{3}{8}y^2.$$

By Art. 132, Rule II, $x = \dfrac{y}{8} \pm \sqrt{\tfrac{3}{8}y^2 + \dfrac{y^2}{64}} = \tfrac{25}{64}y^2$,

or $x = \dfrac{y \pm 5y}{8} = \tfrac{3}{4}y$, or $-\tfrac{1}{2}y$.

Take $x = \tfrac{3}{4}y$, and substitute in either (1) or (2), and we have $y^2 = 16$.

$\therefore \quad y = \pm 4$, and $x = \pm 3$.

If we substitute the second value of x, which is $-\tfrac{1}{2}y$, we find an inadmissible result.

Second method. Examples of this class are conveniently solved by substituting for one of the unknown quantities the product of the other and a third unknown quantity, called an *auxiliary* quantity.

2. Solve
$$2y^2 - 4xy + 3x^2 = 17 \quad \ldots \ldots \quad (1)$$
$$y^2 - x^2 = 16 \quad \ldots \ldots \quad (2)$$

Put $y = vx$, and substitute in (1) and (2). Thus
$$x^2(2v^2 - 4v + 3) = 17 \quad \ldots \ldots \quad (3)$$
$$x^2(v^2 - 1) = 16 \quad \ldots \ldots \quad (4)$$

By division, $\qquad \dfrac{2v^2 - 4v + 3}{v^2 - 1} = \frac{17}{16}.$

$$\therefore \quad 32v^2 - 64v + 48 = 17v^2 - 17;$$
or $\qquad 15v^2 - 64v = -65.$

Solving, we obtain $\qquad v = \frac{5}{3},$ or $\frac{13}{5}.$

Take $v = \frac{5}{3}$, and substitute in either (3) or (4).

From (4) $\quad x^2(\frac{25}{9} - 1) = 16;$ $\quad \therefore \quad x^2 = 9.$

$$\therefore \quad x = \pm 3, \quad \text{and} \quad y = vx = \frac{5}{3}x = \pm 5.$$

Again, take $v = \frac{13}{5}, x^2(\frac{169}{25} - 1) = 16;$ $\quad \therefore \quad x^2 = \frac{25}{9}.$

$$\therefore \quad x = \pm\frac{5}{3}, \quad \text{and} \quad y = vx = \pm\frac{13}{3}.$$

Any pair of equations which are *of the second degree and homogeneous*, can be solved by either of these methods, though the second is usually preferred.

Solve the following:

3. $x^2 + 3xy = 28,$ $\qquad$ *Ans.* $x = \pm 4,$ or $\mp 14,$

$\quad xy + 4y^2 = 8.$ $\qquad\qquad\qquad y = \pm 1,$ or $\pm 4.$

4. $x^2 + xy + 2y^2 = 74,$ $\qquad\qquad x = \pm 8,$ or $\pm 3,$

$\quad 2x^2 + 2xy + y^2 = 73.$ $\qquad\qquad y = \mp 5,$ or $\pm 5.$

5. $x^2 + xy - 6y^2 = 24,$ $\qquad\qquad\qquad x = \pm 6,$

$\quad x^2 + 3xy - 10y^2 = 32.$ $\qquad\qquad\qquad y = \pm 2.$

6. $x^2 + xy - 6y^2 = 21,$ $\qquad\qquad\qquad x = \pm 9,$

$\quad xy - 2y^2 = 4.$ $\qquad\qquad\qquad y = \pm 4.$

NOTE. — Many examples of homogeneous equations of the second degree are easily solved by Case II or III. Only those examples of this class are to be solved by Case IV that cannot be solved by either Case II or III.

146. Case V. When the Two Equations are Symmetrical with respect to x and y. — *An expression is said to be symmetrical with respect to two letters when these letters are similarly involved, i.e., when they can be interchanged without altering the expression.* Thus, the expression $a^3 + a^2x + ax^2 + x^3$ is symmetrical with respect to a and x, since if we write x for a, and a for x, we get the same expression. Also $x^4 + 3x^2y + 3xy^2 + y^4$ is symmetrical with respect to x and y.

Examples of this class may frequently be solved by substituting for the unknown quantities, the sum and difference of two others.

1. Solve
$$x^4 + y^4 = 82 \quad \ldots \ldots (1)$$
$$x - y = 2 \quad \ldots \ldots (2)$$

Put $x = u + v$, and $y = u - v$;

(2) becomes $(u + v) - (u - v) = 2;$ $\therefore$ $v = 1.$

(1) becomes $(u + 1)^4 + (u - 1)^4 = 82.$

$\therefore$ $2(u^4 + 6u^2 + 1) = 82;$

or $u^4 + 6u^2 - 40 = 0.$

Hence, Art. 135, $(u^2 + 10)(u^2 - 4) = 0.$

$\therefore$ $u^2 = 4$, or $-10.$

$\therefore$ $u = \pm2$, or $\pm\sqrt{-10}.$

Thus
$$x = 3, -1, \quad 1 \pm \sqrt{-10},$$
$$y = 1, -3, \quad -1 \pm \sqrt{-10}.$$

2. Solve
$$(x^2 + y^2)(x^3 + y^3) = 280 \quad \ldots \ldots (1)$$
$$x + y = 4 \quad \ldots \ldots (2)$$

Put $x = u + v$, and $y = u - v$;

(2) becomes $(u + v) + (u - v) = 4;$ $\therefore$ $u = 2.$

Also $x^2 + y^2 = (2 + v)^2 + (2 - v)^2 = 8 + 2v^2,$

and $x^3 + y^3 = (2 + v)^3 + (2 - v)^3 = 16 + 12v^2.$

Hence (1) becomes $(8 + 2v^2)(16 + 12v^2) = 280$,

or $\qquad\qquad (4 + v^2)(4 + 3v^2) = 35.$

$$\therefore \quad v^4 + \tfrac{16}{3}v^2 = \tfrac{19}{3}.$$

$$\therefore \quad v^2 = -\tfrac{8}{3} \pm \tfrac{11}{3} = 1, \text{ or } -\tfrac{19}{3}.$$

$$\therefore \quad v = \pm 1, \text{ or } \pm\sqrt{-\tfrac{19}{3}}.$$

$$\therefore \quad x = \quad 3, 1, 2 \pm \sqrt{-\tfrac{19}{3}},$$

$$y = \quad 1, 3, 2 \mp \sqrt{-\tfrac{19}{3}}.$$

Solve the following:

3. $x - y = 2$, and $x^5 - y^5 = 242$.

Ans. $x = 3$, or -1; $\ y = 1$, or -3.

4. $x - y = 1$, and $x^5 - y^5 = 781$.

Ans. $x = 4, -3$; $\ y = 3, -4$.

5. $x + y = 3$, and $x^5 + y^5 = 33$.

Ans. $x = 1, 2$; $\ y = 2, 1$.

147. Special Methods. — The preceding cases will be sufficient as a general explanation of the methods to be employed; but in some cases special artifices are necessary. One that is often used with advantage consists in considering the *sum*, *difference*, *product*, or *quotient*, of the two unknown quantities as a *single quantity*, and first finding its value. Other artifices may also be used with advantage, but familiarity with them can be obtained only by experience.

1. Solve $x^2 + 4xy + 3x = 40 - 6y - 4y^2$ . . . . (1)

$\qquad\qquad 2xy - x^2 = 3$ . . . . (2)

From (1) we have $x^2 + 4xy + 4y^2 + 3x + 6y = 40$;

or $\qquad (x + 2y)^2 + 3(x + 2y) - 40 = 0.$

Consider $x + 2y$ as a single unknown quantity, and find its value from this quadratic. Thus,

(Art. 135), $[(x + 2y) + 8][(x + 2y) - 5] = 0.$

$$\therefore \quad x + 2y = -8, \quad \ldots \ldots \ldots (3)$$

or $\qquad\qquad x + 2y = \quad 5 \quad \ldots \ldots \ldots (4)$

From (2) and (4) we obtain
$$x = 1, \text{ or } \tfrac{3}{2};$$
$$y = 2, \text{ or } \tfrac{7}{4}.$$

From (2) and (3) we obtain
$$x = \frac{-4 \pm \sqrt{10}}{2},$$
$$y = \frac{-12 \mp \sqrt{10}}{2}.$$

2. Solve $x^2y^2 - 6x = 34 - 3y$ . . . . (1)
$$3xy + y = 18 + 2x \quad . . . . (2)$$

Multiply (2) by 3 and subtract the result from (1),
$$x^2y^2 - 9xy + 20 = 0.$$
$$\therefore \ (xy - 5)(xy - 4) = 0.$$
$$\therefore \ xy = 5 \quad . . . . . (3)$$
$$xy = 4 \quad . . . . . (4)$$

From (2) and (3) we obtain
$$x = 1, \text{ or } -\tfrac{5}{2},$$
$$y = 5, \text{ or } -2.$$

From (2) and (4) we obtain
$$x = \frac{-3 \pm \sqrt{17}}{2}, \quad \text{and} \quad y = 3 \pm \sqrt{17}.$$

Solve the following:

3. $x^2 + y = 73 - 3x - 2xy,$ *Ans.* $x = 4, 16, -12 \pm \sqrt{58},$
   $y^2 + x = 44 - 3y.$ $y = 5, -7, -1 \mp \sqrt{58}.$

4. $\dfrac{x^2}{y^2} + \dfrac{4x}{y} = 12,$ $x = 6, \tfrac{18}{7},$
   $x - y = 3.$ $y = 3, -\tfrac{3}{7}.$

5. $x^2 + 3xy = 54,$ $x = \pm 3, \pm 36,$
   $xy + 4y^2 = 115.$ $y = \pm 5, \mp \tfrac{23}{2}.$

6. $x^4 - x^2 + y^4 - y^2 = 84,$ $x = \pm 3, \pm 2,$
   $x^2 + x^2y^2 + y^2 = 49.$ $y = \pm 2, \pm 3.$

148. Simultaneous Quadratic Equations with Three Unknown Quantities.

1. Solve

$$xy + xz = 27 \quad \ldots \ldots \ldots (1)$$
$$yz + yx = 32 \quad \ldots \ldots \ldots (2)$$
$$zx + zy = 35 \quad \ldots \ldots \ldots (3)$$

Add (1) and (2) and subtract (3) from the sum,

$$2xy = 24 ; \quad \therefore \quad xy = 12 \quad \ldots \ldots (4)$$

Subtract (4) from (1), $xz = 15 \quad \ldots \ldots \ldots (5)$

Subtract (4) from (2), $yz = 20 \quad \ldots \ldots \ldots (6)$

Multiply (4) and (5), $x^2yz = 180 \quad \ldots \ldots \ldots (7)$

Divide (7) by (6), $x^2 = 9 \quad \therefore \quad x = \pm 3.$

Hence from (4), $y = 12 \div (\pm 3) = \pm 4.$

And from (5), $z = 15 \div (\pm 3) = \pm 5.$

Thus $x = \pm 3, y = \pm 4, z = \pm 5,$

all the upper signs being taken together.

Solve the following:

2. $3yz + 2zx - 4xy = 16,$ *Ans.* $x = \pm 1,$

   $2yz - 3zx + xy = 5,$ $y = \pm 2,$

   $4yz - zx - 3xy = 15.$ $z = \pm 3.$

3. $6(x^2 + y^2 + z^2) = 13(x + y + z) = 4\frac{81}{6},$ $x = \frac{8}{3}, \frac{3}{2},$

   $xy = z^2.$ $y = \frac{3}{2}, \frac{8}{3},$

                                          $z = \pm 2.$

149. Problems Leading to Simultaneous Quadratic Equations.

1. The small wheel of a bicycle makes 135 revolutions more than the large wheel in a distance of 260 yards; if the circumference of each were one foot more, the small wheel would make 27 revolutions more than the large wheel in a distance of 70 yards: find the circumference of each wheel.

Let x = the circumference of the small wheel in feet,

and y = the circumference of the large wheel in feet.

Then the two wheels make $\dfrac{780}{x}$ and $\dfrac{780}{y}$ revolutions respectively in a distance of 260 yards.

Hence $\qquad \dfrac{780}{x} - \dfrac{780}{y} = 135$;

or $\qquad \dfrac{1}{x} - \dfrac{1}{y} = \tfrac{9}{52}$ . . . . . . (1)

Similarly from the second condition,

$$\dfrac{210}{x+1} - \dfrac{210}{y+1} = 27;$$

or $\qquad \dfrac{1}{x+1} - \dfrac{1}{y+1} = \tfrac{9}{70}$ . . . . . (2)

From (1), $x = \dfrac{52y}{9y+52}$; $\quad \therefore \quad x+1 = \dfrac{61y+52}{9y+52}$.

Substituting in (2), $\dfrac{9y+52}{61y+52} - \dfrac{1}{y+1} = \tfrac{9}{70}$.

$$\therefore \quad 9y^2 - 113y = 52.$$

Solving, we obtain $\quad y = 13$, or $-\tfrac{4}{9}$.

Substituting $y = 13$ in (1), we find $x = 4$. The negative value of y is inadmissible.

Hence the small wheel is 4 feet, and the large wheel is 13 feet in circumference.

2. A man starts from the foot of a mountain to walk to its summit; his rate of walking during the second half of the distance is half a mile per hour less than his rate during the first half, and he reaches the summit in $5\tfrac{1}{2}$ hours. He descends in $3\tfrac{3}{4}$ hours by walking at a uniform rate, which is one mile per hour more than his rate during the first half of the ascent: find the distance to the summit, and the rates of walking.

Let $2x =$ the number of miles to the summit,

and $\quad y =$ the rate of walking, in miles per hour, during the first half of the ascent.

Then $\dfrac{x}{y} =$ the time in hours for the first half of the ascent;

and $\dfrac{x}{y - \tfrac{1}{2}} =$ the time in hours for the second half of the ascent.

Hence $\qquad \dfrac{x}{y} + \dfrac{x}{y - \frac{1}{2}} = 5\frac{1}{2}$ . . . . . (1)

Similarly $\qquad \dfrac{2x}{y + 1} = 3\frac{3}{4}$ . . . . . . (2)

From (2) $\qquad x = \tfrac{15}{8}(y + 1)$ . . . . (3)

From (1) $\quad x(2y - \tfrac{1}{2}) = \tfrac{11}{2}y(y - \tfrac{1}{2})$ . . . . . . (4)

Substituting (3) in (4),

$$\tfrac{15}{8}(y + 1)(2y - \tfrac{1}{2}) = \tfrac{11}{2}y(y - \tfrac{1}{2}).$$

$$\therefore \quad 28y^2 - 89y = -15.$$

Solving, we obtain $\qquad y = 3$, or $\tfrac{5}{28}$.

Substituting $y = 3$ in (3), we find $x = \tfrac{15}{2}$. The other value of y is inapplicable, because by supposition y is greater than $\frac{1}{2}$.

Hence the whole distance to the summit is 15 miles, and the rates of walking are 3, $2\frac{1}{2}$, and 4 miles per hour.

3. The sum of the squares of two numbers is 170, and the difference of their squares is 72 : find the numbers.
                                               Ans. 11 ; 7.

4. The product of two numbers is 108, and their sum is twice their difference : find the numbers. *Ans.* 6 ; 18.

5. The product of two numbers is 6 times their sum, and the sum of their squares is 325 : find the numbers.
                                               Ans. 10 ; 15.

6. A certain rectangle contains 300 square feet ; a second rectangle is 8 feet shorter, and 10 feet broader, and also contains 300 square feet : find the length and breadth of the first rectangle. *Ans.* 20 ; 15.

7. Find two numbers such that their sum may be 39, and the sum of their cubes 17199. *Ans.* 15 and 24.

8. The product of two numbers is 750, and the quotient of one divided by the other is $3\frac{1}{3}$: find the numbers.
                                               Ans. 50 and 15.

EXAMPLES OF SIMULTANEOUS QUADRATICS.

NOTE. — In the great variety of simultaneous quadratic equations, it is impossible to give rules for every solution. The artifices employed in Algebraic work are very numerous. The student is cautioned not to go to work upon a pair of equations at random, but to study them until he sees how they can be reduced to a simpler equation by addition, multiplication, factoring, or by some other process, and then to perform the operations thus suggested.

Solve the following :

1. $x + 4y = 14,$
$y^2 + 4x = 2y + 11.$
Ans. $x = 2, -46,$
$y = 3, 15.$

2. $3x + 2y = 16,$
$xy = 10.$
$x = 2, \frac{10}{3},$
$y = 5, 3.$

3. $x + 2y = 9,$
$3y^2 - 5x^2 = 43.$
$x = 1, -\frac{71}{17},$
$y = 4, \frac{112}{17}.$

4. $3x - y = 11,$
$3x^2 - y^2 = 47.$
$x = 4, 7,$
$y = 1, 10.$

5. $4x + 9y = 12,$
$2x^2 + xy = 6y^2.$
$x = -24, \frac{6}{5},$
$y = 12, \frac{4}{5}.$

6. $3x + 2y = 5xy,$
$15x - 4y = 4xy.$
$x = \frac{2}{3}, 0,$
$y = \frac{3}{2}, 0.$

7. $x + y = 51,$
$xy = 518.$
$x = 37, 14,$
$y = 14, 37.$

8. $x - y = 18,$
$xy = 1075.$
$x = 43, -25,$
$y = 25, -43.$

9. $x^2 + y^2 = 89,$
$xy = 40.$
$x = \pm 8, \pm 5,$
$y = \pm 5, \pm 8.$

10. $x^2 + y^2 = 178,$
$x + y = 16.$
$x = 13, 3,$
$y = 3, 13.$

11. $x^2 + y^2 = 185,$
$x - y = 3.$
$x = 11, -8,$
$y = 8, -11.$

12. $x^2 - xy + y^2 = 76$,   *Ans.* $x = \quad 10, \quad 4,$
  $x + y = 14.$ $y = \quad 4, \quad 10.$

13. $x - y = 3,$ $x = \quad 7, - 4,$
  $x^2 - 3xy + y^2 = -19.$ $y = \quad 4, - 7.$

14. $\dfrac{1}{x^2} + \dfrac{1}{y^2} = \frac{481}{576}$, $x = \quad \frac{5}{6}, \quad \frac{3}{8},$
  $\dfrac{1}{x} + \dfrac{1}{y} = \frac{29}{24}.$ $y = \quad \frac{3}{8}, \quad \frac{5}{6}.$

15. $\dfrac{1}{x^2} + \dfrac{1}{y^2} = \frac{61}{900}$, $x = \pm\, 6, \pm\, 5,$
  $xy = 30.$ $y = \pm\, 5, \pm\, 6.$

16. $x^2 + y^2 + xy = 208,$ $x = \quad 12, \quad 4,$
  $x + y = 16.$ $y = \quad 4, \quad 12.$

17. $x^2 - y^2 = 16,$ $x = \quad 5,$
  $x - y = 2.$ $y = \quad 3.$

18. $x^3 - y^3 = 7xy,$ $x = \quad 4, - 2,$
  $x - y = 2.$ $y = \quad 2, - 4.$

19. $x + y = 23,$ $x = \quad 14, \quad 9,$
  $x^3 + y^3 = 3473.$ $y = \quad 9, \quad 14.$

20. $x + y = 35,$ $x = \quad 8, \quad 27,$
  $x^{\frac{1}{3}} + y^{\frac{1}{3}} = 5.$ $y = \quad 27, \quad 8.$

21. $x - y = \sqrt{x} + \sqrt{y},$ $x = \quad 16, \quad 9,$
  $x^{\frac{3}{2}} - y^{\frac{3}{2}} = 37.$ $y = \quad 9, \quad 16.$

22. $x^4 + x^2y^2 + y^4 = 2923,$ $x = \pm\, 7, \pm\, 3,$
  $x^2 - xy + y^2 = 37.$ $y = \pm\, 3, \pm\, 7.$

23. $x^4 + x^2y^2 + y^4 = 9211,$ $x = \pm\, 9, \pm\, 5,$
  $x^2 - xy + y^2 = 61.$ $y = \pm\, 5, \pm\, 9.$

24. $x^3 - y^3 = 56,$ $x = \quad 4, - 2,$
  $x^2 + xy + y^2 = 28.$ $y = \quad 2, - 4.$

25. $x^3 + y^3 = 126,$ $x = \quad 5, \quad 1,$
  $x^2 - xy + y^2 = 21.$ $y = \quad 1, \quad 5.$

26. $\dfrac{1}{x^3} + \dfrac{1}{y^3} = 1\frac{1}{125}$,   *Ans.* $x =$ 5,   1,

$\dfrac{1}{x} + \dfrac{1}{y} = 1\frac{1}{5}$. $y =$ 1,   5.

27. $x^2 + xy = 45,\ y^2 + xy = 36$. $x = \pm\ 5;\ y = \pm 4$.

28. $2x^2 - xy = 56,\ 2xy - y^2 = 48$. $x = \pm\ 7;\ y = \pm 6$.

29. $x^2 - 2xy = 15,\ xy - 2y^2 = 7$. $x = \pm 15;\ y = \pm 7$.

30. $x^2y(x+y) = 80,\ x^2y(2x-3y) = 80$. $x = \pm\ 4;\ y = \pm 1$.

31. $x^3 + 1 = 9y$, $x = 2, \frac{1}{2}, -1$,

$x^2 + x = 6y$. $y = 1, \frac{1}{8},\ \ 0$.

NOTE. — It will be seen that Examples 17 to 31 can be solved by Case III.

32. $x^2 + 3xy = 54$,   *Ans.* $x = \pm 3, \pm 36$,

$xy + 4y^2 = 115$. $y = \pm 5, \mp \frac{23}{2}$.

33. $x^2 + xy = 24$, $x = \pm 4, \pm 6\sqrt{-2}$,

$2y^2 + 3xy = 32$. $y = \pm 2, \mp 8\sqrt{-2}$.

34. $x^2 - 3xy = 10$, $x = \pm 5, \pm 4$,

$4y^2 - xy = -1$. $y = \pm 1, \pm \frac{1}{2}$.

35. $x^2 + xy - 2y^2 = -44$, $x = \pm 14, \pm 1$,

$xy + 3y^2 =\ \ \ 80$. $y = \mp\ 8, \pm 5$.

36. $x^2 + 3xy =\ \ 54$, $x = \pm 3, \pm 36$,

$xy + 4y^2 = 115$. $y = \pm 5, \mp 11\frac{1}{2}$.

37. $x(x + y) = 40$, $x = \pm 5, \pm 4\sqrt{2}$,

$y(x - y) =\ \ 6$. $y = \pm 3, \pm\ \sqrt{2}$.

38. $x(x+y) + y(x-y) = 158$, $x = \pm 9, \pm 8\sqrt{2}$,

$7x(x+y) =\ \ 72y(x-y)$. $y = \pm 7, \pm\ \sqrt{2}$.

39. $x + y = 4,\ x^4 + y^4 =\ \ \ 82$. $x = 3, 1;\ y = 1, 3$.

40. $x - y = 3,\ x^5 - y^5 = 3093$, $x = 5, -2;\ y = 2, -5$.

41. $x^2y^2 + 13xy + 12 = 0$, $x =\ \ \ 4, -3, \dfrac{1 \pm \sqrt{5}}{2}$.

$x + y = 1$. $y = -3,\ \ \ 4, \dfrac{1 \mp \sqrt{5}}{2}$.

42. $x + y = 5,$ Ans. $x = 6, -1, \dfrac{5 \pm \sqrt{17}}{2},$

$\qquad 4xy = 12 - x^2y^2.$ $y = -1, 6, \dfrac{5 \mp \sqrt{17}}{2}.$

43. $x^2y^2 + 5xy = 84,$ $x = 7, 1, 4 \pm \sqrt{28},$

$\qquad x + y = 8.$ $y = 1, 7, 4 \mp \sqrt{28}.$

44. $x^2 + 4y^2 + 80 = 15x + 30y,$ $x = 4, 3, 6, 2,$

$\qquad xy = 6.$ $y = \frac{3}{2}, 2, 1, 3.$

45. $9x^2 + y^2 - 63x - 21y + 128 = 0,$ $x = 2, \frac{2}{3}, 4, \frac{1}{3},$

$\qquad xy = 4.$ $y = 2, 6, 1, 12.$

46. $x^4 + y^4 = 14x^2y^2,$ $x = \dfrac{a}{2}(1 \pm \sqrt{3}), \dfrac{a}{2}\left(1 \pm \dfrac{1}{\sqrt{3}}\right),$

$\qquad x + y = a.$ $y = \dfrac{a}{2}(1 \mp \sqrt{3}), \dfrac{a}{2}\left(1 \mp \dfrac{1}{\sqrt{3}}\right).$

47. Find two numbers whose difference added to the difference of their squares is 14, and whose sum added to the sum of their squares is 26.   *Ans.* 4, 2.

48. Find two numbers such that twice the first with three times the second may make 60, and twice the square of the first with three times the square of the second may make 840.   *Ans.* 18 and 8, or 6 and 16.

49. Find two numbers whose sum is nine times their difference, and whose product diminished by the greater number is equal to twelve times the greater number divided by the less.   *Ans.* 5, 4.

50. Find two numbers whose difference multiplied by the difference of their squares is 32, and whose sum multiplied by the sum of their squares is 272.   *Ans.* 5, 3.

51. Find two numbers whose product is equal to their sum, and whose sum added to the sum of their squares is 12.   *Ans.* 2, 2.

52. Find two numbers whose sum added to their product is 34, and the sum of whose squares diminished by their sum is 42.   *Ans.* 4, 6.

53. A number consisting of two digits has one decimal place; the difference of the squares of the digits is 20, and if the digits be reversed, the sum of the two numbers is 11 : find the number.   *Ans.* 6.4, or 4.6.

54. A man has to travel a certain distance ; and when he has traveled 40 miles he increases his speed 2 miles per hour. If he had traveled with his increased speed during the whole journey he would have arrived 40 minutes earlier ; but if he had continued at his original speed he would have arrived 20 minutes later. Find the whole distance he had to travel, and his original speed.   *Ans.* 60, 10.

55. A and B are two towns situated 18 miles apart on the same bank of a river. A man goes from A to B in 4 hours, by rowing the first half of the distance and walking the second half. In returning he walks the first half at the same rate as before, but the stream being with him, he rows $1\frac{1}{2}$ miles per hour more than in going, and accomplishes the whole distance in $3\frac{1}{2}$ hours. Find his rates of walking and rowing.   *Ans.* $4\frac{1}{2}$ walking, $4\frac{1}{2}$ rowing at first.

56. A and B run a race round a two mile course. In the first heat B reaches the winning post 2 minutes before A. In the second heat A increases his speed 2 miles per hour, and B diminishes his as much ; and A then arrives at the winning post 2 minutes before B. Find at what rate each man ran in the first heat.   *Ans.* 10, 12 miles per hour.

57. Find two numbers whose product is equal to the difference of their squares, and the sum of their squares equal to the difference of their cubes.   *Ans.* $\frac{1}{2}\sqrt{5}$, $\frac{1}{4}(5 + \sqrt{5})$.

58. The fore-wheel of a coach makes 6 revolutions more than the hind-wheel in going 120 yards ; but if the circumference of each wheel be increased 1 yard, the fore-wheel will make only 4 revolutions more than the hind-wheel in the same distance : find the circumference of each wheel.

Ans 4 and 5 yards.

CHAPTER XV.

RATIO—PROPORTION—VARIATION.

150. Ratio — Definitions. — The *relative magnitude* of two quantities, measured by the number of times which the first contains the second, is called their *Ratio*.

The ratio of a to b is usually written $a:b$; a is called the first *term*, and b the second *term* of the ratio. The first term is often called the *antecedent*, and the second term the *consequent*.

Magnitudes must always be expressed by means of *numbers;* and the *number of times* which one number contains the other is found by dividing the one by the other. Hence the ratio $a:b$ may be measured by the fraction $\frac{a}{b}$.

Thus, the ratio $a:b$ is equal to $\frac{a}{b}$, or is $\frac{a}{b}$.

Concrete quantities of different kinds can have no ratio to one another; thus, we cannot compare pounds with yards, or dollars with days.

To compare two quantities, they must be expressed in terms of the same unit. For example, the ratio of 4 yards to 15 inches is measured by the fraction

$$\frac{4 \times 3 \times 12}{15} = \tfrac{48}{5}.$$

A ratio is called a ratio of *greater inequality*, of *less inequality*, or of *equality*, according as the antecedent is *greater than, less than*, or *equal to* the consequent.

Ratios are *compounded* by multiplying together the antecedents of the given ratios for a new antecedent, and the consequents for a new consequent. Thus, the ratio compounded of the three ratios,

$$3a : 2b, \ 4ab : 5c^2, \ c : a,$$

is $3a \times 4ab \times c : 2b \times 5c^2 \times a$, or $6a : 5c.$

When the ratio $a : b$ is compounded with itself, the resulting ratio is $a^2 : b^2$, and is called the *duplicate ratio* of $a : b$. Similarly, the ratio $a^3 : b^3$ is called the *triplicate ratio* of $a : b$. Also the ratio $a^{\frac{1}{2}} : b^{\frac{1}{2}}$ is called the *subduplicate ratio* of $a : b$.

If we interchange the terms of a ratio, the result is called the *inverse ratio*. Thus

$$b : a \text{ is the inverse of } a : b.$$

The inverse ratio is the reciprocal of the direct ratio.

When the ratio of two quantities can be exactly expressed by the ratio of two integers, the quantities are said to be *commensurable;* when the ratio *cannot* be exactly expressed by the ratio of two integers, they are said to be *incommensurable.*

Although we cannot find two integers which will *exactly* measure the ratio of two incommensurable quantities, yet we can always find two integers whose ratio differs from the required ratio by as small a quantity as we please.

For example, the ratio of a diagonal to a side of a square cannot be exactly expressed by the ratio of two whole numbers, for this ratio is $\sqrt{2}$, and we cannot find any fraction which is *exactly* equal to $\sqrt{2}$; but by taking a sufficient number of decimals, we may find $\sqrt{2}$ to any required degree of approximation. Thus

$$\sqrt{2} = 1.4142135 \ \ldots \ \ldots$$

and therefore $\sqrt{2} > \frac{1414213}{1000000}$ and $< \frac{1414214}{1000000}.$

That is, the ratio of a diagonal to a side of a square lies between $\frac{1414213}{1000000}$ and $\frac{1414214}{1000000}$, and therefore differs from either of these ratios by less than one-millionth; and since the decimals may be continued without end in extracting the square root of 2, it is evident that this ratio can be expressed as a fraction with an error *less than any assignable quantity.*

In general. When a and b are incommensurable, divide b into n equal parts each equal to x, so that $b = nx$, where n is a positive integer. Also let $a > mx$, but $< (m + 1)x$; then

$$\frac{a}{b} > \frac{mx}{nx} \text{ and } < \frac{(m + 1)x}{nx};$$

that is, $\frac{a}{b}$ lies between $\frac{m}{n}$ and $\frac{m + 1}{n}$; so that $\frac{a}{b}$ differs from $\frac{m}{n}$ by a quantity less than $\frac{1}{n}$. And since we can choose x (our unit of measure) as small as we please, n can be made as great as we please, and therefore $\frac{1}{n}$ can be made as small as we please. Hence two integers, m and n, can be found whose ratio will express the ratio $a : b$ to any required degree of accuracy.

Note. — The student should observe that the *Algebraic* definition of ratio deals with numbers, or with magnitudes represented by numbers, while the *Geometric* definition of ratio deals with *concrete* magnitudes, such as lines or areas represented Geometrically, but not referred to any common unit of measure.

151. Properties of Ratios. — (1) *If the terms of a ratio be multiplied or divided by the same number the value of the ratio is unaltered.*

For $$\frac{a}{b} = \frac{ma}{mb} \text{ (Art. 79).}$$

Thus the ratios $2 : 3$, $6 : 9$, and $2m : 3m$, are all equal to each other.

Two or more ratios are compared by reducing the fractions which measure them to a common denominator. Thus, suppose $a : b$ and $c : d$ are two ratios. Then $\frac{a}{b} = \frac{ad}{bd}$, $\frac{c}{d} = \frac{bc}{bd}$; hence the ratio $a : b >$, $=$, or $<$ the ratio $c : d$, according as $ad >$, $=$, or $< bc$.

The ratio of two fractions can be expressed as a ratio of two integers.

Thus the ratio $\dfrac{a}{b} : \dfrac{c}{d}$ is measured by the fraction $\dfrac{\frac{a}{b}}{\frac{c}{d}}$ or $\dfrac{ad}{bc}$; and is therefore equivalent to the ratio $ad : bc$.

(2) *A ratio of greater inequality is diminished, and a ratio of less inequality is increased, by adding the same quantity to each of its terms; that is, the ratio is made more nearly equal to unity.*

Let $a : b$ be the ratio, and let $a + x : b + x$ be the new ratio formed by adding x to each of its terms.

Then
$$\frac{a}{b} - \frac{a + x}{b + x} = \frac{x(a - b)}{b(b + x)};$$

and $a - b$ is positive or negative according as a is greater or less than b.

Hence $\dfrac{a + x}{b + x} <$, or $> \dfrac{a}{b}$ according, as $a >$, or $< b$; that is, the resulting ratio is brought nearer to unity.

For example, if to each term of the ratio $3 : 2$ we add 12, the new ratio $15 : 14$ is less than the former, because $\frac{15}{14} = 1\frac{1}{14}$ is clearly less than $\frac{3}{2} = 1\frac{1}{2}$.

Also, if to each term of the ratio $2 : 3$ we add 12, the new ratio $14 : 15$ is greater than the former, since $\frac{14}{15}$ is clearly greater than $\frac{2}{3}$.

(3) Similarly, it can be proved that *a ratio of greater inequality is increased, and a ratio of less inequality is diminished, by taking the same quantity from both its terms.*

(4) *The following is a very important proposition concerning equal ratios.*

If $\dfrac{a}{b} = \dfrac{c}{d} = \dfrac{e}{f} = \ldots \ldots$, then each of these ratios

$$= \left(\frac{pa^n + qc^n + re^n + \ldots}{pb^n + qd^n + rf^n + \ldots} \right)^{\frac{1}{n}},$$

where p, q, r, n are any quantities whatever.

Let $\dfrac{a}{b} = \dfrac{c}{d} = \dfrac{e}{f} = \ldots \ldots = k$;

then $\quad a = bk,\, c = dk,\, e = fk, \ldots \ldots$;

therefore $pa^n + qc^n + re^n + \ldots = pb^n k^n + qd^n k^n + rf^n k^n + \ldots$

$\therefore \left(\dfrac{pa^n + qc^n + re^n + \ldots}{pb^n + qd^n + rf^n + \ldots}\right)^{\frac{1}{n}} = k = \dfrac{a}{b} = \dfrac{c}{d} = \dfrac{e}{f} = \quad (1)$

By giving different values to $p,\, q,\, r,\, n$, many particular cases of this general proposition may be deduced; or they may be proved independently by the above method.

Suppose $n = 1$, then we have from (1)

$$\dfrac{a}{b} = \dfrac{c}{d} = \dfrac{e}{f} = \ldots \ldots = \dfrac{pa + qc + re + \ldots}{pb + qd + rf + \ldots} \quad . \quad (2)$$

Suppose $n = 1$, and $p = q = r = \ldots \ldots$, then (1) becomes

$$\dfrac{a}{b} = \dfrac{c}{d} = \dfrac{e}{f} = \ldots = \dfrac{a + c + e \ldots}{b + d + f \ldots} \quad . \quad (3)$$

That is, *when a series of fractions are equal, each of them is equal to the sum of all the numerators divided by the sum of all the denominators.*

EXAMPLES.

1. If $\dfrac{x}{y} = \frac{3}{4}$, find the value of $\dfrac{5x - 3y}{7x + 2y}$.

$$\dfrac{5x - 3y}{7x + 2y} = \dfrac{\dfrac{5x}{y} - 3}{\dfrac{7x}{y} + 2} = \dfrac{\frac{15}{4} - 3}{\frac{21}{4} + 2} = \tfrac{3}{29}.$$

2. If $a : b$ be in the duplicate ratio of $a + x : b + x$, prove that $x^2 = ab$.

From the condition $\left(\dfrac{a + x}{b + x}\right)^2 = \dfrac{a}{b}$,

$\therefore \quad a^2 b + 2abx + bx^2 = ab^2 + 2abx + ax^2.$

$\therefore \quad x^2 = ab.$

Find the ratio compounded of

3. The ratios $4:15$ and $25:36$.   *Ans.* $5:27$.

4. The ratio $27:8$, and the duplicate ratio of $4:3$. $6:1$.

5. The ratio $169:200$, and the duplicate ratio of $15:26$.

                           Ans. $9:32$.

6. If $4x^2 + y^2 = 4xy$, find the ratio $x:y$. $1:2$.

7. What is the ratio $x:y$, if the ratio $4x + 5y : 3x - y$ is equal to 2?   *Ans.* $7:2$.

8. If $7x - 4y : 3x + y = 5:13$, find the ratio $x:y$.

                           Ans. $3:4$.

PROPORTION.

152. Definitions. — Four quantities are said to be in *proportion* when the ratio of the first to the second is equal to the ratio of the third to the fourth; and the terms of the ratios are said to be *proportionals*.

Thus, if $\dfrac{a}{b} = \dfrac{c}{d}$, then a, b, c, d, are called proportionals, or are said to be in proportion. The proportion is written

$$a:b = c:d,$$

or $$a:b :: c:d,$$

which is read " a is to b as c is to d."

The Algebraic test of a proportion is that the two fractions which represent the ratios shall be equal.

The four terms of the two equal ratios are called the *terms* of the proportion. The first and fourth terms are called the *extremes*, and the second and third, the *means*. Thus, in the above proportion, a and d are the extremes and b and c the means.

Quantities are said to be in *continued proportion* when the first is to the second, as the second is to the third, as the third to the fourth, and so on. Thus a, b, c, d, e, f, . . . are in continued proportion when

$$a:b = b:c = c:d = d:e = e:f = \ldots \ldots$$

If a, b, c, be in continued proportion, b is said to be a *mean proportional* between a and c; and c is said to be a *third proportional* to a and b.

If a, b, c, d be in continued proportion, b and c are said to be *two mean proportionals* between a and d; and so on.

153. Properties of Proportions. — (1) *If four quantities are in proportion, the product of the extremes is equal to the product of the means.*

Let the proportion be $a : b = c : d$.

Then by definition (Art. 156), $\dfrac{a}{b} = \dfrac{c}{d}$.

Multiplying by bd, $\qquad ad = bc$ . . . . (1)

Hence if any three terms of a proportion are given, the fourth may be found from the relation $ad = bc$.

NOTE. — This proposition furnishes a more convenient test of a proportion than the one in Art. 152. Thus, to ascertain whether $2 : 5 :: 6 : 16$, it is only necessary to compare the product of the means and extremes; and since 5×6 is not equal to 2×16, we see that 2, 5, 6, 16, are *not* in proportion.

If $b = c$, we have from (1), $ad = b^2$; $\therefore b = \sqrt{ac}$.

That is, *the mean proportional between two given quantities is equal to the square root of their product.*

(2) Conversely, *If the product of two quantities be equal to the product of two others, two of them may be made the extremes, and the other two the means, of a proportion.*

For let $\qquad\qquad ad = bc$.

Dividing by bd, $\qquad\qquad \dfrac{a}{b} = \dfrac{c}{d}$;

that is, $\qquad\qquad a : b :: c : d$.

In a similar manner it may be shown that the proportions

$$a : c :: b : d,$$
$$b : a :: d : c,$$
$$b : d :: a : c,$$
$$c : d :: a : b, \text{ etc.,}$$

are all true provided that $ad = bc$.

If four quantities are in proportion they will be in proportion by

(3) *Inversion.* — If $a : b :: c : d$, then $b : a :: d : c$.

For $\dfrac{a}{b} = \dfrac{c}{d}$; therefore $1 \div \dfrac{a}{b} = 1 \div \dfrac{c}{d}$;

that is, $\dfrac{b}{a} = \dfrac{d}{c}$; or $b : a :: d : c$.

(4) *Alternation.* — If $a : b :: c : d$, then $a : c :: b : d$.

For $ad = bc$; therefore $\dfrac{ad}{cd} = \dfrac{bc}{cd}$;

that is $\dfrac{a}{c} = \dfrac{b}{d}$; or $a : c :: b : d$.

(5) *Composition.* — If $a : b :: c : d$, then $a+b : b :: c+d : d$.

For $\dfrac{a}{b} = \dfrac{c}{d}$; therefore $\dfrac{a}{b} + 1 = \dfrac{c}{d} + 1$;

that is $\dfrac{a + b}{b} = \dfrac{c + d}{d}$; or $a + b : b :: c + d : d$.

(6) *Division.* — If $a : b :: c : d$, then $a - b : b :: c - d : d$.

For $\dfrac{a}{b} = \dfrac{c}{d}$; therefore $\dfrac{a}{b} - 1 = \dfrac{c}{d} - 1$;

that is $\dfrac{a - b}{b} = \dfrac{c - d}{d}$; or $a - b : b :: c - d : d$.

In a similar manner it may be shown that the sum (or the difference) of the first and second of two quantities is to the *first* as the sum (or the difference) of the third and fourth is to the *third*.

(7) *Composition and Division.* — If $a : b :: c : d$, then

$$a + b : a - b :: c + d : c - d.$$

For by (5) and (6),

$$\frac{a + b}{b} = \frac{c + d}{d}, \qquad \frac{a - b}{b} = \frac{c - d}{d};$$

by division,

$$\frac{a + b}{a - b} = \frac{c + d}{c - d};$$

or

$$a + b : a - b :: c + d : c - d.$$

(8) *If three quantities are in continued proportion, the first is to the third in the duplicate ratio of the first to the second.*

For if $a:b::b:c$, then $\dfrac{a}{b}=\dfrac{b}{c}$.

Now $\dfrac{a}{c}=\dfrac{a}{b}\times\dfrac{b}{c}=\dfrac{a}{b}\times\dfrac{a}{b}=\dfrac{a^2}{b^2}$.

Hence $a:c::a^2:b^2$.

Similarly it may be shown that if $a:b::b:c::c:d$, then $a:d::a^3:b^3$.

(9) *Quantities which are proportional to the same quantities, are proportional to each other.*

If $a:b::e:f$, and $c:d::e:f$, then $a:b::c:d$.

For $\dfrac{a}{b}=\dfrac{e}{f}$, and $\dfrac{c}{d}=\dfrac{e}{f}$; therefore $\dfrac{a}{b}=\dfrac{c}{d}$,

or $a:b::c:d$.

(10) *The products of the corresponding terms of two or more proportions are in proportion.*

For if $a:b::c:d$, and $e:f::g:h$,

then $\dfrac{a}{b}=\dfrac{c}{d}$, and $\dfrac{e}{f}=\dfrac{g}{h}$;

therefore $\dfrac{ae}{bf}=\dfrac{cg}{dh}$; or $ae:bf::cg:dh$.

(11) *When four quantities are in proportion, if the first and second be multiplied, or divided, by any quantity, as also the third and fourth, the resulting quantities will be in proportion.*

For if $a:b::c:d$, then $\dfrac{a}{b}=\dfrac{c}{d}$;

therefore $\dfrac{ma}{mb}=\dfrac{nc}{nd}$; or $ma:mb::nc:nd$.

Similarly it may be shown that $\dfrac{a}{m}:\dfrac{b}{m}::\dfrac{c}{n}:\dfrac{d}{n}$.

(12) In a similar manner it may be shown that *if the first and third terms be multiplied, or divided, by any quantity, and also the second and fourth, the resulting quantities will be in proportion.*

(13) *If four quantities are in proportion, the like powers, or roots, of these quantities will be in proportion.*

For if $a : b :: c : d$, then $\dfrac{a}{b} = \dfrac{c}{d}$;

therefore $\quad \dfrac{a^n}{b^n} = \dfrac{c^n}{d^n}$; $\quad \therefore \ a^n : b^n :: c^n : d^n.$

Also $\quad \dfrac{a^{\frac{1}{n}}}{b^{\frac{1}{n}}} = \dfrac{c^{\frac{1}{n}}}{d^{\frac{1}{n}}}$; $\quad \therefore \ a^{\frac{1}{n}} : b^{\frac{1}{n}} :: c^{\frac{1}{n}} : d^{\frac{1}{n}}.$

(14) *If any number of quantities are in proportion, any antecedent is to its consequent, as the sum of all the antecedents is to the sum of all the consequents.*

For if $a : b :: c : d :: e : f$,

then by (1), $ad = bc$, and $af = be$; also $ab = ba$.

Adding $\quad a(b + d + f) = b(a + c + e)$;

therefore by (2), $a : b :: a + c + e : b + d + f.$

This also follows directly from (3) of Art. 151.

(15) When $\dfrac{a}{b}, \dfrac{c}{d}, \dfrac{e}{f}, \ \ldots \ldots$ are unequal, it follows from Ex. 3 of Art. 106, that

$$\frac{a + c + e + g + \ldots \ldots}{b + d + f + h + \ldots \ldots}$$

is greater than the least, and less than the greatest, of the fractions $\dfrac{a}{b}, \dfrac{c}{d}, \dfrac{e}{f}, \dfrac{g}{h}, \ \ldots \ldots$

It is obvious from the preceding propositions that if four quantities are in proportion, many other proportions may be derived from them. The propositions just proved are often useful in solving problems. In particular, the solution of certain equations is greatly facilitated by a skilful use of the operations of *composition* and *division*.

EXAMPLES.

1. If $a : b :: c : d$,

show that $a^2 + ab : c^2 + cd :: b^2 - 2ab : d^2 - 2cd$.

Let $\dfrac{a}{b} = \dfrac{c}{d} = x$; then $a = bx$, and $c = dx$.

$$\therefore \quad \frac{a^2 + ab}{c^2 + cd} = \frac{b^2x^2 + b^2x}{d^2x^2 + d^2x} = \frac{b^2}{d^2}.$$

Also
$$\frac{b^2 - 2ab}{d^2 - 2cd} = \frac{b^2 - 2b^2x}{d^2 - 2d^2x} = \frac{b^2}{d^2}.$$

Therefore by (9), $a^2 + ab : c^2 + cd :: b^2 - 2ab : d^2 - 2cd$.

2. If $\dfrac{3a + 6b + c + 2d}{3a - 6b + c - 2d} = \dfrac{3a + 6b - c - 2d}{3a - 6b - c + 2d}$,

prove that $\qquad a : b :: c : d$.

By (7) $\qquad \dfrac{2(3a + c)}{2(6b + 2d)} = \dfrac{2(3a - c)}{2(6b - 2d)}$.

By (4) $\qquad \dfrac{3a + c}{3a - c} = \dfrac{6b + 2d}{6b - 2d}$.

Again by (7), $\qquad \dfrac{6a}{2c} = \dfrac{12b}{4d}$; $\quad \therefore \quad a : b :: c : d$.

3. Find a fourth proportional to x^3, xy, $5x^2y$.   *Ans.* $5y^2$.

4. Find a mean proportional between $12ax^2$ and $3a^3$.

       Ans. $6a^2x$.

5. Find a third proportional to x^3 and $2x^2$. $4x$.

6. If $a : b :: c : d$, show that

   (1) $ac : bd :: c^2 : d^2$.

   (2) $ab : cd :: a^2 : c^2$.

   (3) $a^2 : c^2 :: a^2 - b^2 : c^2 - d^2$.

7. If $a : b :: c : d$, prove that

(1) $\qquad ab + cd : ab - cd :: a^2 + c^2 : a^2 - c^2$.

(2) $a^2 + ac + c^2 : a^2 - ac + c^2 :: b^2 + bd + d^2 : b^2 - bd + d^2$.

(3) $\qquad a : b :: \sqrt{3a^2 + 5c^2} : \sqrt{3b^2 + 5d^2}$.

(4) $\qquad a + b : c + d :: \sqrt{a^2 + b^2} : \sqrt{c^2 + d^2}$.

8. If $a : b :: c : d :: e : f$, prove that
$$2a^2 + 3c^2 - 5e^2 : 2b^2 + 3d^2 - 5f^2 :: ae : bf.$$

9. Solve the equation $\dfrac{x^2 + x - 2}{x - 2} = \dfrac{4x^2 + 5x - 6}{5x - 6}$.

Ans. $x = 0, -2$.

10. Find x in terms of y from the proportions $x : y :: a^3 : b^3$, and $a : b :: \sqrt[3]{c + x} : \sqrt[3]{d + y}$.

Ans. $x = \dfrac{c}{d}y$.

11. If a, b, c, d are in continued proportion, prove that
$$a : d :: a^3 + b^3 + c^3 : b^3 + c^3 + d^3.$$

VARIATION.

154. Definition. — One quantity is said to *vary directly* as another when the two quantities depend upon each other in such a manner that if one be changed the other is changed *in the same proportion*.

Thus, if a train moving uniformly, travels 40 miles in an hour, it will travel 80 miles in 2 hours, 120 miles in 3 hours, and so on; the distance in each case being increased or diminished in the same ratio as the time. This is expressed by saying that when the velocity is uniform, *the distance is proportional to the time*, or more briefly, *the distance varies as the time*. We may express this result with Algebraic symbols thus: let A and a be the *numbers* which represent the distances traveled by the train in the times represented by the numbers B and b; that is, when A is changed to any other value a, B must be changed to another value b, so that $A : a :: B : b$; then A is said *to vary directly as B*, or more briefly, *to vary as B*.

Another phrase,* which is also in use, is "*A is proportional to B.*"

---

* Strictly speaking, this phrase is better than the one "varies as," which is somewhat antiquated; but in deference to usage we retain it. The student must not suppose that the variation here considered is the only kind. We are not here concerned with *variation in general*, but merely with the simplest of all the possible kinds of variation.

This relation is sometimes expressed by the symbol $\propto$, so that $A \propto B$ is read "*A* varies as *B*."

It will thus be seen that variation is merely an abridged method of expressing proportion, and that *four* quantities are understood though only *two* are expressed.

If A varies as B, then A is equal to B multiplied by some constant quantity.

For suppose that a, a_1, a_2, ... , b, b_1, b_2, ... are corresponding values of A and B.

Let a and b denote one pair of these values, so that when A has the value a, B has the value b; then we have by the definition, $A : a :: B : b$. Hence

$$A = \frac{a}{b}B = mB,$$

where m is equal to the constant ratio $a : b$.

155. Different Cases of Variation. — *There are four different kinds of variation.*

(1) One quantity is said to vary *Directly* as another when the two increase or decrease together in the same ratio. Thus,

$$A \propto B, \text{ or } A = mB \text{ (Art. 154)}.$$

For example, If a man works for a certain sum per hour, the amount of his wages *varies* as the number of hours during which he works.

(2) One quantity is said to vary *Inversely* as another when the first varies as the *reciprocal* of the other. Thus *A varies inversely as B* is written

$$A \propto \frac{1}{B}, \text{ or } A = \frac{m}{B}, \text{ where } m \text{ is a constant.}$$

For example, If a man has to perform a certain journey, the *time* in which he will perform it varies *inversely* as his speed. If he *doubles* his speed, he will go in *half* the time; and so on.

(3) One quantity is said to vary as two others *Jointly*, when the first varies as the product of the other two. Thus *A varies as B and C jointly* is written

$$A \propto BC, \text{ or } A = mBC, \text{ where } m \text{ is a constant.}$$

For example, The wages to be received by a workman will vary as the *number* of days he has worked and the *wages per day* jointly.

(4) One quantity is said to vary *Directly as a second and Inversely as a third*, when it varies jointly as the second and the reciprocal of the third. Thus *A varies directly as B and inversely as C* is written

$$A \propto \frac{B}{C}, \text{ or } A = m\frac{B}{C}, \text{ where } m \text{ is a constant.}$$

For example, The base of a triangle varies *directly* as the area and *inversely* as the altitude.

In the different cases of variation just defined, to determine the constant *m* it will only be necessary to have given one set of corresponding values.

Example 1. If $A \propto B$, and $A = 3$ when $B = 12$, we have

$$A = mB; \quad \therefore \quad 3 = m \times 12;$$

or $\quad\quad\quad m = \frac{1}{4}; \quad \therefore \quad A = \frac{1}{4}B.$

2. If A varies as B and inversely as C, and $A = 6$ when $B = 2$ and $C = 9$, we have

$$A = m\frac{B}{C}; \quad \therefore \quad 6 = m \times \tfrac{2}{9};$$

or $\quad\quad\quad m = 27; \quad \therefore \quad A = 27\frac{B}{C}.$

156. Propositions in Variation. — The simplest method of treating *variations* is to convert them into *equations*.

(1) If $A \propto B$, and $B \propto C$, then $A \propto C$.

For let $\quad A = mB$, and $B = nC$ (Art. 154),

where *m* and *n* are constants.

Then $$A = mnC;$$

$$\therefore \quad A \propto C, \text{ since } mn \text{ is constant.}$$

In like manner, if $A \propto B$, and $B \propto \dfrac{1}{C}$, then $A \propto \dfrac{1}{C}$.

(2) If $A \propto C$, and $B \propto C$, then $A \pm B \propto C$, and $\sqrt{AB} \propto C$.

For let $$A = mC, \text{ and } B = nC,$$

where m and n are constants.

Then $$A \pm B = (m \pm n)C.$$

$$\therefore \quad A \pm B \propto C, \text{ since } m \pm n \text{ is constant.}$$

Also $$\sqrt{AB} = \sqrt{mnC^2} = C\sqrt{mn}.$$

$$\therefore \quad \sqrt{AB} \propto C, \text{ since } \sqrt{mn} \text{ is constant.}$$

(3) If $A \propto BC$, then $B \propto \dfrac{A}{C}$, and $C \propto \dfrac{A}{B}$.

For let $$A = mBC; \text{ then } B = \frac{1}{m}\frac{A}{C}.$$

$$\therefore \quad B \propto \frac{A}{C}. \quad \text{Similarly } C \propto \frac{A}{B}.$$

(4) If $A \propto B$, and $C \propto D$, then $AC \propto BD$.

For let $$A = mB, \text{ and } C = nD.$$

Then $$AC = mnBD. \quad \therefore \quad AC \propto BD.$$

(5) If $A \propto B$, then $A^n \propto B^n$.

For let $$A = mB; \text{ then } A^n = m^n B^n.$$

$$\therefore \quad A^n \propto B^n.$$

(6) *If $A \propto B$ when C is constant, and $A \propto C$ when B is constant, then $A \propto BC$ when both B and C are variable.*

The variation of A depends on the variations of the two quantities B and C. Suppose these latter variations to take place successively, each in its turn producing its own effect on A.

Let then B be changed to b, and in consequence let A be changed to a', C being constant; then, by supposition,

$$\frac{A}{a'} = \frac{B}{b}.$$

Now let C be changed to c, and in consequence let a' be changed to a, b being constant; then, by supposition,

$$\frac{a'}{a} = \frac{C}{c}.$$

Hence

$$\frac{A}{a'} \times \frac{a'}{a} = \frac{B}{b} \times \frac{C}{c};$$

or

$$\frac{A}{a} = \frac{BC}{bc}. \qquad \therefore \ A \propto BC.$$

The following are illustrations of this proposition.

The amount of work done by *a given number of men* varies directly as the number of days they work, and the amount of work done *in a given time* varies directly as the number of men; therefore when the number of days and the number of men are both variable, the amount of work will vary as the product of the number of men and the number of days.

Again, the area of a triangle varies directly as the base when the height is constant, and directly as the height when the base is constant; hence when both the base and the height are variable, the area will vary as the product of the base and height.

In the same manner, if A varies as each of any number of quantities, B, C, D, . . . when the rest are constant, then when they all vary A varies as their product. Also, the variations may be either direct or inverse.

NOTE. — This principle is interesting because of its frequent occurrence in Physical Science. For example, in the theory of gases it is found by experiment that the pressure p of a gas varies as the "absolute temperature" t when the volume v is constant, and that the pressure varies inversely as the volume when the temperature is constant; that is,

$$p \propto t, \text{ when } v \text{ is constant};$$

and

$$p \propto \frac{1}{v}, \text{ when } t \text{ is constant}.$$

From these results, we should expect that, when both t and v vary, we should have the formula

$$p \propto \frac{t}{v}, \text{ or } \frac{pv}{t} = \text{a constant,}$$

which by actual experiment is found to be the case.

1. If y varies inversely as $x^2 - 1$, and is equal to 24 when $x = 10$, find y when $x = 5$.

Since $y \propto \dfrac{1}{x^2 - 1}$, $y = \dfrac{m}{x^2 - 1}$, by (2) of Art. 159.

As $y = 24$ when $x = 10$, we have

$$24 = \frac{m}{99}. \qquad \therefore \quad m = 24 \times 99.$$

Hence, when $x = 5$, we have

$$y = \frac{24 \times 99}{x^2 - 1} = 99.$$

2. The pressure of a gas varies jointly as its density and its absolute temperature; also when the density is 1 and the temperature 300, the pressure is 15. Find the pressure when the density is 3 and the temperature is 320.

Let $p =$ the pressure, $t =$ the temperature, and $d =$ the density.

Then, since $p \propto td$, we have $p = mtd$, by (3) of Art. 159.

As $p = 15$ when $t = 300$ and $d = 1$, we have

$$15 = m \times 300 \times 1. \qquad \therefore \quad m = \tfrac{1}{20}.$$

Hence, when $d = 3$ and $t = 320$, we have

$$p = \tfrac{1}{20} \times 320 \times 3 = 48.$$

3. The time of a railway journey varies directly as the distance and inversely as the velocity; the velocity varies directly as the square root of the quantity of coal used per mile, and inversely as the number of cars in the train. In a journey of 25 miles in half an hour with 18 cars, 10 cwt. of coal is required: how much coal will be consumed in a journey of 21 miles in 28 minutes with 16 cars?

Let $t =$ the time in hours, $d =$ the distance in miles, $v =$ the velocity in miles per hour, $q =$ the quantity of coal in cwt., and $n =$ the number of cars.

Then we have $t \propto \dfrac{d}{v}$, and $v \propto \dfrac{\sqrt{q}}{n}$.

$$\therefore \quad t \propto \frac{dn}{\sqrt{q}}, \quad \text{or} \quad t = m\frac{dn}{\sqrt{q}}.$$

As $d = 25$ when $t = \frac{1}{2}$, $n = 18$, and $q = 10$, we have

$$\tfrac{1}{2} = m\frac{25 \times 18}{\sqrt{10}}. \quad \therefore \ m = \frac{\sqrt{10}}{25 \times 36}, \text{ and } t = \frac{\sqrt{10}}{25 \times 36}\frac{dn}{\sqrt{q}}.$$

Hence, when $d = 21$, $t = \frac{28}{60}$, and $n = 16$, we have

$$\tfrac{28}{60} = \frac{\sqrt{10} \times 21 \times 16}{25 \times 36\sqrt{q}} = \frac{\sqrt{10} \times 28}{25 \times 3\sqrt{q}}.$$

$$\therefore \ \sqrt{q} = \frac{\sqrt{10} \times 28 \times 60}{25 \times 3 \times 28} = \tfrac{4}{5}\sqrt{10}.$$

$$\therefore \ q = \tfrac{32}{5} = 6\tfrac{2}{5}.$$

Hence the quantity of coal is $6\frac{2}{5}$ cwt.

4. *A* varies as *B*, and *A* is 5 when *B* is 3; what is *A* when *B* is 5?　　　　　　　　　　　　　　　*Ans.* $8\frac{1}{3}$.

5. *A* varies inversely as *B*, and *A* is 4 when *B* is 15; what is *A* when *B* is 12?　　　　　　　　　　　*Ans.* 5.

6. If $x \propto y$ and $y \propto z$, show that $xz \propto y^2$.

7. If $x \propto \dfrac{1}{y}$, and $y \propto \dfrac{1}{z}$, prove that $z \propto x$.

8. If $x \propto z$ and $y \propto z$, prove that $x^2 - y^2 \propto z^2$.

EXAMPLES.

Find the ratio compounded of

1. The ratio $32 : 27$, and the triplicate ratio of $3 : 4$.
　　　　　　　　　　　　　　　　　　　Ans. $1 : 2$.

2. The ratio $6 : 25$, and the subduplicate ratio of $25 : 36$.
　　　　　　　　　　　　　　　　　　　Ans. $1 : 5$.

3. The triplicate ratio of $x : y$, and the ratio $2y^2 : 3x^2$.
　　　　　　　　　　　　　　　　　　　Ans. $2x : 3y$.

4. If $x : y = 3\frac{1}{3}$, find the value of $(x - 3y) : (2x - 5y)$.
　　　　　　　　　　　　　　　　　　　Ans. $1 : 5$.

5. If $\dfrac{a}{b} = \frac{3}{4}$, and $\dfrac{x}{y} = \frac{5}{7}$, find the value of $\dfrac{3ax - by}{4by - 7ax}$.
　　　　　　　　　　　　　　　　　　　Ans. $17 : 7$.

6. Find $x : y$, having given $x^2 + 6y^2 = 5xy$.　　　2 or 3.

7. Find two numbers in the ratio of 5 to 6, and whose sum is 121.   *Ans.* 55 and 66.

8. For what value of x will the ratio $15 + x : 17 + x$ be $\frac{1}{2}$?   *Ans.* -13.

9. Find x in order that $x + 1 : x + 4$ may be the duplicate ratio of $3 : 5$.   *Ans.* $11 : 16$.

10. Two numbers are in the ratio of $4 : 5$, and if 6 be taken from each, the ratio is that of $3 : 4 :$ find the numbers.
                                    Ans. 24, 30.

11. Find two numbers in the ratio of $5 : 6$, such that their sum has to the difference of their squares the ratio of $1 : 7$.
                                    Ans. $35 : 42$.

12. Find x so that $x : 1$ may be the duplicate of the ratio $8 : x$.   *Ans.* 4.

13. If $2x : 3y$ be in the duplicate ratio of $2x - m : 3y - m$, prove that $m^2 = 6xy$.

14. If $A : B$ be the subduplicate ratio of $A - x : B - x$, prove that $x = AB : (A + B)$.

15. Prove that if $\dfrac{a_1 + a_2x}{a_2 + a_3y} = \dfrac{a_2 + a_3x}{a_3 + a_1y} = \dfrac{a_3 + a_1x}{a_1 + a_2y}$, each of these ratios is equal to $1 + x : 1 + y$, supposing $a_1 + a_2 + a_3$ not to be zero.

16. If $\dfrac{a - b}{ay + bx} = \dfrac{b - c}{bz + cx} = \dfrac{c - a}{cy + az} = \dfrac{a + b + c}{ax + by + cz}$, prove that each of these ratios $= 1 : x + y + z$, supposing $a + b + c$ not to be zero.

17. Find a mean proportional between a^3b and ab^3.
                                    Ans. a^2b^2.

18. Find a third proportional to $(a - b)^2$ and $a^2 - b^2$.
                                    Ans. $(a + b)^2$.

19. If $a : b :: c : d$, prove that

   (1) $2a + 3c : 3a + 2c :: 2b + 3d : 3b + 2d$.

   (2) $la + mb : pa + qb :: lc + md : pc + qd$.

   (3) $\sqrt{a^2 + b^2} : \sqrt{c^2 + d^2} :: \sqrt[3]{a^3 + b^3} : \sqrt[3]{c^3 + d^3}$.

   (4) $a^2c + ac^2 : b^2d + bd^2 :: (a + c)^3 : (b + d)^3$.

Find the value of x in each of the proportions:

20. $3x - 1 : 6x - 7 : : 7x - 10 : 9x + 10.$ *Ans.* 8 or $\frac{2}{3}$.

21. $x^2 - 2x + 3 : 2x - 3 : : x^2 - 3x + 5 : 3x - 5.$ 2 or 0.

22. $2x - 1 : x + 4 : : x^2 + 2x - 1 : x^2 + x + 4.$ 5 or 0.

23. $(\sqrt{x+1} + \sqrt{x-1}) : (\sqrt{x+1} - \sqrt{x-1}) : : 4x - 1 : 2.$ $1\frac{1}{4}$.

24. If $a : b : : c : d : : e : f$, prove that

$$a^3 + c^3 + e^3 : b^3 + d^3 + f^3 : : ace : bdf.$$

25. If $a : b : : c : d$, prove that

(1) $a(c + d) = c(a + b).$

(2) $\dfrac{(a + c)(a^2 + c^2)}{(a - c)(a^2 - c^2)} = \dfrac{(b + d)(b^2 + d^2)}{(b - d)(b^2 - d^2)}.$

(3) $\dfrac{pa^2 + qab + rb^2}{la^2 + mab + nb^2} = \dfrac{pc^2 + qcd + rd^2}{lc^2 + mcd + nd^2}.$

26. If x and y be unequal and x have to y the duplicate ratio of $x + z : y + z$, prove that z is a mean proportional between x and y.

27. If $a : b : : p : q$, prove that $a^2 + b^2 : \dfrac{a^3}{a+b} : : p^2 + q^2 : \dfrac{p^3}{p+q}.$

28. If four quantities are in proportion, and the second is a mean proportional between the third and fourth, prove that the third will be a mean proportional between the first and second.

29. If $\dfrac{a + b + c + d}{a - b + c - d} = \dfrac{a + b - c - d}{a - b - c + d}$, prove that a, b, c, d are in proportion.

30. Each of two vessels contains a mixture of wine and water; a mixture consisting of equal measures from the two vessels contains as much wine as water, and another mixture consisting of four measures from the first vessel and one from the second is composed of wine and water in the ratio of 2 : 3. Find the proportion of wine and water in each of the vessels.

Ans. In the first the wine is $\frac{1}{3}$ of the whole; in the second $\frac{2}{3}$.

31) If $x \propto y$, and $y = 7$ when $x = 18$, find x when $y = 21$.
Ans. 54.

32. If $x \propto \dfrac{1}{y}$, and $y = 4$ when $x = 15$, find y when $x = 6$.
Ans. 10.

33. A varies jointly as B and C; and $A = 6$ when $B = 3$ and $C = 2$: find A when $B = 5$ and $C = 7$. *Ans.* 35.

34. A varies jointly as B and C; and $A = 9$ when $B = 5$ and $C = 7$: find B when $A = 54$ and $C = 10$. *Ans.* 21.

35. A varies directly as B and inversely as C; and $A = 10$ when $B = 15$ and $C = 6$: find A when $B = 8$ and $C = 2$.
Ans. 16.

36. If $3a + 7b \propto 3a + 13b$, and $a = 5$ when $b = 3$, find the equation between a and b. *Ans.* $3a = 5b$.

37. $A \propto B$, and $A = 2$ when $B = 1$; find A when $B = 2$.
Ans. 4.

38. If $A^2 + B^2 \propto A^2 - B^2$, prove that $A + B \propto A - B$.

39. $3A + 5B \propto 5A + 3B$; and $A = 5$ when $B = 2$: find the ratio $A : B$. *Ans.* 5 : 2.

40. $A \propto nB + C$; and $A = 4$ when $B = 1$ and $C = 2$; and $A = 7$ when $B = 2$ and $C = 3$: find n. *Ans.* 2.

41. If $x^2 \propto y^3$, and $x = 2$ when $y = 3$, find the equation between x and y. *Ans.* $27x^2 = 4y^3$.

42. If y varies as the sum of two quantities, one of which varies as x directly, the other as x inversely, and if, $y = 4$ when $x = 1$, and $y = 5$ when $x = 2$, find the equation between x and y.
Ans. $y = 2x + \dfrac{2}{x}$.

43. If $y =$ the sum of two quantities, one of which varies directly as x, and the other inversely as x^2; and if $y = 19$ when $x = 2$, or 3; find y in terms of x. *Ans.* $y = 5x + \dfrac{36}{x^2}$.

44. If y varies as the sum of three quantities of which the first is constant, the second varies as x, and the third as x^2; and if $y = 0$ when $x = 1$, $y = 1$ when $x = 2$, and $y = 4$ when $x = 3$; find y when $x = 7$. *Ans.* 36.

CHAPTER XVI.

ARITHMETIC, GEOMETRIC, AND HARMONIC PROGRESSIONS.

ARITHMETIC PROGRESSION.

157. Definitions — Formulæ. — A number of terms formed according to some law is called a *series*. Quantities are said to be in *Arithmetic Progression** when they increase or decrease by a constant difference, called the *common difference*.

Thus, the following series are each in Arithmetic Progression :

$$2, 5, 8, 11, 14, 17, \ldots \ldots \ldots$$
$$9, 7, 5, 3, 1, -1, -3, -5, \ldots \ldots$$
$$a, a + d, a + 2d, a + 3d, a + 4d, \ldots \ldots$$

The letters A. P. are often used for shortness instead of the term Arithmetic Progression.

The common difference is found by subtracting *any* term of the series from that which immediately *follows* it. In the first series above the common difference is 3 ; in the second it is -2 ; in the third it is d.

The series is said to be *increasing* or *decreasing*, according as the common difference is *positive* or *negative*. Thus, the first series above is *increasing*, and the second is *decreasing*.

If we examine the third series above, we see that *the coefficient of d in any term is less by one than the number of the term in the series.*

Thus the   2d term is $a + d$,

3d term is $a + 2d$,

4th term is $a + 3d$,

---

* Called also *Arithmetic Series.*

and so on. Hence if n be the number of terms, and if l denote the last, or n^{th} term, we have

$$l = a + (n - 1)d \quad \ldots \quad \ldots \quad (1)$$

Let S denote the sum of n terms of this series; then we have

$$S = a + (a + d) + (a + 2d) + \ldots + (l - 2d) + (l - d) + l;$$

and, by writing the series in the reverse order, we have

$$S = l + (l - d) + (l - 2d) + \ldots + (a + 2d) + (a + d) + a.$$

Adding together these two equations, we have

$$2S = (a + l) + (a + l) + (a + l) + \ldots \ldots \text{ to } n \text{ terms}$$
$$= n(a + l).$$

$$\therefore \quad S = \frac{n}{2}(a + l) \quad \ldots \quad \ldots \quad (2)$$

By (1) and (2) we have $S = \dfrac{n}{2}[2a + (n - 1)d]$ . . (3)

We have here three useful formulæ, (1), (2), (3), which should be remembered; in each of these any one, of the letters may denote the unknown quantity when the other three are known. For example, in (1), we can write down any term of an A. P. when the first term, the common difference, and the number of the term are given. Thus, if the first term of an A. P. is 5 and the common difference is 3,

the 10th term $= 5 + (10 - 1)3 = 32$,

and the 20th term $= 5 + (20 - 1)3 = 62$.

Also in (2), if we substitute given values for S, n, l, we obtain an equation for finding a; and similarly in (3). Thus,

1. Find the sum of 20 terms of the series 1, 3, 5, 7, . . .

Here $a = 1$, $d = 2$, $n = 20$; therefore by (3)

$$S = \tfrac{20}{2}[2 + 19 \times 2] = 10 \times 40 = 400.$$

2. The first term of a series is 5, the last 45, and the sum 400; find the number of terms, and the common difference.

Here $a = 5, l = 45, S = 400$; therefore by (2)

$$400 = \frac{n}{2}(5 + 45) = 25n. \quad \therefore \quad n = 16.$$

By (1) $45 = 5 + 15d.$ $\therefore$ $d = 2\frac{2}{3}.$

When *any two* terms of an A. P. are given, the series can be completely determined; for the data furnish two simultaneous equations, with two unknown quantities, which may be solved by methods previously given.

3. The 10th and 15th terms of an A. P. are 25 and 5 respectively; find the series.

Here $25 = a + 9d$;

and $5 = a + 14d.$

By subtraction, $20 = -5d.$ $\therefore$ $d = -4.$

Then $a = 5 - 14d = 61.$

Hence the series is 61, 57, 53, ...

4. Find the sum of the first n odd integers.

Here $a = 1$, and $d = 2$; therefore by (3)

$$S = \frac{n}{2}[2 + (n - 1)2] = \frac{n}{2} \times 2n = n^2.$$

Thus the sum of any number of consecutive odd integers beginning with unity, is the square of their number.[*]

Find the last term and sum of the following series:

5. 14, 64, 114, ... to 20 terms.   *Ans.* 964, 9780.

6. 9, 5, 1, ... to 100 terms. $-387, -18900.$

7. $\frac{1}{4}, -\frac{1}{4}, -\frac{3}{4},$ . . . . . . to 21 terms. $-9\frac{3}{4}, -99\frac{3}{4}.$

Find the sum of the following series:

8. 5, 9, 13, ... to 19 terms.   779.

9. 12, 9, 6, ... to 23 terms. $-483.$

Find the series in which

10. The 27th term is 186, and the 45th is 312.

Ans. 4, 11, 18, ...

---

[*] This proposition was known to the Greek geometers.

11. The 9th term is -11, and the 102d is $-150\frac{1}{2}$.

Ans. 1, $-\frac{1}{2}$, -2, ...

12. The 16th term is 214, and the 51st is 739.

Ans. -11, 4, 19, ...

158. Arithmetic Mean. — When three quantities are in Arithmetic Progression, the middle one is called the *Arithmetic Mean* of the other two.

Thus if a, b, c are in A. P., b is the arithmetic mean of a and c; and by the definition of A. P. we have

$$b - a = c - b;$$

$$\therefore \quad b = \tfrac{1}{2}(a + c).$$

Thus *the arithmetic mean of any two quantities is half their sum.*

Between any two given quantities any number of terms may be inserted so that the whole series thus formed shall be in A. P. ; the terms thus inserted are called the *arithmetic means.*

For example, to insert four arithmetic means between 10 and 25.

Here we have to find an A. P. with 4 terms between 10 and 25, so that 10 is the first and 25 is the sixth term.

By (1) of Art. 157,

$$25 = 10 + 5d; \quad \therefore \quad d = 3.$$

Thus the series is 10, 13, 16, 19, 22, 25 ;
and the required arithmetic means between 10 and 25 are

$$13, 16, 19, 22.$$

In general.   To insert n arithmetic means between a and b.

Here we have to find an A. P. with n terms between a and b, so that a is the first and b is the $(n + 2)^{\text{th}}$ term.

By (1) of Art. 161,

$$b = a + (n + 2 - 1)d = a + (n + 1)d$$

$$\therefore \quad d = \frac{b - a}{n + 1}.$$

Thus the required means are

$$a + \frac{b-a}{n+1}, \quad a + 2\frac{b-a}{n+1}, \quad \ldots \ldots \quad a + n\frac{b-a}{n+1}.$$

1. Find the sum of the first p terms of the series whose n^{th} term is $3n - 1$.

By putting $n = 1$, and $n = p$ respectively, we obtain

first term $= 2$, last term $= 3p - 1$.

Hence by (2), Art. 157, $S = \frac{p}{2}(2 + 3p - 1) = \frac{p}{2}(3p + 1)$.

In an Arithmetic Progression when a, S, and d are given, n is to be found by solving the *quadratic* (3), Art. 157. When both roots are positive and integral, there is no difficulty in interpreting the result corresponding to each.

2. How many terms of the series 24, 20, 16, ... must be taken that the sum may be 72?

Here $a = 24$, $d = -4$, $S = 72$. Then from (3), Art. 157, we have

$$72 = \frac{n}{2}[2 \times 24 + (n - 1)(-4)] = 24n - 2n(n - 1).$$

$$\therefore \ n^2 - 13n + 36 = 0, \text{ or } (n - 4)(n - 9) = 0.$$

$$\therefore \ n = 4, \text{ or } 9.$$

Both of these values satisfy the conditions of the question; for if we take the first 4 terms, we get 24, 20, 16, 12; and if we take the first 9 terms, we get 24, 20, 16, 12, 8, 4, 0, −4, −8, in either of which the sum is 72; the last 5 terms of the last series destroy each other, so that the sum of the first 4 terms is the same as the sum of the first 9 terms.

When one of the roots is negative or fractional, it is inapplicable, for a *negative* or a *fractional number of terms* is, strictly speaking, without meaning. In some cases however a suitable interpretation can be given for a negative value of n.

3. How many terms of the series $-9, -6, -3, \ldots \ldots$ must be taken that the sum may be 66?

Here $\qquad 66 = \frac{n}{2}[-18 + (n-1)3].$

$\therefore \; n^2 - 7n - 44 = 0;$ or $(n-11)(n+4) = 0.$

$\therefore \; n = 11,$ or $-4.$

If we take 11 terms of the series, we have

$$-9, -6, -3, 0, 3, 6, 9, 12, 15, 18, 21;$$

the sum of which is 66.

If we begin at the *last* term and *count backwards* four terms, the sum is also 66. From this we see that, although the negative solution does not directly answer the question proposed, we are enabled to give it an intelligible meaning as follows: begin at the *last* term of the series which is furnished by the positive value of n, and *count backwards* for as many terms as the negative value indicates; then the result will be the given sum. We thus see that the *negative* value for n answers a question closely connected with that to which the positive value applies.

4. How many terms of the series $26, 21, 16, \ldots \ldots$ must be taken that the sum may be 74?

Here $\qquad 74 = \frac{n}{2}[52 + (n-1)(-5)].$

Solving, we get $n = 4,$ or $7\frac{2}{5}.$

Thus, the only applicable value of n is 4. We infer that of the two numbers 7 and 8, one corresponds to a sum greater, and the other to a sum less than 74.

5. Insert 3 arithmetic means between 12 and 20.
   Ans. 14, 16, 18.

6. Insert 5 arithmetic means between 14 and 16.
   Ans. $14\frac{1}{3}, 14\frac{2}{3}, \ldots \ldots$

7. Insert 17 arithmetic means between 93 and 69.
   Ans. $91\frac{2}{3}, 90\frac{1}{3}, \ldots \ldots 70\frac{1}{3}.$

How many terms must be taken of the series

8. 42, 39, 36, ... to make 315?   *Ans.* 14, or 15.

9. −16, −15, −14, ... to make −100?   8, or 25.

10. 20, 18¾, 17½, ... to make 162½?   13, or 20.

11. The sum of three numbers in A. P. is 39, and their product is 2184; find them.   *Ans.* 12, 13, 14.

12. The sum of 10 terms of an A. P., whose first term is 2, is 155; what is the common difference?   *Ans.* 3.

GEOMETRIC PROGRESSION.

159. Definition — Formulæ. — Quantities are said to be in *Geometric Progression* when they increase or decrease by a constant factor, called the *common ratio.*

Thus, the following series are each in Geometric Progression (G. P.):

$$3, 6, 12, 24, 48, \ldots \ldots$$
$$3, 1, \tfrac{1}{3}, \tfrac{1}{9}, \tfrac{1}{27}, \ldots \ldots$$
$$a, ar, ar^2, ar^3, ar^4, \ldots \ldots$$

The common ratio is found by dividing *any* term of the series by that which immediately *precedes* it. In the first series above the common ratio is 2; in the second it is $\tfrac{1}{3}$; in the third it is r.

The series is said to be *increasing* or *decreasing*, according as the common ratio is *greater* than 1, or *less* than 1. Thus, the first series above is *increasing*, and the second is *decreasing.*

Note 1. — An Arithmetic Progression is formed by repeated addition or subtraction; a Geometric Progression by repeated multiplication or division.

If we examine the third series above, we see that *the exponent of r in any term is less by one than the number of the term in the series.*

Thus, the   2d term is ar,
    3d term is ar^2,
    4th term is ar^3,

and so on. Hence if n be the number of terms, and if l denote the last, or n^{th} term, we have

$$l = ar^{n-1} \quad . \quad . \quad . \quad . \quad . \quad . \quad (1)$$

Let S denote the sum of n terms of this series ; then we have

$$S = a + ar + ar^2 + \ldots\ldots + ar^{n-2} + ar^{n-1} ;$$

multiplying by r, we have

$$Sr = ar + ar^2 + \ldots\ldots + ar^{n-2} + ar^{n-1} + ar^n.$$

Hence by subtraction, we have

$$Sr - S = ar^n - a \; \underline{\quad} \; \text{or} \quad (r-1)S = a(r^n - 1).$$

$$\therefore \quad S = \frac{a(r^n - 1)}{r - 1}, \text{ or } \frac{a(1 - r^n)}{1 - r} \quad . \quad . \quad . \quad . \quad (2)$$

Multiplying (1) by r, and substituting in (2), we get

$$S = \frac{rl - a}{r - 1}, \text{ or } \frac{a - rl}{1 - r}, \quad . \quad . \quad . \quad (3)$$

a form which is sometimes useful.

Note 2. — It will be found convenient to remember both forms given in (2) for S, and to use the first form in all cases when r is *positive* and > 1, and the second when r is *negative* or < 1.

1. Find the 8th term of the series $-\frac{1}{3}, \frac{1}{2}, -\frac{3}{4}, \ldots\ldots$
Here $a = -\frac{1}{3}, n = 8, r = \frac{1}{2} \div (-\frac{1}{3}) = -\frac{3}{2}$: therefore by (1)

$$l = -\frac{1}{3}(-\frac{3}{2})^7 = -\frac{1}{3}(-\frac{2187}{128})$$
$$= \frac{729}{128} = \text{the 8th term.}$$

2. Sum the series $1, 3, 9, \ldots\ldots$ to 6 terms.
Here $a = 1, n = 6, r = 3$; therefore by the first form of (2),

$$S = \frac{3^6 - 1}{3 - 1} = \frac{729 - 1}{2} = 364.$$

3. Sum the series $81, 54, 36, \ldots\ldots$ to 9 terms.
Here $a = 81, n = 9, r = 54 \div 81 = \frac{2}{3}$; therefore by the second form of (2),

$$S = \frac{81[1 - (\frac{2}{3})^9]}{1 - \frac{2}{3}} = 243[1 - (\frac{2}{3})^9]$$
$$= 243 - \frac{512}{81} = 236\frac{55}{81}.$$

4. Sum the series 2, -3, $\frac{9}{2}$, $-$ . . . . . to 7 terms.

Here $a = 2$, $n = 7$, $r = -\frac{3}{2}$; therefore by the second form of (2),

$$S = \frac{2[1 - (-\frac{3}{2})^7]}{1 - (-\frac{3}{2})} = \frac{2[1 + \frac{2187}{128}]}{\frac{5}{2}}$$
$$= \frac{4}{5} \times \frac{2315}{128} = 14\frac{15}{32}.$$

5. Find the 6th term of each of the following series:

(1) 9, 3, 1, etc. ;   (2) 2, -3, $\frac{9}{2}$, etc. ;   (3) a^2, ab, b^2, etc.

Ans. (1) $\frac{1}{27}$;   (2) $-\frac{243}{16}$;   (3) $\dfrac{b^5}{a^3}$.

Sum the following series:

6. 1,   4, 16, ... to 6 terms.   *Ans.* 1365.

7. 25, 10, 4, ... to 4 terms. $40\frac{3}{5}$.

8. $\frac{2}{3}$, -1, $\frac{3}{2}$, ... to 7 terms. $\frac{463}{96}$.

9. 3, -1, $\frac{1}{3}$, ... to 6 terms. $2\frac{20}{81}$.

160. Geometric Mean. — When three quantities are in Geometric Progression the middle one is called the *Geometric Mean* between the other two.

Thus if a, b, c are in G. P., b is the geometric mean between a and c; and by the definition of G. P., we have

$$\frac{b}{a} = \frac{c}{b};$$
$$\therefore \quad b^2 = ac; \qquad \therefore \quad b = \sqrt{ac}.$$

Thus, *the geometric mean between any two quantities is the square root of their product.*

Quantities which are in G. P. are in continued proportion, and the geometric mean between two quantities is the same as their mean proportional (Art. 152).

Between any two given quantities any number of terms may be inserted so that the whole series thus formed shall be in G. P. ; the terms thus inserted are called the *geometric means*.

For example, to insert three geometric means between 2 and 32.

Here we have to find a G. P. with 3 terms between 2 and 32, so that 2 is the first and 32 is the fifth term.

By (1) of Art. 159, $32 = 2r^4$; $\therefore$ $r = 2$.

Thus the series is   2, 4, 8, 16, 32,   and the required geometric means between 2 and 32 are 4, 8, 16.

In general.   To *insert n geometric means between a and b.*

Here we have to find a G. P. with n terms between a and b, so that a is the first and b is the $(n+2)^{\text{th}}$ term.

By (1) of Art. 159,

$$b = ar^{n+1}; \quad \therefore \ r^{n+1} = \frac{b}{a};$$

$$\therefore \ r = \sqrt[n+1]{\frac{b}{a}} \ \cdot \ \cdot \ \cdot \ \cdot \ \cdot \ \cdot \ (1)$$

Thus the required means are $ar, ar^2, \ldots \ldots ar^n$, where r has the value found in (1).

1. Insert 4 geometric means between 160 and 5.

Ans. 80, 40, 20, 10.

2. Insert 6 geometric means between 56 and $-\frac{7}{16}$.

Ans. $-28, 14, -7, \frac{7}{2}, -\frac{7}{4}, \frac{7}{8}$.

3. Insert 4 geometric means between $5\frac{1}{3}$ and $40\frac{1}{2}$.

Ans. 8, 12, 18, 27.

161. The Sum of an Infinite Number of Terms.— From (2) of Art. 159, we have

$$S = \frac{a(1 - r^n)}{1 - r} = \frac{a}{1 - r} - \frac{ar^n}{1 - r}. \ \cdot \ \cdot \ \cdot \ (1)$$

Now suppose r is a *proper fraction*, positive or negative; then the greater the value of n the smaller is the absolute value of r^n, and consequently of $\dfrac{ar^n}{1 - r}$; and by taking n sufficiently large r^n can be made *as small as we please.* Hence, by taking n large enough, the sum of n terms of the series can be made to differ from $\dfrac{a}{1 - r}$ by as small a quantity as we please.

Thus, *the sum of an infinite number of terms of a decreas-ing Geometric Progression is* $\dfrac{a}{1 - r}$;

or more briefly, *the sum to infinity is* $\dfrac{a}{1 - r}$.

This quantity, $\dfrac{a}{1-r}$, which we call the *sum* of the series, is the *limit* to which the sum approaches, but never actually attains; that is, although no *definite* number of terms will amount to $\dfrac{a}{1-r}$, yet by taking a sufficient number, the sum will reach it as near as we please.

1. Sum the series $\frac{1}{2}$, $\frac{1}{4}$, $\frac{1}{8}$, ...

For n terms we have by (2) of Art. 159,

$$S = \frac{\frac{1}{2}\left(1 - \dfrac{1}{2^n}\right)}{1 - \frac{1}{2}} = 1 - \frac{1}{2^n}.$$

From this result it appears that however many terms be taken, the sum of this series is always less than 1. Also we see that by taking n large enough, the fraction $\dfrac{1}{2^n}$ can be made as small as we please. Hence by taking a sufficient number of terms, the sum can be made to differ from 1 by as little as we please; *and when n is made infinitely great* we have $S = 1$.

This may be illustrated geometrically as follows:

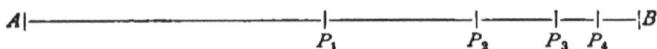

Let AB be a line of unit length. Bisect AB in P_1; bisect P_1B in P_2, P_2B in P_3, P_3B in P_4, and so on indefinitely, always bisecting the remaining distance. It is evident that by a series of such bisections we can never reach B, because we shall always have a distance left equal to half the preceding distance; but by a sufficient number of these bisections we can come nearer to B than any assigned distance, however small, because every bisection carries us over half the remaining distance. That is, if we take a sufficient number of terms of the series

$$AP_1 + P_1P_2 + P_2P_3 + P_3P_4 + \cdots \cdots ,$$

we shall have a result differing from AB, i.e., from unity, by as little as we please. This is simply a geometric way of saying that

$$\frac{1}{2} + \frac{1}{2^2} + \frac{1}{2^3} + \cdots \cdots \text{ to } \infty = 1.$$

Sum the following series to infinity:

2. $1, \frac{1}{2}, \frac{1}{2^2} \ldots \ldots \ldots$ *Ans.* 2.

3. $9, -6, 4, - \ldots \ldots$ $5\frac{2}{5}$.

4. $1, -\frac{1}{3}, \frac{1}{9}, \ldots \ldots$ $\frac{3}{4}$.

5. $1, \frac{1}{4}, \frac{1}{16}, \ldots \ldots$ $\frac{4}{3}$.

162. Value of a Repeating* Decimal. — Repeating decimals furnish a good illustration of infinite Geometric Progressions.

1. Find the value of $.4\overset{..}{23}$.

$$.4\overset{..}{23} = .4232323 \ldots \ldots$$

$$= \frac{4}{10} + \frac{23}{10^3} + \frac{23}{10^5} + \ldots \ldots$$

$$= \frac{4}{10} + \frac{23}{10^3}\left(1 + \frac{1}{10^2} + \frac{1}{10^4} + \ldots \ldots\right)$$

$$= \frac{4}{10} + \frac{23}{10^3}\left(\frac{1}{1 - \frac{1}{10^2}}\right) \text{ [by (1) of Art. 161]}$$

$$= \frac{4}{10} + \frac{23}{10^3} \times \frac{100}{99} = \frac{4}{10} + \frac{23}{990} = \frac{419}{990},$$

which agrees with the value found by the usual rule in Arithmetic.

The value of any repeating decimal may be found by the method employed in the last example; but in practice it may be found more easily by a general rule, which may be proved as follows:

Let P denote the figures which do not repeat, and suppose them p in number; let Q denote the repeating period consisting of q figures. Let S denote the value of the repeating decimal; then

$$S = .PQQQ \ldots \ldots \ldots;$$

$$\therefore \quad 10^p S = P.QQQ \ldots \ldots \ldots;$$

and $\quad 10^{p+q}S = PQ.QQQ \ldots \ldots \ldots;$

---

* Called also *recurring* and *circulating*.

by subtracting,

$$(10^{p+q} - 10^p)S = PQ - P;$$

that is, $10^p(10^q - 1)S = PQ - P;$

$$\therefore \; S = \frac{PQ - P}{(10^q - 1)10^p}.$$

Now $10^q - 1$ is a number consisting of q *nines;* therefore the denominator consists of q nines followed by p ciphers. Hence, for finding the value of a repeating decimal, we have the following

RULE. *Subtract the integral number consisting of the non-repeating figures from the integral number consisting of the non-repeating and repeating figures, and divide by a number consisting of as many nines as there are repeating figures followed by as many ciphers as there are non-repeating figures.*

Find the value of the following repeating decimals:

2. .151515 .... *Ans.* $\frac{5}{33}$.
3. .123123123 .... $\frac{41}{333}$.

4. .1̇6̇ ......   *Ans.* $\frac{1}{6}$.
5. .0̇37̇ ...... $\frac{1}{27}$

HARMONIC PROGRESSION.

163. Definition. — A series of quantities is said to be in *Harmonic Progression* when their reciprocals are in Arithmetic Progression.

Thus, the following series

$$1, \tfrac{1}{3}, \tfrac{1}{5}, \tfrac{1}{7}, \ldots \ldots \text{ and } \tfrac{1}{4}, \tfrac{2}{7}, \tfrac{1}{3}, \tfrac{2}{5}, \ldots \ldots ,$$

are each in Harmonic Progression (H. P.) because their reciprocals,

$$1, 3, 5, 7, \ldots \ldots \text{ and } 4, 3\tfrac{1}{2}, 3, 2\tfrac{1}{2}, \ldots \ldots$$

are in A. P.

The following is therefore a *general form* for a H. P.

$$\frac{1}{a}, \frac{1}{a + d}, \frac{1}{a + 2d}, \ldots \ldots \frac{1}{a + (n - 1)d},$$

because the reciprocals of the terms are in A. P.

From the above definition it follows that all problems relating to quantities in H. P. can be solved by taking the reciprocals of the quantities and using the formulæ relating to A. P. This makes it unnecessary to give any special rules for the solution of problems in H. P.

If a, b, c be three consecutive terms of a H. P., then we have by definition,

$$\frac{1}{b} - \frac{1}{a} = \frac{1}{c} - \frac{1}{b}.$$

$$\therefore \quad \frac{a - b}{ab} = \frac{b - c}{bc}.$$

$$\therefore \quad a - b : b - c :: a : c \quad \dots \dots (1)$$

Thus, *if three quantities are in Harmonic Progression, the difference between the first and the second is to the difference between the second and the third as the first is to the third.*

Sometimes this relation is taken as the definition of Harmonic Progression.

1. The 12th term of a H. P. is $\frac{1}{5}$, and the 19th term is $\frac{3}{22}$; find the series.

Here the 12th and 19th terms of the corresponding A. P. are 5 and $\frac{22}{3}$ respectively. Therefore by

(1), Art. 161, $\qquad 5 = a + 11d$,

and $\qquad\qquad\qquad \frac{22}{3} = a + 18d$.

Solving, we get $\qquad d = \frac{1}{3}, a = \frac{4}{3}$.

Hence the A. P. is $\frac{4}{3}$, $\frac{5}{3}$, 2, $\frac{7}{3}$, $\frac{8}{3}$, ...

and the H. P. is $\quad \frac{3}{4}$, $\frac{3}{5}$, $\frac{1}{2}$, $\frac{3}{7}$, $\frac{3}{8}$, ...

Find the last term of the following harmonic series:

2. 4, 2, $1\frac{1}{3}$, ... to 6 terms. *Ans.* $\frac{2}{3}$.

3. $2\frac{1}{2}$, $1\frac{12}{13}$, $1\frac{9}{16}$, ... to 21 terms. $\frac{5}{14}$.

4. $1\frac{1}{3}$, $1\frac{11}{17}$, $2\frac{2}{13}$, ... to 8 terms. -4.

164. Harmonic Mean. — When three quantities are in Harmonic Progression, the middle one is called the *Harmonic Mean* between the other two.

Thus if a, b, c are in H. P., b is the harmonic mean between a and c; and by the definition of H. P. we have

$$\frac{1}{b} - \frac{1}{a} = \frac{1}{c} - \frac{1}{b};$$

$$\therefore \frac{2}{b} = \frac{1}{a} + \frac{1}{c}.$$

$$\therefore b = \frac{2ac}{a + c}.$$

Thus, *the harmonic mean between any two quantities is twice their product divided by their sum.*

Between any two given quantities any number of terms may be inserted so that the whole series thus formed shall be in H. P.; the terms thus inserted are called the *harmonic means.*

For example, to insert 5 harmonic means between $\frac{2}{3}$ and $\frac{8}{15}$.

Here we have to insert 5 arithmetic means between $\frac{3}{2}$ and $\frac{15}{8}$. Hence, by (1) of Art. 157,

$$\tfrac{15}{8} = \tfrac{3}{2} + 6d; \qquad \therefore \ d = \tfrac{1}{16}.$$

Thus the A. P. is $\frac{3}{2}$, $\frac{25}{16}$, $\frac{26}{16}$, $\frac{27}{16}$, $\frac{28}{16}$, $\frac{29}{16}$, $\frac{15}{8}$;

and therefore the required harmonic means between $\frac{2}{3}$ and $\frac{8}{15}$

are $\qquad \frac{16}{25}, \frac{16}{26}, \frac{16}{27}, \frac{16}{28}, \frac{16}{29}.$

In general. To insert n harmonic means between a and b.

Here we have to insert n arithmetic means between $\dfrac{1}{a}$ and $\dfrac{1}{b}$. By Art. 158, these will be and therefore the required harmonic means between a and b are the reciprocals of these, that is,

$$\frac{(n+1)ab}{(n+1)b+(a-b)}, \frac{(n+1)ab}{(n+1)b+2(a-b)}, \ldots$$

1. Insert 2 harmonic means between 4 and 2. *Ans.* 3, $2\frac{2}{5}$.
2. Insert 3 harmonic means between $\frac{1}{3}$ and $\frac{1}{21}$. $\frac{2}{15}$, $\frac{1}{12}$, $\frac{2}{33}$.
3. Insert 4 harmonic means between 1 and 6. $1\frac{1}{5}$, $1\frac{1}{2}$, 2, 3.

165. Relation between Arithmetic, Geometric, and Harmonic Means.

(1) If A, G, H be the arithmetic, geometric, and harmonic means between a and b, then (Arts. 158, 160, 164),

$$A = \frac{a+b}{2} \quad \ldots \ldots \quad (1)$$

$$G = \sqrt{ab} \quad \ldots \ldots \quad (2)$$

$$H = \frac{2ab}{a+b} \quad \ldots \ldots \quad (3)$$

Therefore $AH = \dfrac{a+b}{2} \times \dfrac{2ab}{a+b} = ab = G^2$;

that is, G is the geometric mean between A and H.

Hence *the geometric mean between any two real positive quantities, a and b, is also the geometric mean between the arithmetic and the harmonic means between a and b.*

(2) From (1) and (2) we have

$$A - G = \frac{a+b}{2} - \sqrt{ab} = \tfrac{1}{2}(\sqrt{a} - \sqrt{b})^2;$$

and from (2) and (3),

$$G - H = \sqrt{ab} - \frac{2ab}{a+b} = \frac{\sqrt{ab}}{a+b}(\sqrt{a} - \sqrt{b})^2.$$

Now if a and b are both positive, $\sqrt{a}$ and $\sqrt{b}$ are both real; therefore $(\sqrt{a} - \sqrt{b})^2$ is positive; also $\sqrt{ab}$ and $a+b$ are both positive. Hence $A - G$ and $G - H$ are positive. Therefore $A > G > H$.

That is, *the arithmetic, geometric, and harmonic means between any two real positive quantities are in descending order of magnitude.**

Three quantities, a, b, c, are in A. P., G. P., or H. P., according as

$$\frac{a-b}{b-c} = \frac{a}{a}, \ \frac{a}{b}, \ \text{or} \ \frac{a}{c}, \ \text{respectively.}$$

The first follows from the definition of A. P. (Art. 157).

In the second, $b(a-b) = a(b-c)$; $\therefore$ $b^2 = ac$. See Art. 160.

The third follows from (1) of Art. 163.

Harmonic properties are interesting chiefly because of their importance in Geometry and in the Theory of Sound. If there be a series of strings of the same substance, the lengths of which are proportional to $1, \frac{1}{2}, \frac{1}{3}, \frac{1}{4}, \frac{1}{5}, \frac{1}{6}$, and if these strings are stretched tight with equal force, and any two of them are sounded together, the effect is found to be harmonious to the ear.

Notwithstanding the comparative simplicity of the law of its formation, there is no general formula for the sum of any number of terms in harmonic progression.

EXAMPLES.

Find the last term and sum of the following series:

1. $1, 1.2, 1.4, \ldots \ldots$ to 12 terms.   *Ans.* 3.2, 25.2.

2. $3\frac{1}{2}, 1, -1\frac{1}{2}, \ldots \ldots$ to 19 terms. $-41\frac{1}{2}, -361$.

3. $64, 96, 128, \ldots \ldots$ to 16 terms.   544, 4864.

Sum the following series:

4. $4, 5\frac{1}{4}, 6\frac{1}{2}, \ldots \ldots$ to 37 terms. $980\frac{1}{2}$.

5. $-3, 1, 5, \ldots \ldots$ to 17 terms.   493.

6. $3a, a, -a, \ldots \ldots$ to a terms. $a^2(4-a)$.

7. $3\frac{1}{3}, 2\frac{1}{2}, 1\frac{2}{3}, \ldots \ldots$ to n terms. $\dfrac{5n(9-n)}{12}$.

---

* These two propositions were known to the Greek geometers.

8. $1\frac{1}{7}$, $1\frac{17}{21}$, $2\frac{10}{21}$, ... to n terms. $\quad$ Ans. $\dfrac{n(17 + 7n)}{21}$.

9. $\dfrac{1}{\sqrt{2} + 1}$, $\sqrt{2}$, $\dfrac{1}{\sqrt{2} - 1}$, ... to 7 terms. $\quad 7(\sqrt{2} + 2)$.

Find the series in which

10. The 15th term is 25, and the 29th term 46.

$\qquad\qquad\qquad$ Ans. 4, 5$\frac{1}{2}$, 7, ...

11. The 15th term is -25, and the 23^{d} term -41.

$\qquad\qquad\qquad$ Ans. 3, 1, -1, ...

12. Insert 14 arithmetic means between $-7\frac{1}{2}$ and $-2\frac{1}{2}$.

$\qquad\qquad$ Ans. $-6\frac{13}{15}$, $-6\frac{8}{15}$, ... $-2\frac{8}{15}$.

13. Insert 36 arithmetic means between $8\frac{1}{2}$ and $2\frac{1}{3}$.

$\qquad\qquad\qquad$ Ans. $8\frac{1}{3}$, $8\frac{1}{6}$, ... $2\frac{1}{2}$.

How many terms must be taken of the series

14. $15\frac{2}{3}$, $15\frac{1}{3}$, 15, ... to make 129? $\quad$ Ans. 9, or 86.

15. $-10\frac{1}{2}$, -9, $-7\frac{1}{2}$, . . . to make -42? $\qquad$ 7, or 8.

16. $-6\frac{4}{5}$, $-6\frac{2}{5}$, -6, ... to make $-52\frac{4}{5}$? $\qquad$ 11, or 24.

17. The sum of three numbers in A. P. is 33, and their product is 792 : find them. $\qquad\qquad$ Ans. 4, 11, 18.

18. An A. P. consists of 21 terms; the sum of the three terms in the middle is 129, and of the last three is 237 : find the series. $\qquad$ Ans. 3, 7, 11, ... 83.

19. The first term of an A. P. is 5, and the fifth term is 11 : find the sum of 8 terms. $\qquad\qquad$ Ans. 82.

20. The sum of four terms in A. P. is 44, and the last term is 17 : find the terms. $\qquad$ Ans. 5, 9, 13, 17.

21. The seventh term of an A. P. is 12, and the twelfth term is 7 ; the sum of the series is 171 : find the number of terms. $\qquad\qquad\qquad$ Ans. 18, or 19.

22. A sets out from a place and travels $2\frac{1}{2}$ miles an hour. B sets out 3 hours after A, and travels in the same direction, 3 miles the first hour, $3\frac{1}{2}$ miles the second, 4 miles the third, and so on. In how many hours will B overtake A? $\quad$ Ans. 5.

23. In the series, 1, 3, 5, etc., the sum of $2n$ terms : the sum of n terms :: x : 1 : find the value of x. $\qquad$ Ans. 4.

24. Find an A. P. such that the sum of the first five terms is one-fourth the sum of the following five terms, the first term being unity.   *Ans.* 1, -2, -5, ... -26.

25. If the sum of m terms of an A. P. be always to the sum of n terms in the ratio of m^2 to n^2, and the first term be unity, find the nth term.   *Ans.* $2n - 1$.

26. If $2n + 1$ terms of the series 1, 3, 5, 7, 9, ... be taken, show that the sum of the alternate terms, 1, 5, 9, ... will be to the sum of the remaining terms 3, 7, 11, ... as $n + 1$ to n.

27. On the ground are placed n stones; the distance between the first and second is one yard, between the second and third three yards, between each of the remaining stones five yards: how far will a person have to travel who shall bring them one by one to a basket placed at the first stone?

Ans. $5n^2 - 17n + 16$ yards.

28. Find a series of arithmetic means between 1 and 21, so that their sum has to the sum of the two greatest of them the ratio of 11 to 4.   *Ans.* 9 means, 3, 5, 7, ... 19.

29. Find the number of arithmetic means between 1 and 19 when the second mean is to the last as 1 to 6.   *Ans.* 17.

30. If the second term of an A. P. be a mean proportional between the first and the fourth, show that the sixth term will be a mean proportional between the fourth and the ninth.

Find the last term of each of the following geometric series:

31. 2, -6, 18, ... to 8 terms.   *Ans.* -4374.

32. 2, 3, $4\frac{1}{2}$, ... to 6 terms. $\frac{243}{16}$.

33. 3, -3^2, 3^3, ... to $2n$ terms. -3^{2n}.

Sum the following series:

34. 1, $-\frac{1}{2}$, $\frac{1}{4}$, ... to 12 terms.   *Ans.* $\frac{1365}{2048}$.

35. 9, -6, 4, ... to 7 terms. $5\frac{58}{81}$.

36. $2, -4, 8, \ldots \ldots$ to $2p$ terms.   *Ans.* $\frac{2}{3}(1 - 2^{2p})$.

37. $\sqrt{2}, \sqrt{6}, 3\sqrt{2}, \ldots \ldots$ to 12 terms. $364(\sqrt{6} + \sqrt{2})$.

38. Insert 3 geometric means between 486 and 6.

Ans. 162, 54, 18.

39. Insert 4 geometric means between $\frac{1}{8}$ and 128.

Ans. $\frac{1}{2}$, 2, 8, 32.

40. Insert 3 geometric means between 1 and 256.

Ans. 4, 16, 64.

41. Insert 4 geometric means between 3 and -729.

Ans. $-9, 27, -81, 243$.

Sum the following series:

42. $\frac{3}{8}, \quad \frac{1}{4}, \quad \frac{1}{6}, \ldots \ldots$ to 6 terms.   *Ans.* $\frac{665}{648}$.

43. $1, \quad -\frac{1}{2}, \quad \frac{1}{4}, \ldots \ldots$ to infinity. $\frac{2}{3}$.

44. $6, \quad -2, \quad \frac{2}{3}, \ldots \ldots$ to infinity. $4\frac{1}{2}$.

45. $\frac{1}{3}, \quad \frac{2}{9}, \quad \frac{4}{27}, \ldots \ldots$ to infinity.   1.

46. $\frac{8}{5}, \quad -1, \quad \frac{5}{8}, \ldots \ldots$ to infinity. $\frac{64}{65}$.

47. $.9, .03, .001, \ldots \ldots$ to infinity. $\frac{27}{29}$.

Find the value of the following repeating decimals:

48. $.4282828 \ldots$ *Ans.* $\frac{212}{495}$. | 50. $.1\dot{6}$.   *Ans.* $\frac{1}{6}$.

49. $.28131313 \ldots$ $\frac{557}{1980}$. | 51. $.3\dot{7}\dot{8}$. $\frac{25}{66}$.

52. The sum of three terms in G. P. is 63, and the difference of the first and third terms is 45: find the terms.

Ans. 3, 12, 48; or 36, -54, 81.

Let a, ar, ar^2 denote the numbers.

53. The sum of the first four terms of a G. P. is 40. and the sum of the first eight terms is 3280: find the series.

Ans. 1, 3, 9, $\ldots \ldots$

54. The sum of three terms in G. P. is 21, and the sum of their squares is 189: find the terms.   *Ans.* 3, 6, 12.

55. A person who saved every year half as much again as he saved the previous year had in seven years saved \$102.95: how much did he save the first year?   *Ans.* \$3.20.

CHAPTER XVII.

PERMUTATIONS AND COMBINATIONS—BINOMIAL THEOREM.

PERMUTATIONS AND COMBINATIONS.

166. Definitions. — The different orders in which a number of things can be arranged, either by taking some or all of them, are called their *Permutations.*

Thus, the permutations of the letters a, b, c, taken one at a time are three, viz., a, b, c; taken two a time, are six, viz., ab, ba, ac, ca, bc, cb; and taken three at a time, are also six, viz.,

$$abc, \ acb, \ bca, \ bac, \ cab, \ cba.$$

The *Combinations* of things are the different groups or collections which can be made, either by taking a part or all of them, without reference to the order in which the things are placed.

Thus, the combinations of the letters a, b, c, taken two at a time are three, viz., ab, ac, bc; ab and ba, though different *permutations*, form the same *combination*, both consisting simply of a and b grouped together.

It appears from this that in forming *combinations* we are concerned only with the *number* of things each group contains; while in forming *permutations* we have also to consider the *order* of the things which make up each group; thus the above six permutations of the letters a, b, c, taken three at a time, form but one combination.

167. The Number of Permutations. — *To find the number of permutations of n different things, taken r at a time.*

Let the different things be represented by n letters, a, b,

r, ... Set a aside; write down the other $n - 1$ letters
in a line; put a before each of them in succession; we thus
obtain ab, ac, ad, etc., or $n - 1$ permutations, each of two
letters in which a stands first. In the same manner there
are $n - 1$ permutations, each of two letters in which b
stands first. Similarly there are $n - 1$ permutations,
each of two letters in which c stands first; and so on for each
of the other letters; and as there are n of them, the whole
number of permutations of the n letters, two together, is
$n(n - 1)$.

Again, set a aside, and group the other $n - 1$ letters, two
and two; as has just been shown, there are $(n - 1)(n - 2)$
such groups. Put a before each of them, and we have
$(n - 1)(n - 2)$ permutations, each of three letters in which
a stands first. Similarly there are $(n - 1)(n - 2)$ permutations, each of three letters in which b stands first; and
so on for each of the other letters. Therefore the whole
number of permutations of n letters taken three at a time is
$n(n - 1)(n - 2)$.

Proceeding thus, and noticing that at any stage, the
number of factors is the same as the number of letters in
each permutation, and that the negative number in the last
factor is one less than the number of letters in each permutation, we shall have the number of permutations of n things,
r together equal to $n(n - 1)(n - 2) \ldots \ldots$ to r factors;
and the rth factor is $n - (r - 1)$ or $n - r + 1$.

Hence, *the whole number of permutations of n things taken
r at a time is*

$$n(n - 1)(n - 2) \ldots \ldots (n - r + 1) \quad . \quad . \ (1)$$

If all the letters are taken together, $r = n$, and (1)
becomes

$$n(n - 1)(n - 2) \ldots \ldots 3 \cdot 2 \cdot 1 \quad . \quad . \quad . \ (2)$$

Hence, *the number of permutations of n things taken all at
a time is equal to the product of the natural numbers from 1
up to n.*

It is usual to denote this product by the symbol $\lfloor n$, which is read " factorial n." * Thus,

Factorial 6, or $\lfloor 6$, means $6 \cdot 5 \cdot 4 \cdot 3 \cdot 2 \cdot 1$, or 720 ;

factorial 5, or $\lfloor 5$, means $5 \cdot 4 \cdot 3 \cdot 2 \cdot 1$, or 120.

From the law of formation it is clear that $\lfloor 7 = 7\lfloor 6$.

More generally, $\lfloor n + 1 = (n + 1)\lfloor n$.

Thus $\lfloor n + 1$ contains all the factors of $\lfloor n$, and one factor, $n + 1$, additional.

Denoting the number of permutations of n things taken r at a time by the symbol $_nP_r$, we have from (1) and (2)

$$_nP_r = n(n-1)(n-2)\ldots\ldots(n-r+1); \quad \text{and} \quad _nP_n = \lfloor n.$$

Thus, $_nP_4 = n(n-1)(n-2)(n-3)$.

Also $_nP_5 = {_nP_4}(n-4) = n(n-1)(n-2)(n-3)(n-4)$;

and so on.

1. Four persons enter a railway carriage in which there are six seats ; in how many ways can they take their places?

Here $n = 6$, and $r = 4$; then by (1) we have

$$_6P_4 = 6 \cdot 5 \cdot 4 \cdot 3 = 360.$$

2. Required the number of changes which can be rung, (1) with 5 bells out of 8, and (2) with the whole peal.

Ans. (1) 6720 ; (2) 40320.

3. Required the number of different ways in which 6 persons can be seated at a dinner table. *Ans.* 720.

168. The Number of Combinations. — *To find the number of combinations of n different things taken r at a time.*

The number of permutations of n things taken r at a time is

$$n(n - 1)(n - 2)\ldots\ldots(n - r + 1). \quad \text{(Art. 167)}.$$

But each combination of r things taken r at a time will

---

* It is also sometimes denoted by $n!$.

make $\lfloor r$ permutations, by (2) of Art. 167; therefore there are $\lfloor r$ times as many permutations as combinations. Hence, calling $_nC_r$ the required number of combinations, we have

$$_nC_r = \frac{n(n-1)(n-2)\ldots\ldots(n-r+1)}{\lfloor r}. \tag{1}$$

This formula for $_nC_r$ may also be written in a different form; for if we multiply the numerator and the denominator by the product of the natural numbers from 1 up to $n - r$, it becomes

$$_nC_r = \frac{n(n-1)(n-2)\ldots\ldots(n-r+1)(n-r)\ldots\ldots 2\cdot 1}{\lfloor r \cdot (n-r)(n-r-1)\ldots\ldots 2\cdot 1}.$$

The numerator is now the product of the natural numbers from n to 1, or is $\lfloor n$ (Art. 167); the denominator is the product of the natural numbers from r to 1, and from $n - r$ to 1. Hence we have

$$_nC_r = \frac{\lfloor n}{\lfloor r \; \lfloor n - r} \quad \cdot \quad \cdot \quad \cdot \quad \cdot \quad \cdot \tag{2}$$

It will be convenient to use (1) for $_nC_r$ in all cases where a numerical result is required, and (2) when it is sufficient to leave it in an Algebraic shape.

NOTE 1. — If in (2) we put $r = n$, we have

$$_nC_n = \frac{\lfloor n}{\lfloor n \; \lfloor 0} = \frac{1}{\lfloor 0};$$

but $_nC_n = 1$, so that if the formula is to be true for $r = n$, the symbol $\lfloor 0$ must be considered as equivalent to 1.

The number of combinations of n things taken r at a time is the same as the number of them taken $n - r$ at a time.

For the number taken $n - r$ at a time is, from (2),

$$_nC_{n-r} = \frac{\lfloor n}{\lfloor n-r \; \lfloor n-(n-r)} = \frac{\lfloor n}{\lfloor n-r \; \lfloor r}, \quad \cdot \tag{3}$$

which $= \; _nC_r$, from (2).

The truth of this proposition is also evident from the consideration that for every different group of r things taken out of n things there is always left a different group of

$n - r$ things. Hence the number of groups of r things out of n must be the same as the number of groups of $n - r$ things. Such combinations are called *complementary.*

NOTE 2. — Put $r = n$; then from (2) and (3), $_nC_n = _nC_0 = 1.$

The proposition just proved is useful in enabling us to abridge Arithmetic work. Thus,

1. Required the number of combinations of 20 things taken 18 together.

The required number is the same as the number taken 2 together.

$$\therefore \quad _{20}C_2 = \frac{20 \times 19}{1 \times 2} = 190.$$

If we had used the formula $_{20}C_{18}$, we should have had to reduce an expression whose numerator and denominator each contained 18 factors.

2. From 12 books, in how many ways can a selection of 5 be made when one specified book is always excluded?

Since the specified book is always to be excluded, we have to select the 5 books out of the remaining 11. Hence

$$_{11}C_5 = \frac{11 \cdot 10 \cdot 9 \cdot 8 \cdot 7}{1 \cdot 2 \cdot 3 \cdot 4 \cdot 5} = 462.$$

3. How many combinations may be made of 10 letters taken 6 at a time? *Ans.* 210.

4. From 11 books, in how many ways can a selection of 4 be made? *Ans.* 330.

169. To Divide m + n Things into Two Classes. — *To find the number of ways in which* m + n *different things can be divided into two classes, so that one may contain* m *and the other* n *things.*

This is equivalent to finding the number of combinations of $m + n$ things m at a time, for every time we select one group of m things, we leave a group of n things behind. Hence by (2) of Art. 168,

$$\text{The required number} = \frac{\lfloor m + n}{\lfloor m \lfloor n}.$$

In a similar manner it may be shown that the number of ways in which $m + n + p$ different things can be divided into three classes containing m, n, p things respectively is

$$\frac{\lfloor m + n + p}{\lfloor m \ \lfloor n \ \lfloor p}.$$

1. There are three bookshelves capable of containing 14, 22, and 24 books; in how many ways can 60 books be allotted to the shelves?

Here we have to divide 60 things into groups of 14, 22, and 24 things.

Hence the required number $= \dfrac{\lfloor 60}{\lfloor 14 \ \lfloor 22 \ \lfloor 24}.$

2. From 7 Englishmen and 4 Americans a committee of 6 is to be formed, containing 2 Americans; in how many ways can this be done?

Here we have to choose 2 Americans out of 4, and 4 Englishmen out of 7. The number of ways in which the Americans can be chosen is $_4C_2$; and the number of ways in which the Englishmen can be chosen is $_7C_4$. Each of the first groups can be associated with each of the second. Hence the required number of ways is

$$_4C_2 \times _7C_4 = \frac{\lfloor 4}{\lfloor 2 \ \lfloor 2} \times \frac{\lfloor 7}{\lfloor 4 \ \lfloor 3} = \frac{\lfloor 7}{\lfloor 2 \ \lfloor 2 \ \lfloor 3} = 210.$$

3. In how many ways can the 52 cards in a pack be divided among 4 players, each to have 13?   *Ans.* $\dfrac{\lfloor 52}{[\lfloor 13]^4}.$

170. Permutations of *n* Things not all Different.

— *To find the number of permutations of n things taken all at a time, when they are not all different.*

Let there be n letters; and suppose p of them to be a, q of them to be b, r of them to be c, and the rest to be unlike.

344 PERMUTATIONS OF n THINGS NOT ALL DIFFERENT.

Let P be the required number of permutations. If in *any one* of the actual permutations, the p letters a were all changed into p letters different from each other and from all the rest, then from this single permutation, without altering the position of any of the remaining letters, we could form $\lfloor p$ new permutations. Hence if this change were made in each of the P permutations, there would be $P \times \lfloor p$ permutations.

Similarly, if in any one of these new permutations, the q letters b were changed into q letters different from each other and from all the rest, then from this single permutation we could form $\lfloor q$ new permutations. Hence the whole number of permutations would now be $P \times \lfloor p \times \lfloor q$.

In like manner, if the r letters c were also changed so that no two were alike, the total number of permutations would be $P \times \lfloor p \times \lfloor q \times \lfloor r$. But this number must be equal to the number of permutations of n different things taken all together, which is $\lfloor n$. Hence

$$P \times \lfloor p \times \lfloor q \times \lfloor r = \lfloor n.$$

$$\therefore \quad P = \frac{\lfloor n}{\lfloor p \ \lfloor q \ \lfloor r}.$$

And similarly any other case may be treated.

1. How many different permutations can be formed **out of** the letters of the word *Mississippi* taken all together?

Here we have 11 letters, of which 4 are i, 4 are s, and 2 are p.

$$\therefore \quad P = \frac{\lfloor 11}{\lfloor 4 \ \lfloor 4 \ \lfloor 2} = 11 \cdot 10 \cdot 9 \cdot 7 \cdot 5 = 34650.$$

2. How many different permutations can be made out of the letters of the word *assassination* taken all together?

Ans. 10810800.

3. How many different permutations can be made out of the letters of the word *Heliopolis?*   *Ans.* 453600.

171. Positive Integral Exponent. — The method of raising a binomial to any power by repeated multiplication has been explained in Art. 104. We shall now prove a formula known as the *Binomial Theorem*,* by which any binomial can be raised to any power without the labor of actual multiplication.

To prove the Binomial Theorem for a positive integral exponent.

By actual multiplication we obtain

$$(x+a)(x+b) = x^2 + (a+b)x + ab,$$
$$(x+a)(x+b)(x+c) = x^3 + (a+b+c)x^2 + (ab+ac+bc)x + abc.$$

In these r~ .!ts we see that the following laws hold:

1. *The number of terms on the right side is one more than the number of the binomial factors on the left side.*

2. *The exponent of x in the first term is the same as the number of binomial factors, and decreases by one in each successive term.*

3. *The coefficient of the first term is unity; of the second term, the sum of the letters a, b, c; of the third term, the sum of the products of the letters a, b, c, taken two at a time; and the fourth term is the product of all the letters.*

We shall now prove that these laws always hold whatever be the number of binomial factors.

Suppose these laws to hold for $n-1$ binomial factors, so that

$$(x+a)(x+b)\ldots(x+k) = x^{n-1} + Ax^{n-2} + Bx^{n-3} + Cx^{n-4} + \ldots K, \quad (1)$$

where $A = a+b+c+\ldots+k$, the sum of the second erms,

$B = ab+ac+bc+\ldots\ldots$, the sum of the products of these terms taken two at a time.

$C = abc+abd+\ldots\ldots$, the sum of the products of these terms taken three at a time.

$\ldots\ldots\ldots\ldots\ldots\ldots\ldots\ldots\ldots\ldots$

$K = abcd \ldots\ldots k$, the product of all these terms.

---

* This theorem was discovered by Newton.

Multiply both sides of (1) by another factor $(x + l)$; thus,

$$(x+a)(x+b) \ldots (x+k)(x+l) = x^n + (A+l)x^{n-1}$$
$$+ (B+Al)x^{n-2} + (C+Bl)x^{n-3} + \ldots + Kl \ldots (2)$$

Now $A + l = a + b + c + \ldots + k + l$
$= $ the sum of all the terms $a, b, c, \ldots l.$

$B + Al = ab + ac + bc + \ldots + al + bl + cl + \ldots + kl.$
$= $ the sum of the products taken two at a time.

$C + Bl = abc + abd + \ldots + abl + acl + bcl + \ldots$
$= $ the sum of the products taken three at a time.

$\ldots \ldots \ldots \ldots \ldots \ldots \ldots$

$Kl = abcd \ldots kl = $ the product of all the terms
$a, b, c, \ldots l.$

Also the exponent of x in the first term is the same as the number of binomial factors, and decreases by 1 in each successive term. ·

Hence if the laws hold when $n - 1$ factors are multiplied together, they hold when n factors are multiplied together; but they have been proved to hold for 3 factors, therefore they hold for 4 factors, and therefore for 5 factors, and so on, generally, for any number whatever.[*]

Now let $b, c, d, \ldots l$, each $= a$; then the binomial factors are all equal, and the first member of (2) becomes

$(x+a)(x+a) \ldots = (x+a)$ taken n times as a factor $= (x+a)^n$;

and the second member becomes

$A + l = a + a + a + \ldots = a$ taken n times $= na.$

$B + Al = aa + aa + \ldots = a^2$ taken as many times as there are combinations of n letters taken 2 at a time $= \dfrac{n(n-1)}{\underline{|2}}$
(Art. 181).

$C + Bl = aaa + aaa + \ldots = a^3$ taken as many times as there are combinations of n letters taken 3 at a time $= \dfrac{n(n-1)(n-2)}{\underline{|3}}$; and so on.

---

[*] This method of proof is called *Mathematical Induction.*

$Kl = aaaa \ldots = a$ taken n times as a factor $= a^n$.
Substituting in (2), we obtain

$$(x + a)^n = x^n + nax^{n-1} + \frac{n(n-1)}{\lfloor 2} a^2 x^{n-2}$$
$$+ \frac{n(n-1)(n-2)}{\lfloor 3} a^3 x^{n-3} + \ldots a^n \quad . \quad . \quad (3)$$

This formula is called the *Binomial Theorem;* the series
in the second member is called the *expansion* of $(x + a)^n$.
In this expansion we observe the following

RULE.

(1) *The exponent of x in the first term is the same as the
exponent of the power, and decreases by unity in each succeeding term; the exponent of a begins with one in the second
term, and increases by unity in each succeeding term.*

(2) *The coefficient of the first term is 1, that of the second
is the exponent of the power, and if the coefficient of any
term be multiplied by the exponent of x in that term, and the
product be divided by the number of the term, the quotient
will be the coefficient of the next term.*

By changing x to a and a to x, we have

$$(a + x)^n = a^n + na^{n-1}x + \frac{n(n-1)}{\lfloor 2} a^{n-2}x^2$$
$$+ \frac{n(n-1)(n-2)}{\lfloor 3} a^{n-3}x^3 + \ldots x^n \quad . \quad . \quad (4)$$

If we write $-a$ for a in (3), we obtain

$$(x - a)^n = x^n - nax^{n-1} + \frac{n(n-1)}{\lfloor 2} a^2 x^{n-2} - \ldots \quad (5)$$

Thus the odd powers of a are *negative* and the even powers
positive, and the last term is positive or negative according
as n is even or odd.

Suppose $a = 1$, then (4) becomes

$$(1+x)^n = 1 + nx + \frac{n(n-1)}{\lfloor 2}x^2 + \frac{n(n-1)(n-2)}{\lfloor 3}x^3 + \ldots x^n, \quad (6)$$

which is the simplest form of the binomial theorem.

1. Expand $(x + a)^5$.

By the rule, we have

$$(x + a)^5 = x^5 + 5x^4a + 10x^3a^2 + 10x^2a^3 + 5xa^4 + a^5.$$

Similarly

2. $(a-2x)^7 = a^7 - 7a^6(2x) + 21a^5(2x)^2 - 35a^4(2x)^3 + 35a^3(2x)^4$
$$- 21a^2(2x)^5 + 7a(2x)^6 - (2x)^7.$$
$$= a^7 - 14a^6x + 84a^5x^2 - 280a^4x^3 + 560a^3x^4$$
$$- 672a^2x^5 + 448ax^6 - 128x^7.$$

Expand the following by the Binomial Theorem :

3. $(x - 3)^5$.   *Ans.* $x^5 - 15x^4 + 90x^3 - 270x^2 + 405x - 243$.

4. $(3x + 2y)^4$. $81x^4 + 216x^3y + 216x^2y^2 + 96xy^3 + 16y^4$.

5. $(x^2 + x)^5$. $x^{10} + 5x^9 + 10x^8 + 10x^7 + 5x^6 + x^5$.

6. $(2 - \frac{3}{2}x^2)^4$. $16 - 48x^2 + 54x^4 - 27x^6 + \frac{81}{16}x^8$.

The sum of the coefficients in the expansion of $(1 + x)^n$ is 2^n. For put $x = 1$; then

$$(1 + x)^n = (1 + 1)^n = 2^n = 1 + n + \frac{n(n - 1)}{\lfloor 2} + \text{etc.}$$
$$= \text{sum of the coefficients.}$$

Also, by putting $x = -1$, we have

$$(1 - 1)^n = 1 - n + \frac{n(n - 1)}{\lfloor 2} - \text{etc.};$$

$\therefore$ $0 =$ sum of the odd coefficients — the sum of the even ones; i.e., the sums of the odd and even coefficients are equal, and therefore each $= \frac{1}{2} \times 2^n = 2^{n-1}$.

172. The rth or General Term of the Expansion. — In the expansion of $(x + a)^n$, we see that the second term is $nx^{n-1}a$; the third term is $\dfrac{n(n - 1)}{\lfloor 2}x^{n-2}a^2$; and so on; the last factor in the denominator of each term being one less than the number of the term to which it applies, one greater than the negative number in the last factor of the numerator, and the same as the exponent of a; and also that the exponent of x is found by subtracting the exponent of a from n. Hence the

$$r^{\text{th}} \text{ term } = \frac{n(n-1)(n-2)\ldots\ldots(n-r+2)x^{n-r+1}a^{r-1}}{\lfloor r-1}.$$

This is called the *general term*, because by giving to r different numerical values, any assigned term may be obtained.

The coefficient of the rth term from the beginning is equal to the coefficient of the rth term from the end.

The coefficient of the rth term from the beginning is

$$\frac{n(n-1)(n-2)\ldots\ldots(n-r+2)}{\lfloor r-1}.$$

By multiplying both terms by $\lfloor n-r+1$, this becomes

$$\frac{\lfloor n}{\lfloor r-1 \lfloor n-r+1}. \quad \text{See (1) and (2) of Art. 168.}$$

The rth term from the end is the $(n-r+2)^{\text{th}}$ term from the beginning, and its coefficient is

$$\frac{n(n-1)\ldots\ldots r}{\lfloor n-r+1}, \text{ which also } = \frac{\lfloor n}{\lfloor r-1 \lfloor n-r+1}.$$

Therefore the coefficients of the latter half of an expansion may be taken from the first half.

1. Find the fifth term of $(a + 2x^3)^{17}$.

Here $n = 17, r = 5$; therefore the

$$5\text{th term } = \frac{17 \cdot 16 \cdot 15 \cdot 14}{1 \cdot 2 \cdot 3 \cdot 4} a^{13} \times 16x^{12} = 38080a^{13}x^{12}.$$

2. Find the 14th term of $(3 - a)^{15}$.   *Ans.* $-945a^{13}$.

3. Find the 7th term of $(a^3 + 3ab)^9$. $61236a^{15}b^6$.

4. Find the 5th term of $(a^2 - b^2)^{12}$. $495a^{16}b^8$.

5. Find the 5th term of $(3x^{\frac{1}{2}} - 4y^{\frac{1}{2}})^9$. $126 \times 3^5 x^{\frac{5}{2}} 4^4 y^2$.

REM. — In the demonstration (Art. 171) we assumed n to denote a *positive integer.* But the Binomial Theorem is also true when n is a positive fraction, or a negative quantity whole or fractional. For the proof of the Binomial Theorem for fractional or negative values of n, the student is referred to the College Algebra (Art. 192).

EXAMPLES.

1. How many different numbers can be formed by using six out of the nine digits, 1, 2, 3, ... 9?   *Ans.* 60480.

2. Required the number of changes which can be rung upon 12 bells taken all together.   *Ans.* 479001600.

3. Required the number of combinations of 24 different letters taken 4 at a time.   *Ans.* 10626.

4. Out of 14 men, in how many ways can 11 be chosen?   *Ans.* 364.

5. How many different products can be formed with any three of the figures 1, 3, 5, 7, 9?   *Ans.* 10.

6. In how many ways can 6 copies of Horace, 4 of Virgil, and 3 of Homer be given to 13 boys, so that each boy may receive a book?   *Ans.* 60060.

7. Out of 7 consonants and 4 vowels, how many words can be made each containing 3 consonants and 2 vowels?   *Ans.* 25200.

8. How many parties of 12 men each can be formed from a company of 60 men?   *Ans.* $\dfrac{\lfloor 60}{\lfloor 12 \; \lfloor 48}$.

9. Out of 12 Republicans and 16 Democrats, how many different committees could be formed, each consisting of 3 Republicans and 4 Democrats?   *Ans.* $\dfrac{\lfloor 12}{\lfloor 3 \; \lfloor 9} \times \dfrac{\lfloor 16}{\lfloor 4 \; \lfloor 12}$.

10. Out of 10 consonants and 4 vowels, how many words can be formed, each containing 3 consonants and 2 vowels?   *Ans.* 86400.

11. There are 10 candidates for 6 vacancies in a committee: in how many ways can a person vote for 6 of the candidates?   *Ans.* 210.

12. In how many ways can a cricket eleven be chosen out of fourteen players? *Ans.* 364.

13. In how many ways could 2 ladies and 2 gentlemen be chosen to make a set at tennis from a party of 4 ladies and 6 gentlemen? *Ans.* 90.

14. In how many ways could 2 ladies and 2 gentlemen be chosen to make a set at tennis from a party of 6 ladies and 8 gentlemen? *Ans.* 420.

15. From 6 ladies and 5 gentlemen, in how many ways could you arrange sides for a game of croquet, so that there would be 2 ladies and one gentleman on each side?

$$\text{\textit{Ans.} } \frac{\lfloor 6 \ \lfloor 5}{(\lfloor 2)^3 \lfloor 3}, \text{ or } 1800.$$

16. Out of 6 ladies and 8 gentlemen, how many different parties can be formed, each consisting of 3 ladies and 4 gentlemen?

$$\text{\textit{Ans.} } \frac{\lfloor 6}{\lfloor 3 \lfloor 3} \times \frac{\lfloor 8}{\lfloor 4 \lfloor 4}.$$

17. If the number of permutations of n things taken 4 together is equal to 12 times the number of permutations of n things taken 2 together: find n. *Ans.* 6.

18. In how many ways can a party of 6 take their places at a round table? *Ans.* 60.

19. How many words of 6 letters may be formed with 3 vowels and 3 consonants, the vowels always having the even places? *Ans.* 36.

Expand the following by the Binomial Theorem.

20. $(2x - y)^5$.
Ans. $32x^5 - 80x^4y + 80x^3y^2 - 40x^2y^3 + 10xy^4 - y^5$.

21. $(3a - \frac{2}{3})^6$.
Ans. $729a^6 - 972a^5 + 540a^4 - 160a^3 + \dfrac{80a^2}{3} - \dfrac{64a}{27} + \dfrac{64}{729}$.

22. $(1 + 2x - x^2)^4$.
Ans. $1 + 8x + 20x^2 + 8x^3 - 26x^4 - 8x^5 + 20x^6 - 8x^7 + x^3$.

23. $(3x^2 - 2ax + 3a^2)^3$.
Ans. $27x^6 - 54ax^5 + 117a^2x^4 - 116a^3x^3 + 117a^4x^2 - 54a^5x + 27a^6$.

Expand to 4 terms :

24. $(1-x)^{\frac{2}{5}}$.   *Ans.* $1 - \frac{2}{5}x - \frac{3}{25}x^2 - \frac{8}{125}x^3$.

25. $(1-3x)^{\frac{1}{3}}$. $1 - x - x^2 - \frac{5}{3}x^3$.

26. $(1-3x)^{-\frac{1}{3}}$. $1 + x + 2x^2 + \frac{14}{3}x^3$.

27. $\left(1+\dfrac{x}{3}\right)^{-3}$. $1 - x + \frac{2}{3}x^2 - \frac{10}{27}x^3$.

28. $(1+\tfrac{1}{2}a)^{-4}$. $1 - 2a + \frac{5}{2}a^2 - \frac{5}{2}a^3$.

29. $(8+12a)^{\frac{1}{3}}$. $4(1 + a - \frac{1}{4}a^2 + \frac{1}{8}a^3)$.

30. $(9-6x)^{-\frac{3}{2}}$. $\frac{1}{27}(1 + x + \frac{5}{6}x^2 + \frac{35}{54}x^3)$.

31. $(4a-8x)^{-\frac{1}{2}}$. $\dfrac{1}{2a^{\frac{1}{2}}}\left(1 + \dfrac{x}{a} + \dfrac{3}{2}\cdot\dfrac{x^2}{a^2} + \dfrac{5}{2}\cdot\dfrac{x^3}{a^3}\right)$.

Write down and simplify :

32. The 4th term of $(x-5)^{13}$.   *Ans.* $-35750x^{10}$.

33. The 10th term of $(1-2x)^{12}$. $-112640x^9$.

34. The 4th term of $\left(\dfrac{a}{3}+9b\right)^{10}$. $40a^7b^3$.

35. The 7th term of $\left(\dfrac{4x}{5}-\dfrac{5}{2x}\right)^9$. $\dfrac{10500}{x^3}$.

36. The 5th term of $(x^{\frac{3}{2}}a^{-\frac{1}{2}} - y^{\frac{5}{2}}b^{-\frac{3}{2}})^8$. $70x^6y^{10}a^{-2}b^{-6}$.

37. The 8th term of $(1+2x)^{-\frac{1}{2}}$. $-\frac{429}{16}x^7$.

38. The 5th term of $(3a-2b)^{-1}$. $\dfrac{16b^4}{243a^5}$.

39. The 14th term of $(2^{10}-2^7x)^{\frac{4}{3}}$. $-1848x^{13}$.

www.ingramcontent.com/pod-product-compliance
Lightning Source LLC
Chambersburg PA
CBHW021401210326
41599CB00011B/964